SEAFOOD QUALITY AND SAFETY
Advances in the New Millennium

Edited by

Fereidoon Shahidi
Department of Biochemistry
Memorial University of Newfoundland
St. John's, Newfoundland, Canada

Benjamin K. Simpson
Department of Food Science and Agricultural Chemistry
MacDonald Campus of McGill University
Ste. Anne de Bellevue, Quebec, Canada

ScienceTech Publishing Company
St. John's, NL, Canada

SEAFOOD QUALITY AND SAFETY
Advances in the New Millennium

Copyright © 2004 by ScienceTech Publishing Co.

All rights reserves. No part of this work covered by the copyright heron may be reproduced or used in any form or by any means - graphic, electronic, or mechanical, including photocopying, recording, taping, or information storage and retrieval systems - without written permission of the publisher.

Published by ScienceTech Publishing Company
St. John's, NL, Canada

Includes bibliographical references and index
ISBN No. 0-9683220-2-6

PRINTED IN THE UNITED STATES OF AMERICA

FOREWORD

The publication of this monograph was made possible through contributions made originally at the Atlantic Fisheries Technology Conference (AFTC 2001) in Rimouski, Quebec. The conference was sponsored and financed by generous donations of the following organizations.

- Ministère de l'Agriculture, des Pêcheries et de l'Alimentation du Québec
- Université du Québec à Rimouski
- Heritage Canada
- Canada Economic Development

The conference was chaired by Lucien Adambounou and Luc Leclerc acted as the executive secretary for the event. The scientific program committee, directed by Benjamin K. Simpson, also included Pierre Blier, Michel Desbiens and James R. Wilson. The other committee members were Alain Samuel, Julie Boyer, Marcel Lévesque, Julie Guillemot and Jocelyne Pellerin.

ScienceTech Publishing Company

CONTENTS

Preface ix

1. SEAFOOD QUALITY AND SAFETY: AN OVERVIEW 1
   Fereidoon Shahidi

2. EMERGING FOOD PROCESSING TECHNOLOGIES:
   FOOD SAFETY AND QUALITY 7
   Michael Ngadi and Maksym Bazhal

3. TECHNICAL POSSIBILITIES FOR INTRODUCING
   TRACEABILITY IN THE SEAFOOD INDUSTRY 31
   T. Børresen and M. Frederiksen

4. FOOD QUALITY AND GLOBAL TRACEABILITY FOR
   E-COMMERCE THROUGH AUTOMATIC DATA
   CAPTURE PROCESS: NEW BAR CODE AND RFID
   TECHNOLOGIES 39
   G. Doyon, P. Veilleux, A. Clement

5. ENHANCEMENT OF PROFITABILITY PERSPECTIVES OF
   AN AQUACULTURE PRODUCTION BY THE EXTRACTION
   OF HIGH COMMERCIAL VALUE BIOMOLECULES:
   EVALUATION OF THE POTENTIAL OF WOLFFISH
   CULTIVATION 61
   Nathalie R. Le François, Mariève Desjardins, Pierre U. Blier

6. COMPARISON OF DEHEADED GUTTED AND UNGUTTED
   MACKEREL (*SCOMBER SCOMBRUS*) PRESERVED AT -30°C
   UNDER OPTIMAL CONDITIONS 71
   Nathalie Fillion and Marie-Élise Carbonneau

7. QIM - A TOOL FOR DETERMINATION OF FISH FRESHNESS 81
   Grethe Hyldig and Jette Nielsen

8. **HIGH PRESSURE PROCESSING OF SEAFOOD** 91
Lebail A, de-Lamballerie-Anton M., Hayert M. and Chevalier D.

9. **INTERACTION BETWEEN RAW MATERIAL CHARACTERISTICS AND SMOKING PROCESS ON SMOKED SALMON QUALITY** 111
Mireille Cardinal, Camille Knockaert, Ole Torrissen, Sjofn Sigurgisladottier, Turid Mørkøre, Magny Thomassen, Jean Luc Vallet

10. **MODELING HEAT AND MOISTURE TRANSFER DURING SHRIMP COOKING** 129
Michael Ngadi

11. **EFFECT OF PRESSURE-SHIFT FREEZING OF CARP (*Cyprinus carpio*) VS AIR-BLAST FREEZING: A STORAGE STUDY** 139
A. Sequeira-Munoz, D. Chevalier, B.K. Simpson, A. Le Bail, H.S. Ramaswamy

12. **BIOGENIC AMINES AND E-NOSE DEVICES: INDICATORS OF SEAFOOD QUALITY** 155
Amaral Sequeira-Munoz, Wen-Xian Du, Francis R. Antoine, Maurice R. Marshall and Cheng-I Wei

13. **STUDY OF COLD-SOLUBLE METHYLATED AGARS** 177
Abdelwahab Riad, Andre Begin, Marie-Rose Van Calsteren and Rachid Lebar

14. **USE OF CHITOSAN FOR STORAGE OF MANGOES: COMPARISON WITH OTHER FUNGICIDES** 187
André Bégin, Isabelle Dupuis, Marc Dufaux and Gonzague Leroux

15. **SHRIMP ENHANCEMENT POTENTIAL** 195
Zine-Al-Abidine Gadhi, Lucien Adambounou, Alain Guillou

16. **HEAVY METAL CONTAMINATED LEACHATE REGENERATION WITH CHITOSAN FLAKES** 203
Benedictus Hope Soga, Shiv O. Prasher and Benjamin Kofi Simpson

17. **CHITOSAN FILM IN SEAFOOD QUALITY PRESERVATION** 223
Fereidoon Shahidi

18. **EMULSION STABILIZING PROPERTIES OF CHITOSAN** 233
Serge Laplante, Sylvie L. Turgeon and Paul Paquin

19. **CHITOSAN MODIFICATIONS FOR PHARMACEUTICAL APPLICATIONS** 251
Le Tien Canh, Monique Lacroix, Pompilia Ispas Szabo, Mircea-Alexandru Mateescu

20. ECONOMICS OF CHITIN, CHITOSAN AND CAROTENO-PROTEIN PRODUCTION FROM SHRIMP WASTE 259
K. Gunjal, H. Teftal, B.K. Simpson and P. Goldsmith

21. CHROMATOGRAPHIC METHOD FOR SIMULTANEOUS QUANTITATION OF N-ACETYLGLUCOSAMINE, GLUCOSAMINE AND ITS OLIGOMERS 275
Éric Demers, Diane Ouellet, Marie-Élise Carbonneau

22. MODELLING THE EFFECT OF TEMPERATURE ON SHELF-LIFE AND ON THE INTERACTION BETWEEN THE SPOILAGE MICROFLORA AND *LISTERIA MONOCYTOGENES* IN COLD-SMOKED SALMON 281
Paw Dalgaard, Esther Murillo Iturrate, Lasse Vigel Jørgensen

23. AN OVERVIEW OF POSTHARVEST TREATMENTS TO REDUCE *VIBRIO* IN OYSTERS 303
Michael Morrissey, Hakan Calik and Shiu-er Shiu

24. MICROBIAL QUALITY OF CULTURED NEWFOUNDLAND MUSSELS AND SCALLOPS 317
M.A. Khan, C. Parrish and F. Shahidi

25. HIGH PRESSURE DESTRUCTION KINETICS OF MICROORGANISMS IN TROUT AND SHRIMP 327
N. Houjaij, H.S. Ramaswamy and E. Idziak

26. IDENTIFICATION OF SPOILAGE MICRO-ORGANISMS AND RESEARCH ON QUALITY INDICES FOR COLD-SMOKED SALMON 337
Françoise Leroi, Jean Jacques Joffraud, Frédérique Chevalier, Mireille Cardinal, Valérie Stohr, Jean-Louis Berdague, Jean-Luc Vallet

27. ISOLATION OF BACTERIOCIN-PRODUCING LACTIC ACID BACTERIA FROM REFRIGERATED SMOKED SALMON, MUSSELS AND SHRIMP 355
M. Desbiens, S. Thibault, G. Imbeault

28. AQUA FEED: RESEARCH CHALLENGES AND FUTURE TRENDS 369
Fernando Luis García-Carreño

INDEX 377

PREFACE

Aquatic species provide an important source of nutrients for human use. Consumption of seafoods and marine oils is actively encouraged for the prevention and possible treatment of cardiovascular diseases, diabetes and rheumatoid arthritis as well as for cancer prevention and mental health. In addition, by-products from the seafood processing industry could serve as a rich source of a variety of biomolecules with potential health benefits. In this regard, enzymes, squalene, carotenoids, chitinous materials, N-acetylglucosamine and glucosamine are of considerable importance. Many studies have shown the effect of marine oils on triacylglycerol and low density lipoprotein (LDL) cholesterol in the human body. Studies on the effects of polyunsaturated fatty acids on modulation of signal transduction and mechanism of action at the cellular and subcellular levels may provide important insight about health benefits of specific fatty acids. In addition, guidelines for managing product quality based on hazard analysis critical control point (HACCP) and ISO 9000 principles have been provided. Many of the attributes related to safety and wholesomeness of seafoods relate to natural and environmental effects as well as processing technologies employed. Thus monitoring of the raw material for possible presence of toxins and environmental pollutants and devising appropriate processing strategies to ensure safety of products is most essential.

We are grateful to the authors who provided a state-of-the-art contributions for inclusion in this monograph. We are also grateful to Ms. Peggy-Ann Parsons for her expert assistance in preparation of the manuscripts.

Fereidoon Shahidi
Benjamin K. Simpson

CHAPTER 1

SEAFOOD QUALITY AND SAFETY: AN OVERVIEW

Fereidoon Shahidi

Department of Biochemistry, Memorial University of Newfoundland, St. John's, NL A1B 3X9 Canada

The consumption and popularity of seafoods has increased steadily in recent years. The beneficial health effects of seafoods, marine oils and their constituent long-chain polyunsaturated fatty acids (LC PUFA) has been particularly important for the growth of the industry. In addition, value-added use of by-products has successfully provided the market with a wide range of products that include chitosan, chitosan oligomer, glucosamine and squalene, among others. However, production of high quality products requires particular attention to their safety and wholesomeness. Thus, devising of innovative technologies for value-added use of source materials as well as by-products would contribute to full utilization of the resources and provide means for disease risk reduction.

INTRODUCTION

The world's annual catch of fish and marine invertebrates has been around 100 MMT in recent years. However, aquaculture developments have led to production of high quality products that have also assisted conservation strategies to be implemented. Proteins and lipids from seafoods have unique features that are quite different from those of the land-based animals. These features may, in part, arise from aquatic environmental factors such as temperature, pressure, salt concentration, availability of oxygen and the presence and concentration of different nutrients and environmental pollutants. For cold water species, the low temperature of the habitat has resulted in their special adaptation and production of antifreeze proteins/glycoproteins and adjustment in blood hemoglobin concentration (Davis *et al.*, 1989). In addition, the proteolytic activity of several species of fish, crustaceans and molluscs is often extremely high due to a need for high cellular turnover in the periods of intensive synthesis of genetic materials.

Seafood protein are well-balanced in their amino acid composition. These proteins are highly sensitive to deteriorative changes due to the presence of enzymes that catabolize different meat constituents. However, under appropriate conditions, loss of fresh quality due to hydrolysis, polymerization, deamination, decarboxylation, oxidation and other deteriorative changes could be prevented or delayed. This would in turn protect products from loss of texture, color and flavor (Shahidi, 1994).

Seafood lipids are quite distinct in their compositional characteristics and contain long-chain omega-3 polyunsaturated fatty acids (PUFA). The potential benefits of omega-3 fatty acids in prevention and possible treatment of cardiovascular diseases, arthritis, mental disorders and diabetes as well as their essential role in fetus development have received considerable attention in recent years. These PUFA are highly prone to oxidation and off-flavor development. Therefore, their protection against oxidation by use of appropriate antioxidants and novel packaging technologies as well as storage conditions is essential.

Of the total amount of harvest, a major portion remains unused or used for production of fish meal and fish oil. This is due to the fact that certain species might suffer for small size, high bone, skin and fat content, as well as unappealing shape. In addition, several species of fish may be used for their roe and production of caviar. The leftover carcass following roe extraction as well as those of their male counterparts may be discarded. Furthermore, processing discards from many species of fish and shellfish could be successfully processed for production of specialty enzymes, xanthophylls, chitin/chitosan, glucosamine and other value-added products.

Quality of seafoods may be compromised during post-harvest period due to poor handling of raw material or as a result of improper storage or processing. Thus, following international standards such as hazard analysis critical control point (HACCP) and ISO 9000 may improve the global marketability of seafoods (e.g. Notermans and Joune, 1995). Furthermore, predictive microbiology may serve as an important tool in seafood handling and processing (Buchanan and Whiting, 1996). Detection of environmental pollutants as well as toxins also deserve adequate attention (e.g. Quilliam and Wright, 1989).

The challenges facing the seafood and aquaculture industry are numerous and include sustainable level of catch, regard for environmental issues, by-product utilization and offering of high quality products to the market without fear arising from safety factors. Thus, hygienic health and processing conditions are important. Special role of aquaculture, formulation of aquaculture feed with nutritional and safety considerations related to prevention of disease must be highlighted. Strategies for full utilization of the catch and processing discards for production of novel products is warranted (Shahidi, 2000).

Seafood Proteins

Seafoods constitute a major source of animal protein in the diet of many populations in the coastal areas. A variety of fresh, dried, fermented and salted fish products supply a major part of the essential amino acid requirements of humans. The modern fish processing methods take advantage of unique functional properties of fish proteins such as their water holding capacity and the gelling ability. The crude protein content of fish, crustaceans and molluscs is 11-27%. However, contribution of non-protein nitrogenous (NPN) compounds to this value may reach up to a quarter of this amount. Free amino acids, peptide, polyamines, nucleotides and their breakdown products, urea and nucleic acids contribute to the overall content of NPN in seafoods. Seafood proteins include sarcoplasmic, myofibrillar and stroma. A large proportion of sarcoplasmic proteins may be composed of albumins (~30%) as well as hemoproteins. The myofibrillar proteins are myosin, actin, actomyosin and troponin, accounting for 40-60% of the total crude protein, in fish. The rest of the muscle proteins are classified as stroma and include collagenous matters, among others.

The presence and action of proteinases in fish tissue may contribute to quality deterioration of products during post-harvest storage, but extensive hydrolysis, such as that in fermented products, may contribute to the pleasing aroma, taste, color and texture of certain other products. In addition, texture toughening of frozen fish arise from cross-linking of its protein chains as a result of glutaminase activity.

In recent years, there has been an interest in recovering digestive enzymes from fish and shellfish processing discards for use as industrial processing aids (Shahidi and Kamil, 2001). Suggested uses of digestive proteases from fish include acids for cheese making, herring fermentation, fish skinning, roe processing and production of specialty kits, as well as medical applications.

Seafood Lipids and Marine Oils

Marine oils provide unique health benefits to the consumers, but also present a difficult challenge to scientists and technologists in delivering their highly unsaturated fatty acids (HUFA) in an odor-free and appealing form. These oils originate from the body of fatty fish, liver of lean white fish and blubber of marine mammals. The fatty acids of marine lipids include eicosapentaenoic acid (EPA), docosahexaenoic acid (DHA) and to a lesser extent docosapentaenoic acid (DPA). There is a rapidly growing body of literature illustrating the health benefits of HUFA. These effects are protection against cardiovascular disease, autoimmune and mental disorders, diabetes, arthritis and arrhythmia, among others (Shahidi and Finley, 2001).

The highly unsaturated nature of the oils containing EPA, DHA and DPA brings about oxidation and rancidity problems. Thus, use of antioxidants and novel processing techniques is necessary in order to allow their use in food applications. It appears that products may gain some stability when used in emulsion systems. However, the oils as such need to be initially refined, bleached and deodorize in order to eliminate the contaminants, impurities and odoriferous materials. Since the refining process also leads to the removal of endogenous antioxidants from the oil, use of appropriate antioxidants at optimum concentrations is required.

For therapeutic purposes, however, marine oils may be subjected to further processing to enrich their omega-3 content. The enrichment process reduces the levels of more saturated fatty acids in the oil. There are different methodologies for production of omega-3 concentrates and the resultant products may be in the free fatty acid, simple alkyl ester or acylglycerol forms. Due to the enrichment, the concentrates so produced are highly prone to oxidation and as such must be protected by addition of antioxidant, together with encapsulation or microencapsulation.

Chitinous Materials and other Value-Added By-Products

Chitosan is a partially deacetylated polymer of glucosamine, produced by deacetylation of chitin. Chitin was first described in 1811 by Professor Braconnot at the Academy of Sciences in Nancy, France. It is found primarily in the shells of shrimp, crab, lobster and crayfish. Recent research on chitin was first stimulated by the need to find a means for disposal of seafood processing discards. Chitosan has a strong positive charge and hence can bind to proteins, metals and other negatively changed molecules. Chitosan is also biocompatible and bioactive, making it useful in both medical and non-medical applications.

Glucosamine, the monomer of chitosan, has been reported to possess benefits for joint health and build up. The product is sold as glucosamine sulfate, but this is often a mixture of glucosamine hydrochloride and sodium or protassium-sulfate. Furthermore, glucosamine may also be sold in conjuction with chondroitins (chondroitin 4-sulfate and chondroitin 6-sulfate). Chondroitins are mucopolysaccharides (MPS) with molecular weights of up to 50,000 Da and could be prepared from connective tissues of slaughtered animals. In combination, while glucosamine helps to form proteoglycans that sit within the space in the cartiledge, chondroitin sulfate acts as a liquid magnet (Shahidi and Kim, 2002).

The by-products in the extraction of chitin from shellfish include minerals, proteins, carotenoproteins and enzymes, among others. These components may also be isolated for further value-added utilization. In particular, isolation and application of enzymes as well as carotenoproteins, free xanthophylls and proteins is of interest. By-products from fish processing industries also include lipid, protein and enzymes. Further extraction and utilization of these by-products is warranted. In all such preparations, high quality of products is only possible if the raw material is of good quality (Shahidi and Synowiecki, 1991).

Cultured Products and Feed

Cultured fish and shellfish may be produced under controlled conditions, often in cages close to shore or where water current is minimal. The quality of products from cultured species may be affected by the type and quality of the feed used for this purpose. Thus, formulation of feed using alternative sources of protein and oil from vegetable sources may result in high fat content of fish, but with a compromised quality. The content of omega-3 fatty acids in cultured fish may be reduced proportionally in favor of an increase in the content of omega-6 fatty acids. Thus, nutritional value and high quality of products are, to a large extent, dependent on the constituents present in the feed.

REFERENCES

Buchanan, R.L. and Whiting R.C. 1996. Risk assessment and predictive microbiology. *J. Food Protec. Supplement.* 31-36.

Davies, P.; Fletcher, G.L. and Hew, C.L. 1989. Fish antifreeze protein genes and their use in transgenic studies. In *Oxford Surveys on Eukaryotic Genes*, Vol. 6. MacLean N (ed.) pp. 85-109. Oxford University Press, Oxford.

Notermans, S. and Joune, J.Z. 1995. Quantitative risk analysis and HACCP: Some remarks. *Food Microbiol.* 12: 425-429.

Quilliam, M.A. and Wright, J.L.C. 1989. The amnesiac shellfish poisoning mystery. *Anal. Chem.* 61: 1053A-1059A.

Shahidi, F. (ed). (2000). Seafood in Health and Nutrition – Transformation in Fisheries and Aquaculture: Global Perspectives p. 552. ScienceTech Publishing Co., St. John's, Canada.

Shahidi, F. 1994. Proteins from seafood processing discards. In *Seafood Proteins*. Z.E. Sikorski, B.S. Pan, F. Shahidi (eds.) pp. 171-193. Chapman & Hall, New York.

Shahidi, F. and Kamil, Y.V.A.J. 2001. Enzymes from fish and aquatic invertebrates and their application in the food industry. *Trends Food Sci. Technol.* 12: 435-464.

Shahidi, F. and Finley, J.W. (eds). 2001. Omega-3 Fatty Acids: Chemistry, Nutrition and Health Effects. ACS Symposium Series 788. American Chemical Society, p. 330. Washington, DC.

Shahidi, F. and Kim S.K. 2002. Quality management of marine nutraceuticals. In *Quality Management of Nutraceuticals*. ACS Symposium Series 803. C.T. HO, Q,T, Zheng (eds.). American Chemical Society, pp. 76-87. Washington, DC.

Shahidi, F. and Synowiecki, J. 1991. Isolation and characterization of nutrients and value-added products from snow crab (*Chinoectes opilio*) and shrimp (*Pandalus borealis*) processing discards. *J. Agric. Food Chem.* 39: 1527-1532.

CHAPTER 2

EMERGING FOOD PROCESSING TECHNOLOGIES: FOOD SAFETY AND QUALITY

Michael Ngadi and Maksym Bazhal

Agricultural and Biosystems Engineering Department
McGill University, Macdonald Campus
Ste-Anne-de-Bellevue, Quebec

Increasing consumer demand for fresh-like and safe products is a limiting factor on selection of food processing technologies. The well established thermal based methods are available and provide a high degree of microbial safety. However, they also degrade product quality. Non-thermal technologies as alternatives or complementary to thermal processes are being sought. Although non-thermal technologies including high electric field pulses, high hydrostatic pressure, ultraviolet light have long been used, they are only recently gaining recognition in the food industry. Application of these non-thermal technologies offers interesting opportunities for mildly processed safe products with preserved sensory and nutritional qualities. Current advances on non-thermal processing technologies with emphasis on pulsed electric field (PEF) is reviewed here. Some critical parameters of PEF processing are identified and described. Potential applications in processing.

INTRODUCTION

Thermal processing has been used in the food industry for several decades with various degrees of success to inactivate pathogenic microbial load in food products. Heat effectively destroys microorganisms. However, it is the balancing act of preserving food quality and maintaining safety that makes thermal processing not very attractive for some products. The current trend in consumers' preference is for fresh-like, minimally processed and high quality products. Emerging technologies for non-thermal (or minimal thermal) processing include high pressure processing, UV light irradiation, magnetic fields, electron irradiation, ozonation and pulsed electric fields. Interest in these technologies has arisen from the desire to overcome the problems of traditional thermal processing. The technologies enable unique modes of energy transfer to foods and target biological cells to achieve inactivation or modification without significantly heating the products. These methods have their strengths and weaknesses depending on process objectives and the type of food to be processed. Among the various emerging technologies, high pressure, pulsed electric field (PEF) and UV light methods seem to be most promising. Studies are on-going to improve understanding of these technologies. This contribution provides an up-to date description of the technologies, review of developments, trends and applications of the emerging technologies with emphasis on pulsed electric fields.

PULSED ELECTRIC FIELDS (PEF)

PEF processing involves the application of externally generated electric field across a food product with the intent of inactivating pathogenic microorganisms, modifying enzymes, intensifying some processes or achieving some specific transformation in the product. The technology has long been used for cell hybridization and electrofusion in genetic engineering and biotechnology. Its application is based on the transformation or rupture of cells under a sufficiently high external electric field, resulting in increased permeability and electrical conductivity of the cellular material. This effect named dielectric breakdown (Zimmermann *et al.*, 1976) or electroplasmolysis (McLellan *et al.*, 1991) can be explained by two main factors: (1) electroporation; that is electroinduced formation and growth of pores in biomembranes as a result of their polarization; and (2) denaturation of cell membranes as a result of their ohmic heating caused by the electric resistance of membranes which is typically much higher than that of cell sap content. Beside these, the physiological impact and electroosmotic effect may also influence electroplasmolysis efficiency (Weaver and Chizmadzhev, 1996).

PEF technologies have been demonstrated to provide a viable alternative to high temperature inactivation of microbial load in liquid foods such as fruit juices and milk (Barbosa-Canova *et al.*, 1999; Knorr, 1999). The majority of research effort on PEF has been on liquid food pasteurization. However, these technologies have also been shown to be applicable for microstructural modification of vegetable, fish and meat tissues (Gadmundsson and Hafsteinsson, 2001; Angersbach *et al.*, 2000; Wu *et al.*, 1998), for intensification of juice yield and for increasing product quality in juice production (Bazhal, 2001), processing vegetable raw materials (Papchenko *et al.*, 1988), winemaking and sugar production (Gulyi *et al.*, 1994). PEF treatment significantly enhances certain food processing unit operations such as pressing (Bazhal and Vorobiev, 2000), diffusion (Jemai, 1997), osmotic dehydration (Rastogi, 1999), and drying (Ade-Omowaye *et al.*, 2001). These processes generally impact quality of foods.

There are various designs of PEF systems depending on pulse characteristics and applications. In general, a PEF system includes the following components: (1) A high voltage DC generator to supply electrical energy, (2) bank of capacitors to store the energy generated by the generator, (3) a high voltage switch to deliver energy to electrodes, and (4) treatment cell to hold and contain samples. The generation of electric pulses across a treatment cell requires relatively slow charging and subsequent fast (in the order of nanoseconds) discharging of electrical energy stored in capacitors. This may present interesting challenges depending on the desired pulse characteristics. Treatment chambers could be static or continuous. Parallel plates and co-axial treatment chambers have been used widely by various workers. Jeyamkondan *et al.* (1999) have provided good descriptions of various treatment chambers.

PEF Waveform Characteristics

Pulse waveform is an important design consideration for a PEF application. Different types of waveform have been used for electroporation and disruption of biological cells (Jemai, 1997; Zhang *et al.*, 1995; Bazhal, 2001). These include continuous, sinusoidal, exponential and square waveforms as illustrated in Figure 1.

Biological cells respond differently to different pulse waveforms. The exponential decay and rectangular pulses are widely used in most PEF applications (Barbosa-Canova *et al.*, 1999). Exponential decay pulses are easier to generate due to the simplicity of their circuit (Zhang *et al.*, 1995). They involves charging a capacitor bank connected in series with a charging resistor. The energy stored in the capacitor is then rapidly switched across a treatment chamber.

Exponential pulse waveforms are generated independent of the electrical resistance of the treatment chamber. Rectangular pulse waveforms are more difficult to generate and normally require a pulse-forming network consisting an array of capacitors, inductors and switching device. Effective generation of rectangular or square pulses depends not only on the generator but also on the material to be treated. Rectangular waveforms are realized when both the treatment chamber and the pulse-forming network have matching impedance.

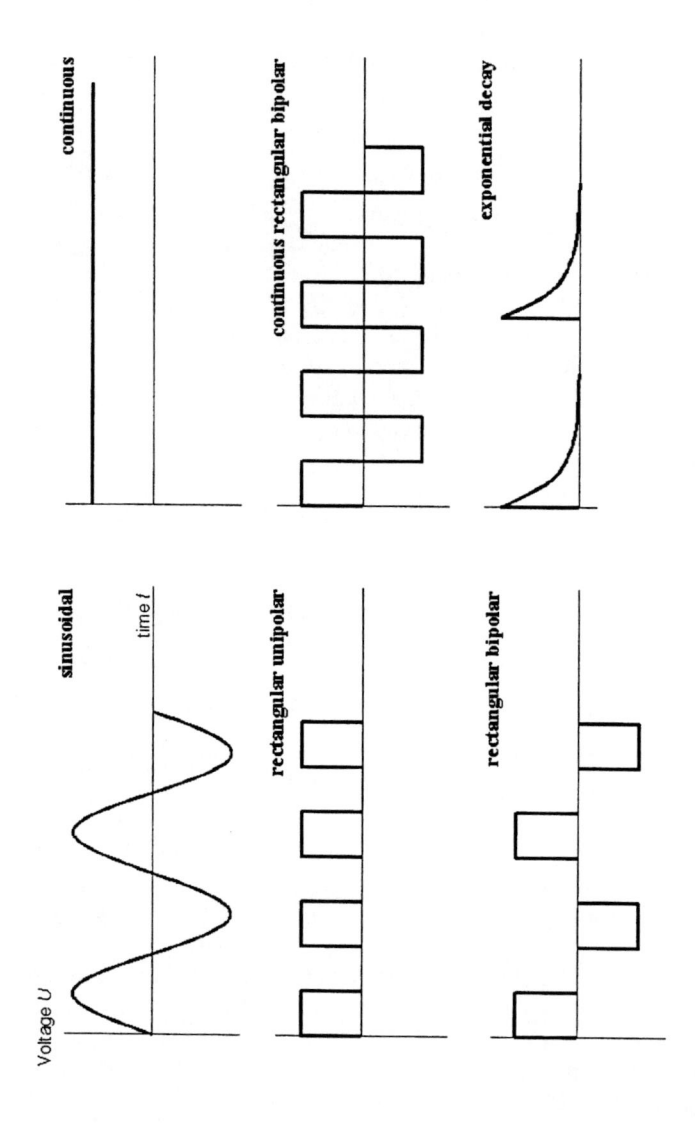

Figure 1. Different types of electric field waveforms.

Rectangular type pulses can be transformed to exponential decay type if the characteristic time for electrical discharge through sample is less than the time constant for charging the total capacitors. High power accumulation in the capacitors is needed to support rectangular shape of the applied pulses. This may present engineering design difficulties in some applications.

Qin *et al.* (1996) studied the effect of using different pulse waveforms on microbial inactivation. Square pulses were reported to be more effective than either exponential or oscillatory pulses. This was attributed to the wider pulse width at higher electric fields obtained using square pulses. Square pulses maintain high electric fields with very high rise and fall times whereas exponential pulses have long tail of low electric fields. Thus electrical energy is dissipated more effectively with square pulses than with exponential pulses. Bi-polar pulses have been reported to be more efficient than mono-polar waveforms for cellular rupture (Kinosita *et al.*, 1992; Bazhal, 1995). Figure 2 shows a typical bi-polar instant reversal pulse waveform used at the Food Engineering Laboratory, McGill University. The effectiveness of bi-polar pulses is due to elimination of the asymmetric properties of cells under bi-polar pulses (Kinosita *et al.*, 1992).

Microbial Inactivation Kinetics

Inactivation of microorganisms by PEF depends on many factors including electric field strength, treatment time, number of pulses, pulse characteristics, medium characteristics as well as type and growth stage of microorganism. Microbial inactivation in liquid medium has been reported to follow the first order kinetics (Hulsheger *et al.*, 1981; Grahl and Markl, 1996) as given in Eq. 1.

$$\log S = B_E (E - E_c) \tag{1}$$

where S is microbial survival rate in fraction given as the ratio of microbial count after treatment, N, to microbial count before treatment, N_o; B_E is electric field constant obtained as the coefficient of regression of the straight survival curves; E is applied electric field and E_c is critical value of electric field below which there will be no inactivation (that is 100% survival). Treatment time can be calculated as product of number of pulses and pulse width. Inactivation kinetics in terms of treatment time may then be calculated as given in Eq. 2 (Hulsheger *et al.*, 1981; Grahl and Markl, 1996):

$$\log S = B_t \frac{t}{t_c} \tag{2}$$

where B_t is electric field constant or coefficient of the regression of the straight survival curves; t is treatment time and t_c is the critical time below which there will be no inactivation. Combining eq. 1 and eq. 2 yields eq. 3.

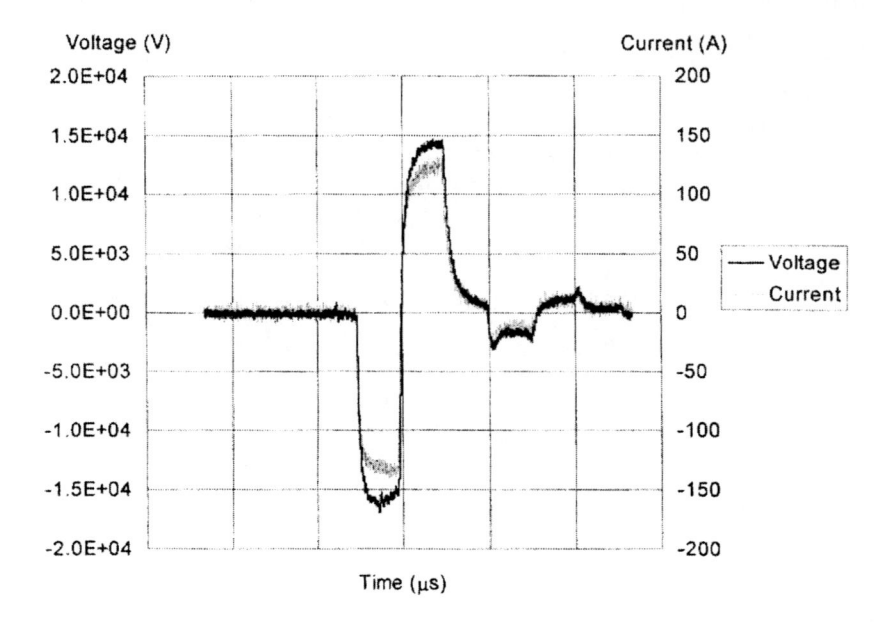

Figure 2. Typical instant charge reversal bi-polar PEF waveform.

$$S = \left(\frac{t}{t_c}\right)^{\frac{E-E_c}{k}} \qquad (3)$$

where k is a constant factor as can be expressed as given in Eq. 4:

$$k = \frac{E - E_c}{B_{t(E)}} = \frac{\log\left(\frac{t}{t_c}\right)}{B_{E(t)}} \qquad (4)$$

Considering the kinetic equations, smaller values of E_c or t_c and higher values of B_E and B_t indicate greater susceptibility of the particular microorganism to PEF processing. The above first order kinetics may be simplistic for most PEF applications. Tailing phenomenon in microbial inactivation kinetics has been observed (Sensoy *et al.*, 1997). A 2 phase kinetic model may be required to adequately describe PEF inactivation kinetics. Other more complicated inactivation kinetics have been reported (Raso *et al.*, 2000, Anderson *et al.*, 1996). The search for the most robust kinetic model for PEF inactivation of microorganisms is currently on-going. Bacterial cells are generally more resistant to PEF inactivation than yeast cells. However, spores are most resistant to PEF. Vegetative cells at the logarithmic growth stage are more susceptible to PEF inactivation. Increasing the conductivity of the medium generally decreases effectiveness of PEF for microbial inactivation since the higher the conductivity of a medium, the lower the achievable peak electric field at a constant input energy.

Quality Issues in PEF Processing

PEF treatment may influence physical and chemical properties of products. The nature and extent of PEF influence on quality changes are still being actively examined. PEF treatment of various liquid foods including apple juice, orange juice, and milk has not shown any significant physio-chemical changes. PEF processed cranberry juice and chocolate milk retained their physical and chemical characteristics (Evrendilek *et al.*, 2001). There was a slight decrease in vitamin C content in PEF treated orange juice compared to heat treated orange juice (Zhang *et al.*, 1997). Qin *et al.* (1995) reported no apparent change in the physical and chemical attributes of PEF processed milk. Barsotti *et al.* (2002) indicated that PEF treatment of model emulsions and liquid dairy cream may result in dispersal of oil droplets and dissociation of fat globule aggregates. It is uncertain if all necessary chemical analysis has been performed to fully ascertain the effect of PEF on the quality of processed foods. However, the clear consensus is that liquid food products generally retain their fresh-like quality after PEF treatment.

Solid food products undergo significant changes when treated with PEF. Changes in electrical conductivity of the treated vegetable samples indicated increasing cell permeability (Lebovka *et al.*, 2000, 2001). The diffusion coefficient of sugar from beetroot increased from 0.68×10^{-9} up to 1.2×10^{-9} m^2/s after PEF treatment (Jemai, 1997; Gulyi *et al.*, 1994). Elastic modulus of sugar beet decreased after PEF treatment (Bazhal, 2001). The microstructure of salmon and chicken changed considerably due to PEF treatment as the muscle cells decreased in size and gaping occurred (Gudmundsson and Hafsteinsson, 2001). Electric field treatment generally affects biological cell membranes whereas heating destructs the cell walls (Calderon-Miranda *et al.*, 1999a). There is the potential of inducing rheological changes in a product as a result of PEF treatment. This phenomenon depends on the type of product involved and requires detailed investigations.

Optimization of PEF Parameters

Pertinent parameters of PEF from the processing point of view include pulse characteristics, electric field strength and treatment time. The pulse characteristic factors include waveform, frequency, width, and number. Also microbial and product characteristics and design of treatment chamber may influence effectiveness of PEF treatment since it determines distribution of electric fields across the product. Successful application of PEF for a product depends on rational and innovative optimization of these parameters. PEF applications allow for better control of electric power input and effective permeability of cellular membranes (Weaver and Chizmadzhev, 1996) without significant temperature elevation (Knorr, 1999). In general, the transmembrane voltage u_m induced on the cell membrane due to an external electric field is given as (Zimmerman, 1976):

$$u_m \approx \alpha d_c E \cos\theta \qquad (5)$$

where d_c is cell diameter; E is electric field strength; θ is angle between a point on membrane surface and direction of electric field strength; α is parameter depending on the cell shape, $\alpha = 0.75$ for spherical cell and $\alpha = 1$ for rectangular cell.

The smaller the exposed cells in the electric field, the higher the field strength required for creating a critical trans-membrane potential needed for the cell membrane's plasmolysis would be. Mean diameters of microorganisms and biological tissue cells are in the ranges of 10 nm-1 µm and 10 µm-1 mm, respectively (Aguilera and Stanley, 1999). Therefore, understandably high electric field pulses with voltages in the range of 20 - 50 kV/cm are used to kill microorganisms for PEF pasteurization. However, for solid materials such as vegetables tissues, for which cells are usually larger than microbial cells, electroplasmolysis can be obtained at much lower electric field strengths. A number of publications report that electroplasmolysis of vegetable cells may be achieved at moderate electric field pulses with voltages in the range of 0.3 - 3 kV/cm (Papchenko *et al.*, 1988; Ngadi *et al.*, 2001). Use of lower electric field pulses in the range of 0.1 - 0.3 kV/cm have also been reported (Kupchik *et al.*, 1998).

It has been established experimentally that the electrical treatment time needed for electroplasmolysis is inversely proportional to electric field strength; the higher the field strength, the less specific energy consumption needed for achieving the same degree of plasmolysis (Lebovka et al., 2000). In liquids, the extent to which electric field strength can be increased is limited by dielectric breakdown of the products and uncontrolled temperature increase in the products. Therefore it is vital to balance the need for higher electric field with product response. Tables 1, 2 and 3 show typical PEF parameters that have been reported in the literature for yeast, bacteria and enzyme inactivation, respectively, in various liquid foods. Different authors have reported different effectiveness of PEF application for the various products. Electric fields in the range from 12 to 80 kV/cm have been used and up to 7 log reduction has been reported. There are currently inherent practical difficulties involved in optimizing PEF processing of foods. It is difficult to compare treatment results obtained by different authors using different PEF systems. PEF systems are very expensive and it has not been possible to build systems that allow wide variation of pulse parameters. Therefore, studies have been conducted using equipment with restricted range of parameters.

Two experimental methods have been proposed for determination of the optimal field strength E for PEF treatment of different solid food tissues. One of the methods is based on estimation of the characteristic damage time τ and the total energy consumption factor during treatment τE^2 as a function of electric field strength (Lebovka et al., 2002). The other approach is based on estimation of the maximal change of sample disintegration index caused by the energy input during each pulse (Bazhal et al., 2002). The optimal E value for electroplasmolysis depends on the type of tissue and is higher for cells with developed secondary cell wall. The overall goal of PEF treatment objective (for instance microbial inactivation in liquid medium versus textural enhancement in solid medium), the material to be treated and all pulse parameters must be taken into consideration in order to optimise PEF treatment for a product.

Prospects of PEF Processing

PEF processing offers great potential for inactivation of pathogenic microorganisms, enzyme inactivation (or activation), and modification of structural and textural characteristics of foods. Extensive studies have been conducted for liquid products and less for solid foods. There are current and on-going projects internationally with the focus to improve understanding of the technology. Areas of intense current research include development of appropriate robust kinetics for microbial that will allow meaningful comparison of different PEF systems, effect of PEF on a wide variety of microorganisms and products, and development of control devices for PEF systems. There may be some benefit in using PEF in combination with other non-thermal technologies (Bazhal et al., 2001; Hudmundsson and Hafsteinsson, 2001; Knorr and Heinz,

Table 1. PEF inactivation of yeast suspended in various media

Yeast	Suspension media	Process condition	Treatment Vessel	Log reduction (max)	Reference
Aspergillus niger	Peptone	<25°C, E=30 kV/cm, t_i=0.5 µs, f=5-10 Hz, 3000 sq. pulses	Parallel plates chamber, d=1 cm	2.0	McGregor et al. (2000)
Saccharomyces cerevisiae	Orange juice	E=50 kV/cm, t=2 ms	ColPure™ PEF system (continuous, coaxial), d=5 mm, v=100 l/h	2.5	McDonald et al. (2000)
Saccharomyces cerevisiae	Apple juice	>30°C, E=12 kV/cm, 20 sq. pulses	Bench, parallel plates	4.2	Qin et al. (1994)
Saccharomyces cerevisiae	Apple juice	<30°C, f=1 Hz, 150 exp. d. pulses	Continuous, coaxial chamber, V=29 ml, d=0.6 cm, v=2-10 l/min,	7	Qin et al. (1995)
Saccharomyces cerevisiae	Peptone	<25°C, E=30 kV/cm, t_i=0.5 µs, f=5-10 Hz, 3000 sq. pulses	Parallel plates chamber, d=1cm	5-6	McGregor et al. (2000)
Saccharomyces cerevisiae	Bacto peptone	30°C, E=30 kV/cm, t_i=4 µs, 20 pulses, f=250 Hz, t=13 ms	Bench, continuous, 6 tubular parallel chambers, V=41 ml, d=2.3 mm	6.0	Aronsson et al. (2001)

E-electric field strength, t_i-pulse width, t-electrical treatment time, f-frequency, V-volume of chamber, v-flow rate, d-electrodes gap, 'exp. d.'-exponential decay waveform, 'sq.'- square waveform.

Table 2. PEF inactivation of bacteria suspended in different media

Bacteria	Suspension media	Process condition	Treatment Vessel	Log reduction (max)	Reference
Lactobacillus plantarum	Orange-carrot juice	E=35.8 kV/cm, t_i=10.3 μs, t=46.3 μs	Bench, continuous, coaxial, d=0.632 cm	2.5	Rodrigo et al. (2001)
Lactobacullus leichmannii	0.1% NaCl	<35°C, E=20 kV/cm, t_i=3 μs, t=145.6 μs. f=1000 Hz, 12 bipolar pulses/chamber	4 chambers, continuous, d=0.292 cm, v=1 ml/s	1.3	Unal et al. (2001)
Leuconostoc mesenteroides	Orange juice	E=50 kV/cm, t=2 ms	ColPure™ PEF system (continuous, coaxial), d=5mm, v=100 l/h	5.0	McDonald et al. (2000)
Leuconostoc mesenteroides	Bacto peptone	30°C, E=30 kV/cm, t_i=4 μs, t=13 ms, f=250 Hz, 20 pulses	Bench, continuous, 6 tubular parallel chambers, V=41 ml, d=2.3 mm	3.0	Aronsson et al. (2001)
Listeria innocua	Orange juice	30°C, E=30 kV/cm, t=2 ms	ColPure™ PEF system (continuous, coaxial) d=5 mm, v=100 l/h	5.0	McDonald et al. (2000)
Listeria innocua	Milk	17°C, E=41 kV/cm, t_i=2.5 μs, f=10 Hz, 63 pulses	Continuous, V=28.6 ml, d=0.6 cm, v=120 l/h	3.9	Dutreux et al. (2000b)
Listeria innocua	Phosphate buffer	40°C, E=30 kV/cm, t_i=3.9 μs	ColPure™ PEF system (continuous, coaxial) d=5 mm, v=200 l/h	6.3	Wouters et al. (1999)
Listeria innocua	Raw skim milk	<34°C, E=50 kV/cm, t_i=2 μs, f=3.5 Hz, 32 exp. d. pulses	Continuous, V=25 ml, d=0.6 cm, v=0.5 l/min	2.4	Calderon-Miranda et al. (1999a)
Listeria innocua	Liquid whole egg (LWE)	<36°C, E=50 kV/cm, t_i=2μs, f=3.5 Hz, 32 exp. d. pulses	Continuous, concentric, V=25 ml, d=0.6 cm, v=0.5 l/min	3.5	Calderon-Miranda et al. (1999b)
Listeria innocua	Bacto peptone	30°C, E=30 kV/cm, t_i=4 μs, t=13 ms, f=250 Hz, 20 pulses	Bench, continuous, 6 tubular parallel chamber, V=41 ml, d=2.3 mm	3.0	Aronsson et al. (2001)
Listeria monocytogene	Distillated water	<4°C, E=20 kV/cm, t=50 μs, f=30 Hz, exp. d. pulses	Static chamber	4.0	Russell et al. (2000)

Table 2. Continued...

Bacteria	Suspension media	Process condition	Treatment Vessel	Log reduction (max)	Reference
Listeria monocytogenes	Pasteurized whole milk (3.5% milkfat), 2% milk, skim milk (0.2%)	<50°C, E=30 kV/cm, t_i=1.5 μs, t=600 μs, f=1700 Hz, bipolar pulses	continuous, cofield flow, V=20 ml, v=0.07 l/s	4.0	Reina *et al.* (1998)
Listeria monocytogenes	0.1% NaCl	E=20 kV/cm, t_i=3 μs, t=145.6 μs, f=1000 Hz, 12 bipolar pulses/chamber	continuous, 4 chambers, d=0.292 cm, v=1 ml/s	1.1	Unal *et al.* (2001)
Micrococcus luteus	Na_2HPO_4 + NaH_2PO_4	17°C, V=28.6 ml, E=32 kV/cm, t_i=2 μs, 50 pulses	Continuous, concentric chamber, V=1800 ml, d=0.6 cm, v=0.5 l/min	2.4	Dutreux *et al.* (2000a)
Pseudomonas aerugenosa	Peptone	<25°C, E=30 kV/cm, t_i=0.5 μs, f=5-10 Hz, 3000 sq. pulses	Parallel plates chamber, d=1 cm	3.4	McGregor *et al.* (2000)
Salmonella Dublin	Milk	63°C, E=37 kV/cm, t_i=36 μs, 40 pulses	parallel plates	4.0	Dunn & Pearlman (1987)
Salmonella Interitidis	Egg white	<30°C, E=35 kV/cm, f=900 Hz, monopolar exp. d. pulses	V=400 ml, d=0.2 cm	3.5	Jeantet *et al.* (1999)
Salmonella senftenberg	McIlvein buffer	E=28 kV/cm, t_i=15 μs, f=5 Hz, sq. pulses	Static parallel plate chamber, d=0.25 cm	6.8	Raso *et al.* (2000)
Salmonella typhimurium	Distilled water	<4°C, E=20 kV/cm, t_i=50μs, f=30 Hz, exp. d. pulses	Static chamber	6.0	Russell *et al.* (2000)
Staphylococcus aureus	Peptone	<25°C, E=30 kV/cm, t_i=0.5 μs, f=5-10 Hz, 3000 sq. pulses	Parallel plates chamber, d=1 cm	3.4	McGregor *et al.* (2000)
Zygosaccharomyces Bailii	Apple juice Orange juice Grape juice Pineapple juice Cranberry Juice	<22°C, E=32.2-36.5 kV/cm	PEF + high hydrostatic pressure	4.8 4.7 5.0 4.3 4.6	Raso *et al.* (1998)

Table 2. Continued...

Bacteria	Suspension media	Process condition	Treatment Vessel	Log reduction (max)	Reference
Lactobacillus plantarum	Orange-carrot juice	E=35.8 kV/cm, t_i=10.3 μs, t=46.3 μs	Bench, continuous, coaxial, d=0.632 cm	2.5	Rodrigo et al. (2001)
Lactobacullus leichmannii	0.1% NaCl	<35°C, E=20 kV/cm, t_i=3 μs, t=145.6 μs, f=1000 Hz, 12 bipolar pulses/chamber	4 chambers, continuous, d=0.292 cm, v=1 ml/s	1.3	Unal et al. (2001)
Leuconostoc mesenteroides	Orange juice	E=50 kV/cm, t=2 ms	ColPure™ PEF system (continuous, coaxial), d=5mm, v=100 l/h	5.0	McDonald et al. (2000)
Leuconostoc mesenteroides	Bacto peptone	30°C, E=30 kV/cm, t_i=4 μs, t=13 ms, f=250 Hz, 20 pulses	Bench, continuous, 6 tubular parallel chambers, V=41 ml, d=2.3 mm	3.0	Aronsson et al. (2001)
Listeria innocua	Orange juice	30°C, E=30 kV/cm, t=2 ms	ColPure™ PEF system (continuous, coaxial) d=5 mm, v=100 l/h	5.0	McDonald et al. (2000)
Listeria innocua	Milk	17°C, E=41 kV/cm, t_i=2.5 μs. f=10 Hz, 63 pulses	Continuous, V=28.6 ml, d=0.6 cm, v=120 l/h	3.9	Dutreux et al. (2000b)
Listeria innocua	Phosphate buffer	40°C, E=30 kV/cm, t_i=3.9 μs	ColPure™ PEF system (continuous, coaxial) d=5 mm, v=200 l/h	6.3	Wouters et al. (1999)
Listeria innocua	Raw skim milk	<34°C, E=50 kV/cm, t_i=2 μs, f=3.5 Hz, 32 exp. d. pulses	Continuous, V=25 ml, d=0.6 cm, v=0.5 l/min	2.4	Calderon-Miranda et al. (1999a)
Listeria innocua	Liquid whole egg (LWE)	<36°C, E=50 kV/cm, t_i=2μs, f=3.5 Hz, 32 exp. d. pulses	Continuous, concentric, V=25 ml, d=0.6 cm, v=0.5 l/min	3.5	Calderon-Miranda et al. (1999b)
Listeria innocua	Bacto peptone	30°C, E=30 kV/cm, t_i=4 μs, t=13 ms, f=250 Hz, 20 pulses	Bench, continuous, 6 tubular parallel chamber, V=41 ml, d=2.3 mm	3.0	Aronsson et al. (2001)
Listeria monocytogene	Distillated water	<4°C, E=20 kV/cm, t=50 μs, f=30 Hz, exp. d. pulses	Static chamber	4.0	Russell et al. (2000)

Table 3. PEF inactivation of enzymes in various media

Enzyme	Suspension media	Process condition	Treatment Vessel	% Inactivation	Reference
Plasmin	SMUF 100 µg plasmin/ml	15°C, E=30 kV/cm, t_i=2 µs, t=100 µs	Continuous, parallel electrodes, d=6 mm, v=45 ml/min	90	Qin et al. (1996)
Protease (Pseudomonas fluorescens)	Skim milk	50°C, E=15 kV/cm, t_i=2 µs, t=196 µs	Continuous (0.5 l/min), coaxial electrodes, d=0.6 cm	60	Vega-Mercado et al. (1995)
Alkaline phosphatase	Raw milk	22°C, E=18.8 kV/cm, t_i=400 µs, t=28 ms	Batch, d=0.1 cm	60	Van Loey et al. (2002)
Lactoperoxidase	Raw milk	E=21.5 kV/cm, W<400 kJ/l, exp. d. pulses	Batch, parallel electrodes, d=0.5 cm	0	Grahl & Markl (1996)
Lipase	Raw milk	20°C, E=21.5 kV/cm, W≈400 kJ/l, exp. d. pulses	Batch, parallel electrodes, d=0.3 cm	60	Grahl & Markl (1996)
Glucose-oxidase (Aspergillus niger)	Buffer (pH=5.1) + 50 mM Na-acetate	20°C, E=50 kV/cm, t_i=2 µs, t=60 µs, exp. d. pulses	Batch, parallel electrodes, d=0.3 cm	75	Ho et al. (1997)
α-Amilase (Bacillus licheniformis)	Deionized water	20°C, E=80 kV/cm, t_i=2 µs, t=60 µs, exp. d. pulses	Batch, parallel electrodes, d=0.3 cm.	85	Ho et al. (1997)
Peroxidase (soybean)	Buffer (pH=6.0) + 100 mM K-phosphate	20°C, E=73 kV/cm, t_i=2 µs, t=60 µs, exp. d. pulses	Batch, parallel electrodes, d=0.3 cm	27	Van Loey et al. (2002)
Polyphenoloxidase (mushroom)	Buffer (pH=6.5) + 50 mM K-phosphate	20°C, E=50 kV/cm, t_i=2 µs, t=60 µs, exp. d. pulses	Batch, parallel electrodes, d=0.3 cm	40	Ho et al. (1997)
Polyphenoloxidase (apple)	McIlvaine buffer (pH=6.5) + 1 M NaCl + 5% w/v PVPP	<15°C, E=24.6 kV/cm, t_i=20 µs, t=6 ms, bipolar exp. d. pulses	Batch, parallel electrodes, d=0.1 cm	95	Ginner et al. (2001)

Table 3. Continued…

Enzyme	Suspension media	Process condition	Treatment Vessel	% Inactivation	Reference
Polyphenoloxidase (pear)	McIlvaine buffer (pH=6.5) + 1 M NaCl + 5% w/v PVPP	<15°C, E=24.6 kV/cm, t_t=20 μs, t=6 ms, bipolar exp. d. pulses	Batch, parallel electrodes, d=0.1 cm	60	Ginner et al. (2001)
Papain (papaya)	1 mM EDTA	35°C, E=50 kV/cm, t_t=4 μs, t=2 ms, sq. pulses	Continuous, co-field flow tubular electrodes, v=46.2 ml/min	85	Van Loey et al. (2002)
Pectinmethylesterase	Orange juice	60°C, E=35 kV/cm, t_t=1.4 μs, t=59 μs	Continuous, co-field flow tubular electrodes, d=1 cm	88	Yeom et al. (2000)
Lactate dehydrogenase (beef heart)	Buffer (pH=7.2) + 20 mM K-phosphate	30°C, E=31.6 kV/cm, t_t=0.96 μs, t=200 μs, exp. d. pulses	Batch, parallel electrodes, d=0.5 cm	0	Barsotti et al. (2002)

2000; Barbosa-Canova et al., 1999). Applications of the technology for liquid products are advanced. However, there are possible important niche applications of the technology for some solid products. PEF technologies may open the way for development of new products. Although PEF has not been widely used in seafood processing, potential applications and benefits exist.

HIGH PRESSURE PROCESSING

The effect of pressure on food microorganisms has been known since the 1890s. However, industrial interest in high pressure processing has increased only in the last decade. Today, some high pressure processed products are available in the market. High pressure processing is based on isobaric compression of process fluids to exert pressure on food products. In high pressure processing, foods are subjected to high pressures in the range of 100 to 1000 MPa to inactivate pathogenic microorganisms, inactivate (or activate) specific enzymes, modify bipolymers such as protein, and induce controlled changes in quality, structure and texture of products (Knorr and Heinz, 2000; Lechowich, 1993). Two fundamental principles of high pressure processing are the le Chatelier's and Pascal's principles. According to le Chatelier's principle, if a system at equilibrium is subjected to a stress such as pressure, the equilibrium will shift in attempt to reduce the stress. Therefore, any phenomenon such as phase transition, chemical reactivity, change in molecular configuration and chemical reaction accompanied by a decrease in volume will generally be enhanced by pressure. Pascal's principle establishes that when there is an increase in pressure at any point in a confined fluid, there will be an equal increase in pressure at every other point in the container. Thus, pressure can be applied instantaneously and uniformly independent of product size and geometry. A typical high pressure system consists of a high pressure vessel and its closure, delivery systems, process fluid and control devices (Farkas and Hoover, 2000). High pressure can be generated by direct compression, indirect compression, or heating of the pressure medium (Mertens and Deplace, 1993).

There may be several mechanisms of microbial inactivation by high pressure including mechanical rupture of cell membranes (Hoover, 1993). The effects of pressure on microorganisms in food are determined by the effect of pressure on water, temperature during pressure treatment, constituents of food, and the properties and the physiological state of the microorganism (Smelt, 1998). Excellent reviews of the effect of high pressure on microorganisms are available in the literature (Barbosa-Canovas et al., 1997; Kalichevsky et al., 1995).

Applications of high pressure in seafood processing include microbial inactivation, extension of shelf life, textural changes, rheological and biochemical changes, freezing and thawing and development of new products (Amanatidou et al., 2000). Applications of high pressure processing technology in seafood processing has been discussed in great detail (Morrissey et al., 2001; Sequeira-Munoz et al., 2001; LeBail, 2001).

UV LIGHT PROCESSING

The use of UV and full spectrum light in food processing is on the increase due to recent advances in power light generation technology. UV irradiation at wavelengths shorter than 280 nm (referred to as UV-C) is known to have biocidal effects and destroys microorganisms by degrading their cell walls and DNA (Rosenthal, 1992). The amount of cell damage depends on the UV absorption characteristics of the medium, target microorganism's resistance to UV exposure and the applied UV dose. UV-C radiation can be generated using low pressure, mercury vapor fluorescent lamps with transparent and colorless quartz tubes. These lamps emit about 95% of their energy in the 253.7 nm range. More recently, high intensity lamps have been developed with enhanced potential for UV inactivation of bacteria. These lamps are low-cost and easy to apply, thus making UV irradiation technology relatively cheap and readily available. The relevant parameters for design of UV irradiation systems include UV intensity and dose, penetration depth, UV absorption by food, and target microorganism.

Ultraviolet irradiation is currently used widely to disinfect water and waste water systems as alternative to chlorine based applications (Qualls et al., 1983; Loge et al., 1996; Downey et al., 1998). It has also been used in the food industry including dairy, meat and vegetable processing plants. The successful use of UV irradiation to reduce pathogenic microbial populations in red meats, poultry and fish has also been documented (Huang and Toledo, 1982; Stermer et al., 1987; Sumner et al., 1995). The major advantages of UV irradiation include low operating cost, minimal maintenance and higher operational safety. However, UV light has a poor penetration depth especially in opaque food products. There is also the problem of "shadowing" in which rough surfaces provide shade and protect microorganisms from the UV light. Thus, UV light applications are generally limited to transparent products, surface treatment of smooth products or thin film of products. The use of pulsed light technology may reduce the penetration limitations of UV light. Possible applications of UV light in seafood processing range from washing, inactivation of pathogenic microbial load on surfaces of packaged products, for the extension of shelf life.

FUTURE PROSPECTS

To adopt a technology, the food industry typically consider variety of factors including the added value to the product provided by using the technology, improvement in the performance or efficiency of the process, cost of operation and marketing strategies. The emerging technologies offer unique opportunities for innovation and creativity in food processing. They will significantly extend the bar and expectations of food safety and quality. Some new products have already been brought to the market as a result of these new technologies. It is expected that new products will be developed as a result of the technologies. However more work is required to further develop the technologies. It will important to reduce start-up and operational cost of technologies of interest.

REFERENCES

Ade-Omowaye, B.I.O.; Angersbah, A.; Eshtiaghi, N.M. and Knorr, D. 2001. Impact of high intensity electric field pulses on cell permeabilisation and as pre-processing step in coconut processing. *Innov. Food Sci. Emerg. Technol.* 1: 203-209.

Aguilera, J.M.; and Stanley, D.W. 1999. *Microstructural principles of food processing and engineering*. Aspen Publishers, Gaithersburg.

Amanatidou, A.; Schluter, O.; Lemkau, K.; Gorris, L.G.M.; Smid, E.J. and Knorr, D. 2000. Effect of combined application of high pressure treatment and modified atmospheres on the shelf life of fresh Atlantic salmon. *Innov. Food Sci. Emerg. Technol.* 1: 87-98.

Anderson, W.A.; Mcclure, P.J.; Bird-Parker, A.C. and Cole, M.B. 1996. The application of a log-logistic model to describe the thermal inactivation of Clostridium botulinum 213B at temperatures below 121.1°C. *J. Applied Bacteriol.* 80: 283-290.

Angersbach, A.; Heinz, V. and Knorr, D. 2000. Effects of pulsed electric fields on cell membranes in real food systems. *Innov. Food Sci. Emerg. Technol.* 1: 135 - 149.

Aronsson, K.; Lindgren, M.; Johansson, B.R. and Ronner, U. 2001. Inactivation of microorganisms using pulsed electric fields: the influence of process parameters on *Escherichia coli, Listeria innocua, Leuconostoc mesenteroides* and *Saccharomyces cerevisiae*. *Innov. Food Sci. Emerg. Technol.* 2: 41-45.

Barbosa-Canovas, G.V.; Gongora-Nieto, M.M.; Pothakamury, U.R. and Swanson, B.G. 1999. *Preservation of foods with pulsed electric fields*. Academic Press, San Diego, CA.

Barbosa-Canovas, G.V.; Pothakamury, U.R.; Palou, E. and Swanson, B.G. 1997. Nonthermal preservation of foods. Marcel Dekker, New York, NY.

Barsotti, L.; Dumay, E.; Mu T.H.; Diaz, M.D.F. and Cheftel, J.C. 2002. Effects of high voltage electric pulses on protein-based food constituents and structures. *Trends Food Sci. Technol.* 12: 136-144.

Bazhal, M.I. 2001. Etude du mécanisme d'électroperméabilisation des tissus végétaux. Application à l'extraction du jus des pommes. Thèse de Doctorat, Université de Technologie de Compiègne, France.

Bazhal, M.I. ; Lebovka, N.I. and Vorobiev, E.I. 2001. Pulsed electric field treatment of apple tissue during compression for juice extraction. *J. Food Eng.* 50: 129-139.

Bazhal, M.I.; Lebovka, N.I. and Vorobiev, E. 2003. Optimisation of pulsed electric field strength for electropermeabilization of vegetable tissues. *Biosystems Eng.* 86: 339-346.

Bazhal, M.I. and Vorobiev, E.I. 2000. Electrical treatment of apple slices for intensifying juice pressing. *J. Sci. Food and Agric.* 80: 1668-1674.

Calderon-Miranda, M.L.; Barbosa-Canovas, G.V. and Swanson, B.G. 1999a. Transmission electron microscopy of *Listeria innocua* treated by pulsed electric fields and nisin in skimmed milk. *Int. J. Food Microbiol.* 51: 31-38.

Calderon-Miranda, M.L.; Barbosa-Canovas, G.V. and Swanson, B.G. 1999b. Inactivation of *Listeria innocua* in liquid whole egg by pulsed electric fields and nisin. *Int. J. Food Microbiol.* 51: 7-17.

Castro, A.; Barbosa-Canovas, G.V. and Swason, B.G. 1993. Microbial inactivation of foods by pulsed electric fields. *J. Food Pres.,* 17: 47 -73.

Downey, D.; Giles, D.K.; Delwiche, M.J. and MacDonald, J.D. 1998. Development and validation of a general model for predicting biological efficacy of UV reactors against plant pathogens in irrigation water. *Trans ASAE,* 41: 849-857.

Dunn, J.E. and Pearlman, J.S. 1987. *Methods and apparatus for extending the shelf life of fluid food products.* U.S. Patent 4,695,472.

Dutreux, N.; Notermans, S.; Gongora-Nieto, M.M.; Barbosa-Canovas, G.V. and Swason, B.G. 2000a. Effects of combined exposure of *Micrococcus luteus* to nisin and pulsed electric fields. *Int. J. Food Microbiol.* 60: 147-152.

Dutreux, N.; Notermans, S.; Wijtzes, T.; Gongora-Nieto, M.M.; Barbosa-Canovas, G.V. and Swanson, B.G. 2000b. Pulsed electric fields inactivation of attached and free-living *Escherichia coli* and *Listeria innocua* under several conditions. *Int. J. Food Microbiol.* 54: 91-98.

Evrendilek, G.A.; Jin, Z.T.; Ruhlman, K.T.; Qiu, X.; Zhang, Q.H. and Richter, E.R. 2000. Microbial safety and shelf-life of apple juices and cider processed by bench and pilot scale PEF systems. *Innov. Food Sci. Emerg. Technol.* 1: 77-86.

Evrendilek, G.A.; Zhang, Q.H. and Richter, E. 1999. Inactivation of *Escherichia coli* O157:H7 and *Escherichia coli* 8739 in apple juice by pulsed electric field. *J. Food Protec.* 62: 793-796.

Farkas, D.F. and Hoover, D.G. 2000. High pressure processing. *J. Food Sci.* 65: supp. 47-64.

Giner, J.; Gimeno, V.; Barbosa-Canovas, G.V. and Martin, O. 2001. Effects of pulsed electric field processing on apple and pear polyphenoloxidases. *Food Sci. Technol. Int.* 7: 339-345.

Grahl, T. and Markl, H. 1996. Killing of microorganisms by pulsed electric fields. *Appl. Microbiol. Biotechnol.* 45: 148-157.

Gudmundsson, M. and Mafsteinsson, H. 2001. Effect of electric field pulses on microstructure of muscle foods and roes. *Trends Food Sci. Technol.* 12: 122-128.

Gulyi, I.S.; Lebovka, N.I.; Mank, V.V.; Kupchik, M.P.; Bazhal, M.I.; Matvienko, A.B. and Papchenko, A.Y. 1994. *Scientific and practical principles of electrical treatment of food products and materials.* UkrINTEI: Kiev, (in Russian).

Heinz, V. and Knorr, D. 2000. Effect of pH, ethanol addition and high hydrostatic pressure on the inactivation of *Bacillus subtilis* by pulsed electric fields. *Innov. Food Sci. Emerg. Technol.* 1: 151-159.

Ho, S.Y.; Mittal, G.S. and Cross, J.D. 1997. Effects of high field electric pulses on the activity of selected enzymes. *J. Food Eng.* 31: 69-84.

Hoover, D.G. 1993. Pressure effects on biological systems. *Food Technol.* 47(6): 150-155.

Huang, Y. and Toledo, R. 1982. Effect of high doses of high and low intensity UV irradiation on surface microbiological counts and storage-life of fish. *J. Food Sci.* 47: 1667-1669, 1731.

Hülsheger, H.; Potel, J. and Niemann, E.G. 1981. Killing of bacteria with electric pulses of high field strength. *Rad. Environ. Biophys.* 20: 53-65.

Iu, J.; Mittal, G.S. and Griffiths, M.W. 2001. Reduction in levels of *Escherichia coli* O157:H7 in apple cider by pulsed electric fields. *J. Food Protec.* 64: 964-969.

Jeantet, R. ; Baron, F.; Nau, F.; Roignant, M. and Brule, G. 1999. High intensity pulsed electric fields applied to egg white: Effect on *Salmonella enteritidis* inactivation and protein denaturation. *J. Food Protec.* 62: 1381-1386.

Jemai, A.B. 1997. Contribution a l'etude de l'effet d'un traitement electrique sur les cossettes de betterave a sucre. Incidence sur le procede d'extraction. Thèse de Doctorat, Universite de Technologie de Compiegne, France.

Jeyamkondan, S. ; Jayas, D.S. and Holley, R.A. 1999. Pulsed electric field processing of foods: A review. *J. Food Protect.* 62: 1088-1096.

Kalichevsky, M.T.; Knorr, D. and Lillford, P.J. 1995. Potential food applications of high-pressure effects on ice-water transitions. *Trends Food Sci. Technol.* 6: 253-259.

Kinosita, K.; Hibino, M.; Itoh, H.; Shigemori, M.; Hirano, K.; Kirino, Y. and Hayakawa, T. 1992. Events of Membrane electroporation visualized on a time scale from microsecond to seconds. *In:* D.C. Chang, B. M. Chassy, J. A. Saunders, A. E. Sowers (Eds.). *Guide to electroporation and electrofusion.* AcademicPress, New York, NY, pp. 29-46.

Knorr, D. 1999. Novel approaches in food-processing technology: new technologies for preserving foods and modifying function. *Current Opinion Biotechnol.* 10: 485-491.

Knorr, D. and Heinz, V. 2000. Effect of pH, ethanol addition and high hydrostatic pressure on the inactivation of Bacillus subtilis by pulsed electric fields. *Innov. Food Sci. Emerg. Technol.* 1: 151-159.

Kupchik, M.P.; Bazhal, M.I.; Guliy, I.S.; Lebovka, N.I. and Mank, V.V. 1998. Pulse electric treatment effect on food stuff, in *Physics of Agro and Food Products ICPAFP'98*, 1998, Lublin, Poland, p 60.

LeBail, A. 2001. High pressure processing of seafood: A review of potential applications. *Paper presented at the 46th Atlantic Fisheries Technology Conference,* August 26-29, Rimouski, Quebec, Canada.

Lebovka, N.I.; Bazhal, M.I. and Vorobiev, E.I. 2000. Simulation and Experimental Investigation of Food Material Breakage Using the Pulsed Electric Field Treatment. *J. Food Eng.* 44: 213-223.

Lebovka, N.I.; Bazhal, M.I. and Vorobiev, E.I. 2001. Pulsed electric field breakage of cellular tissues: visualization of percolative properties. *Innov. Food Sci. Emerg. Technol.* 2: 113-125.

Lebovka, N.I.; Bazhal, M.I. and Vorobiev, E.I. 2002. Estimation of characteristic damage time of food cellular materials in pulsed electric fields. *J. Food Eng.* 54: 337-346.

Lechowich, R.V. 1993. Food safety implications of high hydrostatic pressure as a food processing method. *Food Technol.* 47: 170-172.

Liu, X.; Youcef, A.E. and Chism, G.W. 1997. Inactivation of *Escherichia coli* O157:H7 by the combination of organic acids and pulsed electric field. *J. Food Safety* 16: 287-299.

Loge, F.J.; Emerick, R.W.; Heath, M.; Jacangelo, J.; Tchobanoglous, G. and Darby, J.L. 1996. Ultraviolet disinfection of secondary wastewater effluents: prediction of performance and design. *Water Environ. Res.* 68: 900-916.

McDonald, C.J.; Lloyd, S.W.; Vitale, M.A.; Petersson, K. and Innings, E. 2000. Effects of pulsed electric fields on microorganisms in orange juice using electric field strengths of 30 and 50 kV/cm. *J. Food Sci.* 65: 984-989.

Martin-Belloso, O.; Vega-Mercado, H.; Qin, B.L.; Chang, F.J.; Barbosa-Canovas, G.V. and Swason, B.G. 1997. Inactivation of *Escherichia coli* suspended in liquid egg using pulsed electric fields. *J. Food Process. Preserv.* 21: 193-208.

McGregor, S.J.; Farish, O.; Fouracre, R.; Rowan, N.J. and Anderson, J.G. 2000. Inactivation of pathogenic and spoilage microorganisms in test liquid using pulsed electric fields. *IEEE trans. Plasma Sci.* 28: 144-149.

McLellan, M.R.; Kime, R.L. and Lind, L.R. 1991. Electroplasmolysis and other treatments to improve apple juice yield. *J. Sci. Food Agric.* 57: 303-306.

Mertens, B. and Deplace, G. 1993. Engineering aspects of high-pressure technology in the food industry. *Food Technol.* 47: 164-169.

Morrissey, M.T. 2001. Strategies to reduce vibrios in shellfish with a special emphasis on high processing. *Paper presented at the 46th Atlantic Fisheries Technology Conference*, August 26-29, Rimouski, Quebec, Canada.

Ngadi, M.O.; Arevalo, P.; Raghavan, G.S.V. and Nguyen, D.H. 2001. Pulse electric fields in food processing. *Paper presented at the ASAE annual international meeting*, July 29 – August 1, Sacramento, CA.

Papchenko, A.Y.; Bologa, M.K. and Berzoi, S.E. 1988. *Apparatus for processing vegetable raw material*. US patent 4787303.

Qin, B.L.; Barbosa-Canovas, G.V.; Swason, B.G. and Pedrow, P.D. 1994. Inactivation of microorganisms by pulsed electric field of different voltage waveforms. *IEEE Trans. Dielectrics Electrical Insulation* 1: 1047-1057.

Qin, B.L. ; Chang, F.J. ; Barbosa-Canovas, G.V. and Swanson, B.G. 1995. Nonthermal inactivation of *S. cerevisiae* in apple juice using pulsed electric fields. *Lebensm. Wiss. Technol.* 28: 564-568.

Qin, B.L.; Pothakamury, U.R.; Barbosa-Canovas, G.V. and Swanson, B.G. 1996. Nonthermal pasteurization of liquid foods using high-intensity pulsed electric fields. *Crit. Rev. Food Sci. Nutr.* 36: 603-627.

Qualls, R. and Johnson, J.D. 1983. Bioassay and dose measurement in UV disinfection. *Appl. Environ. Microbiol.* 45: 872-877.

Raso, J.; Alvarez, I.; Condónm, S. and Sala, F.J. 2000. Predicting inactivation of salmonella senftenberg by pulsed electric fields. *Innov. Food Sci. Emerg. Technol.* 1: 21-29.

Raso, J.; Calderon, M.L.; Gongora-Nieto, M.; Barbosa-Canovas, G.V. and Swason, B.G. 1998. Inactivation of *Zygosaccharomyces bailii* in fruit juices by heat, hydrostatic pressure and pulsed electric fields. *J. Food Sci.*, 63: 1042-1044.

Rastogi, N.K.; Eshtiaghi, N.M. and Knorr, D. 1999. Accelerated mass transfer during osmotic dehydration of high intensity electrical field pulse pretreated carrots. *J. Food Sci.* 64: 1020-1023.

Reina, L.D.; Jin, Z.T.; Youcef, A.E. and Zhang, Q.H. 1998. Inactivation of *Listeria monocytogenes* in milk by pulsed electric fields. *J. Food Protec.,* 61: 1203-1206.

Rodrigo, D.; Martinez, A., Harte, F.; Barbosa-Canovas, G.V. and Rodrigo, M. 2001. Study of inactivation of Lactobacillus plantarum in orange-carrot juice by means of pulsed electric fields: Comparison of inactivation kinetics models. *J. Food Protec.* 64: 259-263.

Rosenthal, I. 1992. *Electromagnetic Radiations in Food Science.* Springer-Verlag, New York, NY.

Russell, N.J.; Colley, M.; Simpson, R.K.; Trivett, A.I. and Evans, R.I. 2000. Mechanism of action of pulsed high electric fields (PHEF) on the membranes of food-poisoning bacteria is an 'all-or-nothing' effect. *Int. J. Food Microbiol.* 55: 133-136.

Sensoy, I.; Zhang, H. and Sastry, S. 1997. Inactivation kinetics of Salmonella dublin by pulsed electric field. *J. Food Process Eng.* 20: 367 - 381.

Sequeira-Munoz, A.; Chevalier, D.; Simpson, B.K.; Le Bail, A. and Ramaswamy, S.H. 2001. Effect of pressure-shift freezing of carp vs air-blast freezing: a storage study. *Paper presented at the 46th Atlantic Fisheries Technology Conference*, August 26-29, Rimouski, Quebec, Canada.

Smelt, J.P.P.M. 1998. Recent advances in the microbiology of high pressure processing. *Trends Food Sci. Technol.* 9: 152-158.

Stermer, R.; Lasater-Smith, M. and Brasington, C. 1987. Ultraviolet radiation – an effective bactericide for fresh meat. *J. Food Protec.* 50: 108-111.

Sumner, S.S.; Wallner-Pendleton, E.A.; Froning, G.W. and Stetson, V.E. 1995. Inhibition of *Salmonella typhimurium* on agar medium and poultry skin by ultraviolet energy. *J. Food Protec.* 59: 319-321.

Unal, R.; Kim, I.G. and Youcef, A.E. 2001. Inactivation of *Escherichia coli* O157: H7, *Listeria monocytogenes,* and *Lactobacillus leichmannii* by combinations of ozone and pulsed electric field. *J. Food Protec.* 64: 777-782.

Van Loey, A.; Verachtert, B. and Hendrickx, M. 2002. Effects of high electric field pulses on enzymes. *Trends Food Sci. Technol.* 12: 94-102.

Vega-Mercado, H.; Powers, J.R.; Barbosa-Canovas, G.V. and Swanson, B.G. 1995. Plasmin inactivation with pulsed electric fields. *J. Food Sci.* 60: 1143-1146.

Weaver, J.C.; Chizmadzhev, Y.A. 1996. Theory of electroporation: a review. *Bioelectrochem. Bioenerg.* 41: 135-160.

Wouters, P.C.; Dutreux, N.; Smelt, I.P.P.M. and Lelieveld, H.L.M. 1999. Effects of pulsed electric fields on inactivation kinetics of *Listeria innocua*. *Appl. Env. Micro.* 12: 5364-5371.

Wu, H.; Kolbe, E.; Flugstag, B.; Park, J.W. and Yongsawatdigul, J. 1998. Electrical properties of fish mince during multi-frequency ohmic heating. *J. Food Sci.* 63: 1028-1032.

Yeom, H.W.; Streaker, C.B.; Zhang, Q.H. and Min, D.B. 2000. Effects of pulsed electric field on the quality of orange juice and comparison with heat pasteurisation. *J. Agric. Food Chem.* 48: 4597-4605.

Zimmermann, U.; Pilwat, G.; Beckers, F. and Rieman F. 1976. Effects of external electric fields on cell membranes. *Bioelectrochem. Bioenerg.* 3: 58-83.

Zhang, Q.; Barbosa-Canovas, G.V. and Swanson, B.G. 1995. Engineering aspects of pulsed electric field pasterization. *J. Food Eng.* 25: 261-281.

CHAPTER 3

TECHNICAL POSSIBILITIES FOR INTRODUCING TRACEABILITY IN THE SEAFOOD INDUSTRY

T. Børresen and M. Frederiksen

Danish Institute for Fisheries Research,
Department of Seafood Research, DTU, Lyngby, Denmark

It is necessary to trace the identity and any given property of a product back to its origin, and to document what has occurred during production. A 'track' will have to be laid out throughout the production chain, delivering agreed documentation about the product to the next unit in the chain, with the objective that the last unit be able to trace any feature of the product back to its origin. A traceable unit must be defined, and the actual production chain analyzed in order to introduce traceability at an appropriate level. To handle the large amounts of data, advanced information technology must be used. However, given the present day technical possibilities of applying barcodes and information technology based languages for business communication, there will be a question of who shall have access to which data, and who 'owns' the data handled. This may influence the technical solutions selected. This has been tested in a Danish fresh fish chain as to how the concept of traceability may be introduced throughout the various links. Traceability should provide origin and authenticity of any given product, but this may also be achieved by other methods based on sampling and chemical analysis. Biomarkers may similarly be used to trace specific features in a given product.

INTRODUCTION

Fish is a highly perishable commodity and unless it is preserved in some way it easily becomes inedible (Huss, 1995). Fresh fish has still been traded, but it has often been an open question as to how fit it might be for human consumption, particularly in areas where the temperature is high and spoilage is rapid. Understanding the mechanisms of spoilage and introducing ice to lower the temperature in order to prolong shelf life has helped, but created new problems as several days may pass until the obvious signs of spoilage becomes evident. As better inspection and quality assessment methods have become available (Warm *et al.*, 1998), it is now easier to check a given quality at any stage in the spoilage process. However, inspection and analytical investigation is time consuming, and, in some cases, safety aspects and other characteristics of the product can be just as interesting.

The trend today is that consumers want more and more information about the products they purchase. When it comes to food, this not only concerns the composition of the product, but also its origin and how it has been processed and stored. Further, in some cases political considerations are also of interest. If a product has any defect of any kind, the consumer as well as the retailer want to know where the specific item has been produced, and why exactly this item had a defect. All these circumstances mean makes it necessary to be able to trace any product back to its origin, and to document what has occurred to it during production. The term used is 'traceability', and virtually it means that every unit in a production chain delivers documentation about the product to the next unit in the chain, with the objective of the last unit to be able to trace any feature of the product back to its origin (Moe, 1998). There is also a question of authenticity when the product is a filleted fish and it may be difficult to distinguish e.g. a plaice from another flatfish. There are several methods to determine whether a given species is what it is said to be, of which protein composition or the DNA pattern are the most common approaches (Dahle *et al.*, 1999).

Introducing traceability in the fish production chain is highly needed. This includes all parameters encountered from the moment the fish leaves the water, or even before, and until the final product is presented to the consumer. A number of players, from fishermen, traders, producers and retailers, among others, are often involved.

Defining traceability

According to the International Standardization Organization (ISO, 2000), traceability is defined as:
"The ability to trace the history, application or location of that which is under consideration. When considering a product, traceability can relate to:

- *The origin of material and parts*
- *The processing history*
- *The distribution and location of the product after delivery"*.

For fish and seafood as highly perishable commodities, it is important that the time/temperature conditions are controlled and included as part of the history of the product. Further, food safety is an important aspect, so the documentation of a safe origin and that contamination during processing and transport has not occurred, must also be included. Finally, the origin and authenticity of the product must be guaranteed, so these identifications are necessary.

As it is today, food safety may be considered the driving force for introducing traceability, but in the long run, when applicable systems have been introduced, it is more likely that traceability will be a tool for guaranteeing quality and to set the proper relation between price and quality in commercial relations. For a single producing company it will further aim in protecting the brand, and make it possible to deliver required information urgently when needed, e.g. in connection with possible product defects or any filed reclaims from a customer. This may save the image of the producer and avoid unwanted public concern. Finally, proper traceability inside a company can be used to optimize production efficiency and thereby the costs.

In order to be able to trace something, a track is needed. One may say that all the information collected and recorded along the production chain constitutes the track. This forms the basis for the track and trace concept, where one is tracking forwards and tracing backwards. However, in order to establish a proper documentation system, it is necessary to define a traceable unit.

Defining a traceable unit

In a homogeneous production where all lots of raw material pass through the production line in a series of consecutive operations, ending up in products in the same order as when entering production, traceability is done along a single track (Frederiksen and Bremner, 2001). When the production is splitting the raw material up into several fractions, of which some may be mixed with fractions from other or different raw material lots, traceability becomes more complicated. In some cases, involving intermediate storage for some time, due to ripening or other necessary developments, things may become even more complicated. It is thus necessary to analyze the production in question in order to decide how traceability should be dealt with. In the fish processing area this could be the white fish filleting industry, the herring industry, salmon processing, shrimp processing, or any of the other, varied processing operations in the seafood industry.

In the simplest case, it may be sufficient to define the traceable unit as the lot passing through production. In more complicated operations it may be necessary to define production batches as the traceable unit, and keep a track record of how and when the different parts were handled. However, even in the simplest production line, it may be necessary to monitor the production continuously, as variation in processing conditions may lead to products with changed features. In such cases time series may be considered. In extreme cases the traceable unit could be defined as the single product item leaving the production facility. This is usually not necessary, but it is important to make a proper analysis of the situation such that true traceability is possible according to specifications set by the buyer. This will always be the next unit in the product chain, which could be a wholesaler or a retailer, but ultimately the end customer, who is the consumer, always sets the specifications. To sum up, these specifications must be set according to quality, previous history, origin and authenticity.

Traceability and quality assurance

In the same way as the new concepts of quality assurance like e.g. the HACCP principle (Huss *et al.*, 2000) has been introduced to ascertain that all products leaving the production are up to specified standards, the concept of traceability aims at ascertaining that every item produced is traceable back to its origin. This means that it is not only necessary to guarantee that a given product type is up to a given quality standard, but to actually identify where each item originates from and how it has been processed.

In order to do so, it is necessary to focus more on the traceable unit, entering into the production system, and keep track of its movement through the processing. This adds a new dimension to the product, allowing identification of each unit produced. In addition to being a powerful tool for documenting the exact origin of the product, it also allows the processor to more easily identify where in the production any irregularity may have occurred and to determine when it happened. This can be used to prevent similar incidents to occur in future production. It should also be mentioned that it is necessary to include the common principles of quality assurance (Frederiksen *et al.*, 1997).

Using information technology

As is evident from the already described requirements for the introduction of traceability in the fish processing industry, large amounts of information need to be handled. This applies both among different units in the product chain and within each unit in the chain. Advanced methods of information technology related to data retrieval, handling and storage are therefore needed. In addition, transfer of data among units calls for a common language and harmonization of terms used. An example of easy data transfer together with the physical unit being handled among different partners in a production chain is the bar code system. Existing data can thus be easily picked up and transferred to the next link in the chain as required.

Thus, fish processing operations within a given plant should be analyzed using the language and tools commonly applied within information technology. Electronic data processing rely on a digital basis, meaning that all information need to be transferred to digital form. This should be considered already at the production planning stage, such that equipment and other facilities are prepared for digital operation. Further, communication external to the physical processing facility should be made Internet based, in order to be as flexible and versatile as possible. As the development is very fast within this area, any processor wanting to introduce traceability should be prepared to meet the challenge of ever changing systems of information technology.

Systems for handling data between many different dealers and across borders already exist. The global EAN.UCC system (European Article Numbering – Uniform Code Council) has been developed for distributing data all over the world, and uses an agreed bar code labeling system. Further information is given on the Internet at <www.ean-int.org>.

A final possibility of applying information technology concerns the possibility of simulating production at any stage. Future developments may thus be tried out using electronic modeling before actual production starts. This will reduce costs at start up and help decide which information should be retrieved for a given production.

A case study of traceability in a fresh fish chain

A simple system for introducing traceability in a fresh fish chain has been worked out and tested in Denmark (Frederiksen *et al.*, 2002). On board the fishing vessel the fish is sorted by species and iced in boxes. Each box is labeled with information on fish species, catch date, vessel number, and box number, readable as ordinary numbers and in the form of a bar code. The information is entered into a computer, and the data transmitted via a wireless mobile phone system to the computer system at the shore station. When the boat is docking, different species are sorted according to size, taking care of keeping fish from each day of catch separate. The fish is then again iced in boxes, with new labels attached, and the new data entered into the computer system.

The boxes are now ready for auctioning with documented content in each box. In the Danish chain the boxes are distributed through a wholesaler and further on to a retailer. In this case the retailer was a supermarket with fish being sold in a shop in shop concept. This means that the fish were removed from the boxes, the bar codes read and the data entered into the central computer system. The final customer could then when browsing the available fish in the shop get full information of when the fish had been caught, by which boat, and how it had been handled on the way to the fish market. When buying the fish, the customer would get a label with the needed information printed on it.

The experience with the system when tested has been very positive. In the present case all computers were hooked up into one central system, allowing each party to watch the full history of the fish from the catch to the consumer. Not only could the consumer see who caught the fish, and when, but the fisherman could also see where and when his fish was purchased by the customer. This new system of providing easy accessible information is considered to be of even greater importance at the catering and restaurant market in the future. The waiter can then not only tell exactly where the wine being served with the food comes from, but also where the fish originates!

Identification and authentication

In addition to tracing products as described above, it is sometimes desirable to identify fish product as being produced from a specific species. There are a number of techniques available to identify which species has been used in any given product. This is particularly important in products where all external characteristics of the fish have been removed, as in the form of fillets or mince. With some techniques it is possible to determine which individual species have been mixed in the minced products.

A fairly easy technique to use on raw fish material is isoelectric focusing (Rehbein *et al.*, 1999), where the specific band pattern is used to distinguish among the species. This technique may also be applied on cooked products, but for this purpose DNA typing techniques are more powerful (Rehbein *et al.*, 1997). The DNA techniques are also suitable for determining specimens on the subspecies level, which is sometimes desirable such as that for tuna species.

Proving the authenticity of the product in this way is useful for proving that a product is actually what it is said to be, thus having 'traced' it back to its origin. However, nothing is known about the intermediate steps in the production chain, so this form of authentication can only complement the real traceability.

Biomarkers and traceability

In the same way that checking authenticity by various methods is possible for intermediate or end products, it is also possible to analyze other components that may reveal contaminants, traces of unwanted substances, or specific processing operations that have been performed. This is done by identifying one or several biomolecules, which are then called biomarkers. Their presence at various steps in the product chain can be used to 'trace' the product through the chain.

An example of the use of biomarkers is the analysis of gene products that can make it possible to decide whether genetically modified organisms (GMO) have been used or not (Miraglia, 1998). Another example is the production of aquacultured fish, where specific feed components can be identified in the final product. Aquacultured fish are produced tailor made to meet the specifications of the end user. Thus, if it is required that only marine feed ingredients be present the use of any feed of plant origin can be revealed through analysis of components occurring only in plants. The same technique can be considered when introducing organic production, where it can be checked that the required production methods have been used. However, as for methods of authentication, the use of biomarkers will always complement the real principles of traceability.

CONCLUSIONS

Increasing consumer demands for information about the origin and the history of processing require new methods of documentation. The new requirements can be met by introducing traceability in the production chain. The concepts of traceability build on methods of quality assurance, but go further than presently used. In order to handle the large amounts of data involved, it is necessary to use advanced information technology. This can also be applied for modeling production. In addition to introducing traceability, identification and authentication of products can be done by analyzing intermediate or final products, using e.g. DNA typing techniques. Finally, biomarkers can also be used to identify components throughout a given production chain.

ACKNOWLEDGEMENT

The authors want to thank their colleagues at the Danish Institute for Fisheries Research (DIFRES) for valuable background discussions to this paper. In particular should be mentioned Senior Consultant Erling P. Larsen and Chief of Information Technology Group at the institute. Further, Professor Allan Bremner, who visited DIFRES in the period 1996 to 2000, and who prepared the ground for the present research on traceability, is greatly indebted for inspiring the work.

REFERENCES

Dahle, G.; Etienne, M.; Woolfe, M.; Skarpeid, H.-J.; Downey, G.; Hildrum, K.I.; Lumley, I. and Pitcher, B. 1999. Food authenticity. Fish, shellfish, and fish eggs. In *Food Authenticity - Issues and Methodologies*. Lees, M. (eds), Eurofins Scientific Nantes, France, pp. 105-117

Frederiksen, M. and Bremner, H.A. 2001. Fresh fish distribution chains. *Food Australia*, 54: 117-123

Frederiksen, M.; Popescu, V. and Olsen, K.B. 1997. Integrated Quality Assurance of Chilled Food Fish at Sea. In *Seafood from producer to consumer, integrated approach to quality*. J.B. Luten; T. Børresen; and J. Oehlenschläger (eds.) Elsevier, Amsterdam, pp. 87-96

Frederiksen, M.; Østerberg, C.; Silberg, S.; Larsen, E. and Bremner, H.A. 2002. Development and validation of an Internet based traceability system in a Danish domestic fresh fish chain. *J. Aquatic Food Prod. Technol.* 11: 13-34.

Huss, H.H. (Ed.) 1995. *Quality and Quality Changes in Fresh Fish*, FAO Fish. Technol. Pap. No. 348. 195 pp.

Huss, H.H.; Reilly, A. and Ben Embarek, P.K. 2000. Prevention and control of hazards in seafood. *Food Control* 11: 149-156

ISO (2000) *Quality management systems–Fundamentals and vocabulary*, Brussels, Belgium

Miraglia, M.; Onori, R.; Brera, C. and Cava, E. 1998. Safety assessment of genetically modified food products: an evaluation of developed approaches and methodologies. *Microchem. J.* 59: 154-159

Rehbein, H.; Kress, G. and Schmidt, T. 1997. Application of PCR-SSCP to species identification of fishery products. *J. Sci. Food Agric.* 74: 35-41

Rehbein, H.; Kündiger, R.; Malmheden-Yman, I.; Ferm, M.; Etienne, M.; Jerome, M.; Craig, A.; Mackie, I.; Jessen, F.; Martinez, I.; Mendes, R.; Smelt, A.; Luten, J.; Pineiro, C. and Perez-Martin, R. 1999. Species identification of cooked fish by urea isoelectric focusing and sodium dodecylsulfate polyacrylamide gel electrophoresis: a collaborative study. *Food Chem.* 67: 333-339.

Moe, T. 1998. Perspectives on traceability in food manufacture. *Trends Food Sci. Technol.*, 9: 211-214

Warm, K.; Bøknes, N. and Nielsen, J. 1998. Development of Quality Index Method for evaluation of frozen cod (*Gadus morhua*) and cod fillets. *J. Aquatic Food Prod. Technol.* 7: 45-59

CHAPTER 4

FOOD QUALITY AND GLOBAL TRACEABILITY FOR E-COMMERCE THROUGH AUTOMATIC DATA CAPTURE PROCESS: NEW BAR CODE AND RFID TECHNOLOGIES

G. Doyon[1], P. Veilleux[2] and A. Clement[1]

[1] Canada Ministry of Agriculture and Agri-Food
Food Research and Development Centre (FRDC)
St-Hyacinthe, Quebec J2S 8E3 Canada
[2] Electronic Commerce Council of Canada
Anjou, Quebec H1K 1A1 Canada

The efficiency of technologies and engineering in production and processing, the supply and logistic chain, and also the automatic identification systems (AIS) are all vital components for the global traceability of food commodities such as fish, fish products, packages and vehicles. The combined Bar code (BC) and radio frequency identification (RFID) technology systems as a sound base for the development of tracking tools are required for ISO and HACCP standards and certification programs in the food and fisheries industry. Some of the various applications, benefits, difficulties and challenges are discussed. Finally, the importance of e-commerce (EDI) for small and medium enterprises (SME), data repository, alignment, integrity, standards, image capture, reduce surface symbology (RSS), global trade item number (GTIN) and global location number (GLN) are introduced.

INTRODUCTION

Canada has a leading position internationally for export of food commodities and processed products of excellent quality because of the continued consumer or client satisfaction. Also, the food supply chain is generally safe and consistent. This reputation is most precious to professionals and our foods supply chain partners. One out of seven jobs is from the Agri-Food sectors in Canada and the total business value of our food chain is worth $130 billion annually.

Recent health related diseases and contamination undertakings have happened in far away countries (Belgium, France, England) and those close to us (Brazil, Argentina) and in our backyards (Ontario, Saskatchewan) that awake local, Provincial and Federal agencies. Global traceability is now on everybody's mind. It is an urgent matter. How to do it, with whom and what are choices? Is it costly? Still, today, many individuals are mixed up with ISO and HACCP definitions of terms and their applications.

What has helped better quality structure and management, in the last five years, is a renewal of R&D activities and export initiatives. Better products and process controls have been developed and put in place. From farms, fishermen to shippers' containers, quality and environment abuses must be monitored in batch or real time for insurance claims and liability. Raw materials such as seeds, fertilizers, plants, fruits and vegetables, trees and flowers, animals, meat, ingredients, genetically modified foods (GMF), fish stocks, processed products and also liquids (milk, wine, juice, concentrates) are under scrutiny. More complex integration is required considering the material, the package and vehicles for a well-defined shelf life.

Today's quality control (QC) as well as quality assurance (QA) departments are deeply involved with ISO and HACCP standards and procedures and, with regional and federal laws and regulations. They require documentation with often cumbersome paper format. In recent years, automatic identification (auto ID) has become very popular in various service industries, purchasing department, distribution and food chain supply management, manufacturing process control and material flow systems. Also of great importance in agri-food industry, is to have pertinent information about people, animals, goods, product and vehicles on location in storage and in transit.

The traditional method of entering data into a computer or handheld assistant involves manually keying (TMK) the information using the keyboard and in many occasions are gathered on sheets of paper or instruments. Unfortunately, studies show that the error ratio is 1:300 characters entered. Furthermore, it is very difficult to obtain real-time data entry. To offset the disadvantages of the manual methods, various technologies have been designed and are of current use.

A key element to our success is in the selection of the ideal identifier (Tag/Label id) that shall remain intact in the food supply chain and resist multiple abuses. Proper reading materials and software to decode the data, printer, communications methods and protocols, are to be optimized.

SUPPLY CHAIN BAR CODES AND THE ELECTRONIC COMMERCE COUNCIL (ECCC)

Who we are and structure
Electronic Commerce Council of Canada (not for profit corporation)

Board of Governors
↓
ECC Network services (not for profit corporation)
(ECC Net)
Wholly-owned subsidiary

The ECCC is a Standards Organization providing:

- Company Prefixes, GTIN, GLN
- Development and Publication of Standards & Guidelines on EDI and Bar Codes
- Harmonisation with other countries initiatives
- Operates ECC Networks Services (ECC Net)
    - Image and Data capture (ECC Net I&D)
    - Data Synchronization (Alignment)
    - Marketing information

ECCC Network services (ECCnet) has a goal to secure, web-based National Registry of Product & Images where trading partners exchange information about:

- Products
- Services
- Parties or Locations

It also insures Data Integrity needed for e-commerce transactions by providing a single standardized record of:

- Products Listing Data
- Image & Dimension Data
- Marketing & Nutritional Data
- Global Location Data

Finally, ECC net broadcasts new product information and products changes to participant members.

Other services of ECC net I&D include:

- Database of Product Attributes
- Low-resolution and high-resolution Images and Dimension Capture
- Monthly updates
- Store audits
- Bar Code verification service
- On-line access (coming soon)
- Expanding to include new channels

Joint EAN.UCC system and ECCC initiative

What is the EAN.UCC system? It is an internationally compatible numbering and barcode system for identifying items, processes and services.

It provides a common language of communication for trade and electronic commerce.

The goal of the EAN.UCC (ECCC) initiative is to improve supply chain management and other business processes that reduce costs and/or add value for both goods and services. It also develops, establishes and promotes global, open STANDARDS for identification and communication for the benefit of the user involved and ultimately the consumer.

The Global Trade Item Number (GTIN) is the new term for the 14 digit data structure to uniquely identify trade items (products & services) within the EAN.UCC system. This is a new term only; not a change in standards.

Globally unique 14 digit number to identify trade Items (products and services) are the data structure within GTIN:

- EAN.UCC-14
- EAN.UCC-13
- UCC-12
- EAN.UCC-8

Symbologies

    UPC version A

    UPC version E

Interleave two of five (ITF – 14)

    UAN.UCC - 128

The Global Location Number (GLN) is an EAN.UCC Standard for uniquely identifying organization location and sub-location. For example:

- Company Head Office
- Division within the company
- Each Warehouse and Supermarket
- Farm & Barn and sub-units
- Slaughterhouse
- Etc

The Reduced Surface Symbology (RSS) family contains three linear symbologies to be used with the EAN.UCC system.

RSS-14 encodes the full 14-digit EAN.UCC item identification in a linear symbol that can be scanned omni-directionally by suitable programmed point-of-sale scanners. It also has truncated and stacked format.

The potential benefits are:

- Increased front-end accuracy and speed of transactions
    o Reduced keying errors & inaccurate identification
    o Right price change for the right product
    o Increased consumer confidence
- Enables reporting of item level sales with company and item numbers
- Ability to trace products through the entire supply chain
- Improve ordering process
- Promote the use of EDI

45

RSS-14 (01) 00614141000012

RSS-14 truncated (01) 00614141000012

RSS-14 stacked omni-directional (01) 00614141000012

RSS Limited (01) 00614141000012

RSS Expanded (01) 00614141000012

RSS Expanded Stacked (01) 00614141000012

The RSS family of symbols

Where we are:

1) Retailers, suppliers and vendor participation continues to grow in the User Group
2) RSS has become known within the Grocery, Meat and Produce industry. Awareness has risen dramatically!
3) A pilot project is under way (lab testing)
4) Equipment providers are beginning to step-up their RSS initiatives to meet growing retailer demands
5) RSS initiatives have begun in Europe
6) A cost/benefits analysis was sent out to consulting firms and universities

Upon exhaustion of the UPC system, the numbering organizations: UCC (US) & ECCC (Canada) will start the assignment of Company Prefixes with EAN/UCC-13's.

The retail POS will scan and store the EAN.UCC-13.

There will be expansion of your database GTIN field to 14-digits.

One must remember that:

- GTIN is a global name for trade item (product or service) identification
- GTIN is a data structure, not symbology
- GTIN is the output from the scanners
- GTIN requires a 14-digit file structure
- Company prefix can be of variable length

How we can collaborate in Traceability

The EAN.UCC System with Global Trade Item Number (GTIN) & Global Location Number (GLN)
Images & Bar codes
Farm – Slaughterhouse – processing plant – Distributor and Consumer integration
Reduces surface symbology
Lot and Herd Number

Data hosting and synchronization, including tags
Development of standards on Traceability
Harmonisation with other countries' initiatives

Success factors

Success will be achievable with:

Industrial Sectors Endorsement & Credibility
Support from all major producers, manufacturers and distributors
Data Integrity and quality of images
Product updates
Level of confidentiality and accessibility of data
Creation of users' committees to address important issues
On-going combination and visibility with our partners
Minimal fee structures
Providing custom-made solutions

AUTOMATIC IDENTIFICATION SYSTEM (AIS)

Choices and interconnectivity

One element of success in our quality and logistic system is in the choice of the automatic identification system (AIS) built in with interconnectivity with instruments, devices and networks. Our main focus today being bar coding (BC) and Radio Frequency (RF) identification systems implementation with electronic business (e-commerce) or electronic data interchange (EDI). The importance of the Electronic Commerce Council of Canada (ECCC) as a partner to implement today's and tomorrow's supply chain technologies and traceability for our food industry is also here with documented.

❖ Automatic Identification and Capture: "Can be referred to as systematic (AIC) data collection by means of other than manual method or keyboard."

❖ Our goal is: "A reliable identification of physical objects (entity) in batch or real-time with some meaningful details."

❖ System "A system is a group of components organized in such a way as to produce specific and measurable results on a constant basis."

Important ISO-8402 Definitions

ENTITY
Item which can be individually described and considered.
Note: An **entity** may be, for example: - an activity or a process, - a **product**, - an **organization**, a system or a person, or – any combination thereof.

QUALITY
Totality of characteristics of an **entity** that bear on its ability to satisfy stated and implied needs.
Notes: 1. In a contractual environment, or in a regulated environment, such as the nuclear **safety** field, needs are specified, whereas in other environments, implied needs should be identified and defined. 2. In many instances, needs can change with time; this implies a periodic review of **requirements for quality**. 3. Needs are usually translated into characteristics with specified criteria (see **requirements for quality**). Needs may include, for example, aspects of performance, usability, dependability (availability, reliability, maintainability), **safety**, environment (requirements of society), economics and aesthetics. 4. The term "**quality**" should not be used as a single term to express a degree of excellence in a comparative sense, nor should it be used in a quantitative sense for technical evaluations. To express these meanings, a qualifying adjective should be used. For example, use can be made of the following terms: a) "relative **quality**" where **entities** are ranked on a relative basis in the degree of excellence or comparative sense (not to be confused with **grade**); b)"**quality** level" in a quantitative sense (as used in acceptance sampling) and "**quality** measure" where precise technical evaluations are carried out. 5. The achievement of satisfactory **quality** involves all stages of **quality loop** as a whole. The contributions to **quality** of these various stages are sometimes identified separately for emphasis; for example, **quality** due to definition of needs, **quality** due to product design, quality due to conformance, quality due to referred to as "fitness for use" or "fitness for purpose" or "customer satisfaction" or conformance to the requirements". These represent only certain facets of **quality**, as defined above.

TRACEABILITY
Ability to trace the history, application or location of an **entity** by means of recorded identifications.
Notes: 1. The term "traceability" may have one of three main meanings: a) in a product sense, it may relate to – the origin of materials and parts, -- the product processing history, -- the distribution and location of the product after delivery; b) in a calibration sense, it relates measuring equipment to national or international standards, primary standards, basic physical constants or properties, or reference materials; c) in a data-collection sense, it relates

calculations and data generated throughout the **quality loop** sometimes back to the requirements for quality for an entity. 2. All aspects of traceability requirements, if any, should be clearly specified, for example, in terms of periods of time, point of origin or identification.

Planning and designing for a logistic system

One must consider four (4) principal components such as:

The physical environment
Control and process
Staff and training
Storage and distribution (individually and with the clients)

The analysis of the current operation

Facility layout: size, height, material and data flow
Critical objectives: congestion, damages, noise and safety
Storage locations: floors, pallets, racks, carton flow, bins, shelving and carousels
Movement: lift trucks, pallet jacks and conveyors...
Product profile: number of units, classification
Other elements: shelf life, lot, codes, security number, seasons and readers ...

General description of the operations

- o Various activities
    - Reception, collection
    - Preparation, storage
    - Expedition, shipping

- o Resources
    - Human and physical
    - Management/computer/maintenance

- o Data collection
    - How, when, ... cost
    - Client needs

Execution time ...

- o Priorize tasks per employee and per unit of process

- Equipment profile
- Material profile
- Location profile
- Operator/equipment (choice)
- Operator/tasks priority
- Work standardization/differentiation

Labels and tags (coding)

Types of identifiers
Location
Plate number
Data collection
Action taken/correction
Transfer/expedition

Do we have to choose a Code or/and Tags Today?

1. The general purpose of the code?
2. How much space on it for the producer, manufacturer and product itself?
3. Label use or a direct print?
4. Combination with RFID or not?
5. Utilisation of coding?
   - Internal use?
   - External use also? Must use a standard?
6. Data to be encoded?
   - Alpha, numeric, alphanumeric, constant field length, variable
   - Logos, photos

RFID Principle

7. Printing results
   - Batch, on demand, random, sequential, colour?
8. Selecting a symbol?
   - Industrial specifications
   - Customer specs
   - My internal specs only
   - What is the rest of the industry doing?
   - What does the customer want?

51

RFID Principle
Principe RFID

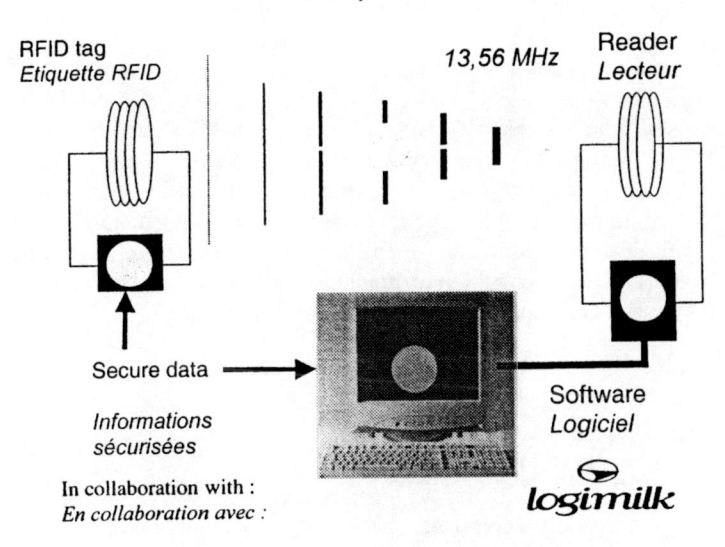

RFID tag / *Etiquette RFID*   13,56 MHz   Reader / *Lecteur*

Secure data
Informations sécurisées

Software / *Logiciel*

In collaboration with :
En collaboration avec :

logimilk

Overview of ID systems

The Auto ID systems (6)

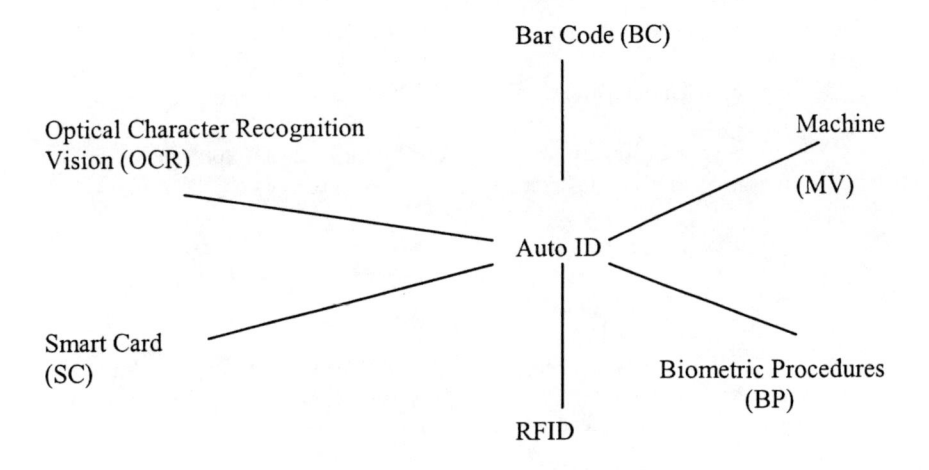

Bar Code (BC)

Optical Character Recognition Vision (OCR)

Machine (MV)

Auto ID

Smart Card (SC)

Biometric Procedures (BP)

RFID

1. Two bar codes basics (linear and 2-dimentional) (BC)

One-dimensional

This is a machine-readable symbol consisting of a series of parallels, adjacent bars and spaces. Predetermined width patterns are used to represent data in a symbol. The narrow bars representing dots and wide bars representing dashes. A scanning device – light pen – or camera "take a picture of". We need a reading and decoding software to analyze the information. We have one-dimensional character recognition technique since data need to be acquired from the widths on lines and spaces along the light-scanning path.

U.P.C. (Universal Product Code)   1973
First 6 Manufacturer
12 characters
    Next 5 Product

Twelfth character: check

Two-dimensional (2D)

Will combine widths of lines with a parallel arrangement of varying height bars. In this case, there are multiple rows of width-modulated bars and spaces. The encoding is into a 2D pattern of data cells (X's and Y's). The cells have different colors (black and white) and shapes (squares, dots or polygons).

2. Optical Character Recognition (OCR)

It is also a 2-dimensional technology with both horizontal and vertical feature of a printed character.

Eg., Passport, bumpy bar code or embossed.

3. Biometric Procedures (BP)

It is the science of counting and body measurement procedures that involves human beings. We compare unmistakable and individual physical characteristics.

In more recent years, genetic identification, i.e. DNA fingerprinting, antibody fingerprinting, various makers and instrumentations were developed to identify or trace humans, animals, plants.

Ex: Fingerprint, handprint, retina (iris) identification, and, more popular are voice recognition, animal, race and herd identification.

Smart Cards (SC)

It is an electronic data storage system, possibly with additional computing capacity (microprocessor card) that is incorporated into a plastic card like a credit card.

Eg., Prepaid telephones smart cards, medical data and security access (doors).

4. Machine Vision (MV)

It is used in manufacturing to sort, to inspect or to measure products automatically. They consist of high-resolution camera (CCD) interfaced with software and computer.

5. Radio Frequency Identification (RFID)

It can read data from tags that are not optically visible to humans. A radio signal is transmitted toward the tag. It responds with a radio signal that is modulated with the instructions stored in the tag, for example, livestock, packages, manufacturing, warehousing and transportation.

Tags are either called "active" or "passive" depending on whether or not they have a battery.

Some frequency ranges are either very low (VLF <= 300KHz; navigation, animal tags), very high (VLF >= 200MHz; radio, TV, FM) and ultra high (UHF; >300 MHz & < 1 GHz; data, communications) or >= 1 GHz & < 30 GHz, microwave. A read/write tag containing a large amount of data can be considered a database. It can be used to find items that may be difficult to locate.

Operation and target frequencies

This is to make sure that no system interferes with nearby radio and television, mobile radio services (police, security services), marine aeronautical radio and mobile phones.

There are frequencies classified worldwide as ISM frequencies ranges (Industrial-Scientific-Medical) and can be used for RFID applications.

Most important ranges for RFID are <= 135 KHz, 13,56 MHz, 27,125 MHz, 40,68 MHz, 433,92 MHz, 869,0 MHz, 915,0 MHz, 2,45 GHz, 5,8 GHz and 24,125 GHz.

The most commonly used by the industries are 125 – 134,2 KHz (animal identification, gas bottle) from ISO 11784-5; 13,56 MHz (baggage, parcels, documents, tickets), 888-889 and 902-908 MHz, 860-950 (long range needs, containers, supply chain), 915,0 MHz (metal), 2,45 GHz (laundry, baggage, metal) and 5,8 GHz.

Challenges and benefits in implementing the Agri-Food and Fisheries Industries

o Some global and costly elements for all users, if we don't

- Theft
- Multiple interests/users
- Counterfeiting
- Trade (regional, international and global demands)
- Adulteration/authentication
- Automatic information (data) exchange
- Return on investments (ROI) per sector
- Traceability at many levels
- Multiple technologies vs choices
- Country's laws and regulations

RFID for Item Management
Target Frequencies

125kHz, 133 kHz	6.8 MHz	13.56 MHz	433 MHz	915 MHz	2.45 GHz	5.8 GHz
10 kHz	100 kHz	1 MHz	10 MHz	100 MHz	1 GHz	10 GHz

Solutions in mind:

- Flexible tags/code and readers, i.e., GTIN, GLN
- Common goals well presented
- Discrete tag and coding
- Reduced human interaction

Benefits of multi-technologies systems per se Bar code and RFID combination

Greater flexibility

Manufacturer can select tag technology compatible with their products and packages

Retailers can optimize their in-store system

Farm and producers adjust to manufacturer coding/tag write/read data

Economics

Reduce inventories, bring tag/coding cost low ($)

New opportunities and alliances

Creating an open/competitive market for technology providers

Partnership

Interoperability of systems, producers, manufacturers and distribution in concert with the global consumer!

Benefits from the use of RFID systems without Bar coding

Coordination of data and material flow

Quality control:

Multiple point controls on line and final quality events

System security:
    If software and computer crash, object character can be retrieved

Data security:
    Data integrity

Flexibility:
    A writeable tag brings more flexible control on the manufacturer or processor

Harsh environment:
    Resistant to dust, moisture, oils, coolants, and gases
    For 100 MHz to 1 GHz damping effect of water and non-conductive substances

Improve or reduce operator time (storage and warehouse) by 55% and movement by 15%

Operator cost reduced by 55% just for picking

Better global internal tracking of assets, staff and products

Comparison of Bar Code and RFID Technologies

Technology	Strength	Weakness
Barcode	Mature technology Established standards Ubiquitous Low implementation cost Human readable	Need clean line of sight Orientation sensitive Human intervention Sensitive to printing and abrasion Static data content No memory
RFID	Line of sight not required Passive or active data collection Not sensitive to environment (generally) Contents changeable Has memory	Emerging technology Lack of standards (in progress) Cost moderate to high today Human readable code "extra"

GENERAL CONCLUSIONS

Beyond quality control and quality assurance programs from the industry: producers, processors, distribution chain and exporters, more open and tailor-made integrated traceability systems have to be developed and implemented with the various Industries, Associations and Federations closely related to the Agri-Food sector.

Traceability tools and logistic are critical in gaining continued success in global food trade. The ECCC is an ideal and unique structure to facilitate and implement EDI, has a representative on international standards and committees, and also is sole distributor of Universal Product Code (UPC) and repository. The actors are in place for global commerce.

- We need to focus, develop solution and testing
- Need time and resources
- Input of all partners and real cooperation
- Evaluation/training

At risks are:

- Loss of consumer/client confidence
- Company and country's supply chain reputation
- Liability
- Loss of job, income

Winning enterprises and organizations must show adaptation and master ISO, HACCP, local and international regulations for international trade. Furthermore give proof of traceability.

Our goals in R & D are to put the right information, technology and know how:

- In the right format
- In the right hand
- At the right time
- And making sure it is accurate, safe and intact

The appropriate coding and tagging technologies can be profitable for the Agriculture and Agri-food and Fisheries systems.

ACKNOWLEDGMENTS

Special thanks to our industrial partners in R&D, sharing knowledge on technologies: T-Log, Logmilk, IdentiGEN, R. Moroz and Deuteron acquisition.

REFERENCES

Frederiksen, M. and Bremner, A. 2001. Fresh Fish Distribution Chains. An Analysis of Three Danish and Three Australian Chains. *Food Australia* 54: 119-123.

Anon, 1999. Salmon Tag for Trademark and Traceability. *Eurofish Magazine* 3: 74-76.

Finkenzeller, K. 1999. RFID Handbook. Radio-Frenquency Identification Fundamentals and Applications. John Wiley and Son Ltd, Chichester, England, 304 p.

Moe, T. 1998. Perspectives on Traceability in Food Manufacture. *Trends Food Sci. Technol.* 9: 211-214.

Palmer, R.C. 1995. The Bar Code Book. Helmers Pub. Inc. New Hampshire, 386 p.

CHAPTER 5

ENHANCEMENT OF PROFITABILITY PERSPECTIVES OF WOLFFISH CULTIVATION BY THE EXTRACTION OF HIGH VALUE BIOMOLECULES

Le François Nathalie R.[1,2], Mariève Desjardins[1] and Pierre U. Blier[1]

[1]Université du Québec à Rimouski, Département de Biologie, Chimie et Sciences de la Santé, Rimouski, Québec, G5L 3A1, Canada
[2]Ministère de l'Agriculture, des Pêcheries et de l'Alimentation du Québec (MAPAQ), Centre Aquacole Marin de Grande-Rivière, Grande-Rivière, Québec, G0C 1V0, Canada

The profitability margin of most aquaculture production (based exclusively on flesh production) is heavily vulnerable to fluctuations in market prices, feed, labor and/or energy costs, setting obstacles to the emergence of finfish mariculture initiatives. This study proposes to evaluate the economical potential of high value biomolecules extraction from cultivated fish to enhance profitability. This could promote private investors interest in developing Québec's maritime regions under-exploited mariculture potential. The Atlantic and spotted wolffish (Anarhichas lupus and A. minor) have been identified as high potential marine fish species for cultivation under Quebec's environmental conditions. Evaluation and comparison of both species for seasonal and size variations in 1) antifreeze proteins production, 2) content and properties of antimicrobial polypeptides and 3) content and kinetic properties of digestive enzymes is proposed.

INTRODUCTION

In the province of Québec, there is a growing interest for marine fish farming, partly related to coastal environment which offers plenty of good quality rearing sites. Considering this natural potential, provincial and federal governments are actively supporting the development of mariculture (scallop, mussels, finfish, etc.). However, no marine-based production of fish is in operation in Québec. The spotted and Atlantic wolffish (*Anarhichas minor* and *A. lupus*, respectively) have been identified as high potential species under Québec's environmental conditions. These fish are cold, disease and stress resistant, can be reared in very high densities and are able to withstand low oxygen as well as low salinity conditions (Le François *et al.*, 2002 ; Le François *et al.*, 2001a; Blier *et al.*, 2001). They hatch fully developed and are readily fed on commercial feed formulations (no live prey required) (Halfyard *et al.*, 2000) and convert most of their energy in growth (sedentary habits), which is considerably rapid in captivity (Falk-Petersen *et al.*, 1999; Moksness and Pavlov, 1996). Unlike most marine fish species, for example Atlantic cod (*Gadus morhua*) and Atlantic halibut (*Hippoglossus hippoglossus*), wolffish cultivation is very much similar to salmoniculture (size of eggs and larvae, feed formulation acceptance, survival, rearing units, fecundity, etc.), making technology transfer toward the industry easily conceivable.

The wolffish is a species that is not targeted by commercial fisheries (by-catch) in Canada (Scott and Scott, 1988) and the wolffish fisheries (mainly Icelandic) can be deficient during the year, thus providing good opportunities to establish a niche market supplied by aquaculture production year-round. Popular in Europe, this quality product has never been strongly targeted in North America. Recently, wolffish was given the Gold Award for the best new product at the 2000 International Boston Seafood Show. It features long, thick, white fleshed filets, delicate in taste, firm in texture, boneless, parasites free and rich in omega-3 fatty acids. Until now, Norway is the only producer of reared wolffish and they are actually running pre-commercial spotted wolffish farms in the region of Tromsø. Newfoundland and Québec (Canada) are presently involved in R&D activities for these species, in collaboration with Norway. Iceland and Chile are also interested in these novel mariculture species.

In the aquaculture sector, a company's profitability depends on fish production costs and flesh market price, which is subject to regular fluctuations. Therefore, a production diversification with by-products exploitation could stabilize and potentially increase benefits. This mini-review focuses on three different biomolecules that could be extracted from wastes of aquaculture industry (which are largely under-used (Haard, 1998)) and sold separately, mainly antifreeze proteins (AFP) (Fletcher *et al.*, 1999), antimicrobial polypeptides (AMP) (Hancock, 2000) and digestive enzymes (DE) (Simpson *et al.*, 1999; Haard, 1992, 1998). Given their numerous applications in diverse biotechnological fields with consequent high market prices, these biomolecules could confer a substantial added value to the basic product.

The originality of this approach lies in the integration of the diversification possibilities early in the process of establishing a novel and sustainable mariculture production (Figure 1).

BIOMOLECULES FROM FISH AND SOME OF THEIR APPLICATIONS

Antifreeze proteins (AFP)

Several species of cold water marine fishes produce antifreeze proteins (AFP) to avoid freezing of their tissues. Until now, four AFP forms have been characterized (http://www.afprotein.com). These proteins are over 500X more effective at depressing tissue freezing point than all the other blood osmolytes. They can also modify or suppress growth of ice crystals, inhibit recrystallization and protect membranes against cold induced damage (Fletcher *et al.*, 1999).

AFP show plenty of future applications and are presently actively studied. Possible uses for AFP include incorporation in frozen foods (e.g. dairy products) to prevent recrystallization of ice crystals already present in the product (in order to preserve its original texture) and cryoprotectants in cell cultures, gametes and organs cryopreservation (Fletcher *et al.*, 1999). Another potential biomedical application lies in cryosurgery (Pham *et al.*, 1999). This special technique makes the AFP kill the cancerous cells by affecting their action on crystal growth.

The principal source of AFP is plasma from some cold marine fish species. Such plasma can contain AFP concentrations varying between 1 and 30 g/L of plasma for an average price of $500/mg (Feeney and Yey, 1998). According to provisions from A/F Protein Canada (http://www.afprotein.com), demands for this resource will soon exceed production capacities of fisheries supply activity. So, several researchers are presently working in order to permit a future large-scale AFP production by using genetic methods like recombinant DNA (Solomon and Appels, 1999; Tong *et al.*, 2000). However, the reticence of people towards genetically modified organisms (GMO) and the limit imposed by the natural resource itself could provide an exclusive room to a marketing niche for naturally derived AFP and other biomolecules from the aquaculture industry.

Atlantic wolffish has been found to synthetize a type III AFP in its blood (Scott *et al.*, 1988). This AFP type have been proven the best of the three types tested in protecting bovine oocytes from cold-induced damages (Rubinsky *et al.*, 1991). Such AFP production by wolffish could allow the rearing of this species in natural conditions, in sea cages or net-pens all year long. In most fish that have been found to synthetize AFP, two abiotic factors lead to the production of

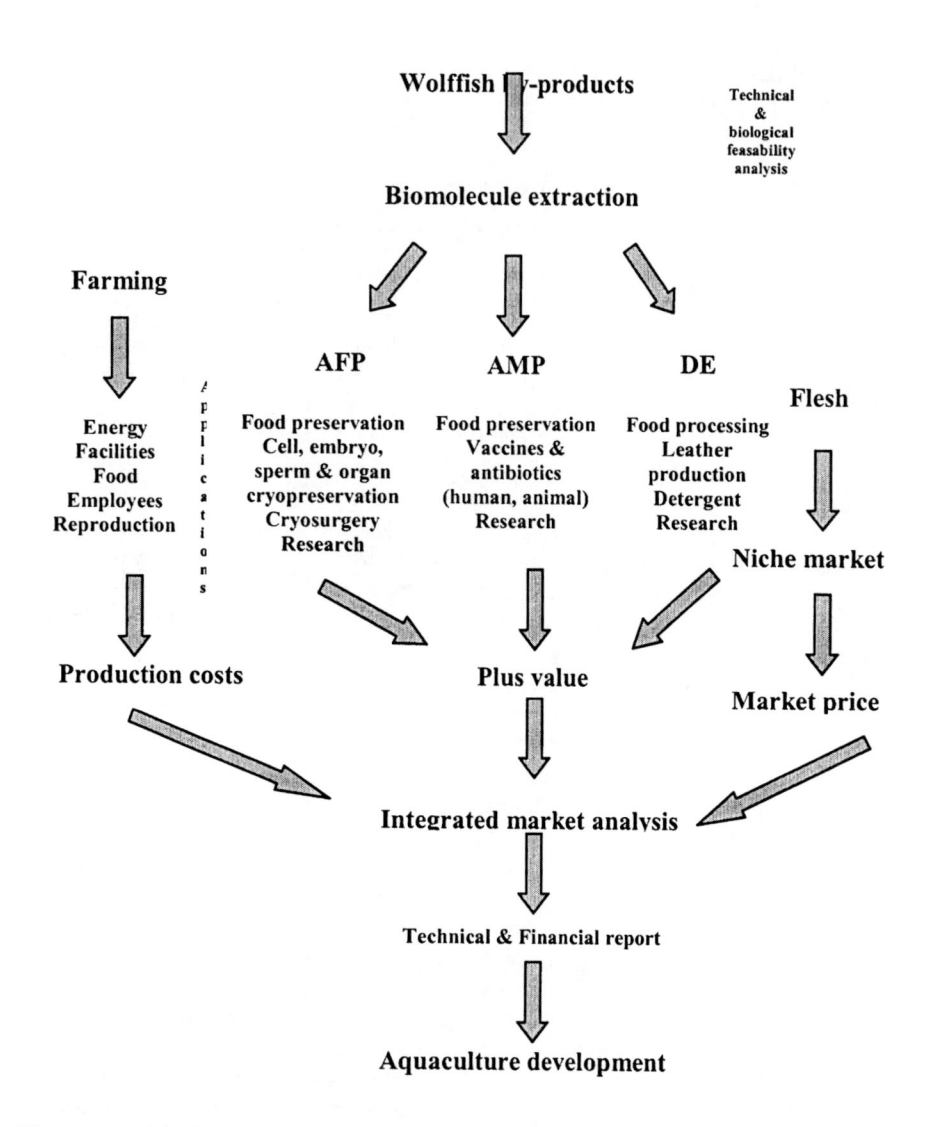

Figure 1. Global study planning.

these biomolecules: water temperature and, principally, photoperiod (Fletcher, 1981). Species like the Atlantic cod (*Gadus morhua*) and the winter flounder (*Pseudopleuronectes americanus*) experience seasonal AFP production (Fletcher, 1981; Fletcher *et al.*, 1982), while others, like Ocean pout (*Macrozoarces americanus*), produce it year-round, with higher concentrations during winter (Fletcher *et al.*, 1985). There is a lack of knowledge concerning both species of wolffish's AFP production pattern (Scott *et al.*, 1988). To assess the potential of AFP production by a wolffish aquaculture operation aimed at the biomolecule market, this production should be characterised by a monitoring on a seasonal cycle and, considering that intra-species AFP production variations exist (Goddard *et al.*, 1994), the effect of age and size should also be evaluated.

Antimicrobial polypeptides (AMP)

Amphipathic cationic antimicrobial polypeptides (AMP) have recently been discovered in mucus, skin and viscera of numerous fish species (Richards *et al.*, 2001). These peptides, present in all species of life, are important components of the first, non-specific, innate immune response (Hancock and Lehrer, 1998). The better known among these natural antibiotics are cecropins, melittins, magainins and defensins (Hancock and Lehrer, 1998). AMP have been proven to act in a broad-spectrum fashion against bacteria, parasites, fungi and encapsuled viruses (Cole *et al.*, 1997).

Applications for AMP are equally promising. For instance, it is a public health fact that over-utilization of traditional vaccines and antibiotics (of synthetic or chemical origins), increase the microbial resistance and encourage the acceleration of new strain's apparition (Epand and Vogel, 1999). Moreover, the use of classic antibiotics in farmed animal foods is proven to be persistent in derived products, with consequent allergy development occurrences in humans. The method by which AMP destroy pathogens (direct contact with microbial membranes leading to its lysis) (Yang *et al.*, 2000) precludes resistance acquisition (Hancock, 2000). Furthermore, AMP are expected to be harmlessly degraded in the gastrointestinal tract because of their protein nature (avoiding cross-resistance observed with traditional medical antibiotics). The public could be more favorable towards the use of the less aggressive and more natural defence tool. Experiments have already shown than some of these AMP show activity against microorganisms resistant to conventional antibiotics (Hancock and Lehrer, 1998). Presently, there are several studies in the medical field to assess the potential of AMP as therapeutic aids, and some results suggest that these polypeptides could be of substantial efficiency in topical treatments (Hancock, 2000). The animal farming industry, including fish farms (Jia *et al.*, 2000), could equally benefit from AMP use as antibiotics. These peptides could also be incorporated in foods to improve their preservation quality and safety. Because of these numerous potential applications, there is actually important world-wide research efforts directed toward natural antibiotic production such as AMP.

Because of the abrasive nature of its diet (mostly hard-bodied invertebrates), the wolffish's digestive tract produces important quantities of mucus (Barsukov, 1972). The skin also shows substantial mucus production (personal observation). We therefore suspect that this mucus contains important proportions of AMP (yet to be characterized) with elevated antimicrobial activity, which have to be assessed.

Digestive enzymes (DE)

The principal proteases produced by fish digestive tract are trypsin, chymotrypsin and pepsin (Lemieux et al., 1999). Pancreatic lipases and chitinases are also abundant (Moe and Place, 1999). An important point in favour of digestive enzymes (DE) from marine organisms are the high catalytic capacities they display at cold temperatures.

Digestive enzymes are widely used in industry (especially proteases), mostly in detergent, leather production and food processing (Haard, 1992). Proteases are used as processing aids for many products like cheese, chocolate, eggs, meat, fish and vegetables, as well as for the production of protein hydrolyzates and flavor extracts. The general capacity of marine DE to get active at cold temperatures offers very interesting opportunities at the level of the food processing industry, where a cold chain must be maintained. These kind of DE are also recognized to display relatively high molecular activity, physiological efficiency and low free energy of activation, compared to homologous DE from endothermic organisms (Haard, 1998). Their low thermal stability raise the possibility to easily inactivate them in a product by a lower thermal treatment, which does not alter the quality of the product treated (Haard and Simpson, 2000). Furthermore, some of these DE are salt activated, contrary to those of mammalian origin. Such a property can be advantageous in food treatments including significant amount of salts, like fermentations (Haard, 1998).

With so many applications, there is an expanding market for DE from marine organisms (reaching 1.5 billion US dollars in 1998 and expected to double in the next years) (Haard, 1998)) and their extraction from marine organisms is a growing field of interest (Haard, 1997, 1998; Simpson et al., 1999; Haard & Simpson, 2000). Until now, the DE market have been dominated by proteases. However, lipases could eventually become the most used industrial DE in the near future (Haard, 1998). An important factor in favor of DE extraction from marine fishes resides in the confidence crisis involving the bovine industry, which forces food processing industry to seek for other supplying sources than butchery by-products (info. AquaBiokem^{BSL} inc.).

Wolffishes are carnivorous and thus, display important protease and lipase activities. These fishes could consist a good natural source of varied DE for the industry. The kinetic properties of these DE have to be assessed.

PERSPECTIVES FOR QUÉBEC'S AQUACULTURE

With recent development in marine biotechnologies and identification of this development as a priority for the east of Québec, the efficient use of fish by-products originating from the fisheries and aquaculture activities has to be assessed. However, before entrepreneurs could include extraction and purification of high value biomolecules in their business plans, they should be aware of the potential production of these biomolecules from their installations, as well as access to the market and the value of these biomolecules. Therefore, technical and biological feasability studies, plus an integrated market analysis, should be initiated.

REFERENCES

Barsukov, V.V. 1972. The wolffish (Anarhichadidae). Smithsonian Institute, Springfield, MA, 292 p.

Blier, P.; Le François, N.R.; Lemieux, H. and Paradis, C. 2001. Potentiel biologique de différentes espèces de poissons marins à des fins de développement de la mariculture dans l'est du Québec. UQAR, Rimouski, Canada, p. 283.

Chakrabartty, A. and Hew, C.L. 1991. The effect of enhanced alpha-helicity on the activity of a Winter flounder antifreeze polypeptide. *Eur. J. Biochem.* 202: 1057-1063.

Cole, A.M.; Weis, P. and Diamond, G. 1997. Isolation and characterization of pleurocidin, an antimicrobial peptide in the skin secretions of Winter flounder. *J. Biol. Chem.* 272: 12008-12013.

Epand, R.M. and Vogel, H.J. 1999. Diversity of antimicrobial peptides and their mechanisms of action. *Biochem. Biophys. Acta* 1462: 11-28.

Falk-Petersen, I.-B.; Hansen, T.K.; Fieler, R. and Sunde, L.M. 1999. Cultivation of the spotted wolffish Anarhichas minor (Olafsen) – a new candidate for cold-water fish farming. *Aquac. Res.* 30: 711-718.

Feeney, R.F. and Yey, Y. 1998. Antifreeze proteins: current status and possible food uses. *Trends Food Sci. Technol.* 9: 1-15.

Fletcher, G.L. 1981. Effects of temperature and photoperiod on the plasma freezing point depression, Cl⁻ concentration and protein "antifreeze" in Winter flounder. *Can. J. Zool.* 59: 193-201.

Fletcher, G.L.; Goddard, S.V. and Wu, Y. 1999. Antifreeze proteins and their genes: from basic research to business opportunity. *Chemtech.* 30: 17-28.

Fletcher, G.L.; Hew, C.L.; Li, X.; Haya, K. and Kao, M.H. 1985. Year-round presence of high levels of plasma antifreeze peptides in a temperate fish, Ocean pout (*Macrozoarces americanus*). *Can. J. Zool.* 63: 488-493.

Fletcher, G.L.; Slaughter, D. and Hew, C.L. 1982. Seasonal changes in the plasma levels of glycoprotein antifreeze, Na^+, Cl^- and glucose in Newfoundland Atlantic cod (*Gadus morhua*). *Can. J. Zool.* 60: 1851-1854.

Goddard, S.V.; Wroblewski, J.S.; Taggart, C.; Howse, K.A.; Bailey, W.L. and Fletcher, G.L. 1994. Overwintering of adult Northern Atlantic cod (*Gadus morhua*) in cold inshore waters as evidenced by plasma antifreeze glycoprotein levels. *Can. J. Fish. Aquat. Sci.* 51: 2834-2842.

Haard, N.F. 1992. A review of proteolytic enzymes from marine organisms and their application in the food industry. *J. Aquatic Food Prod. Technol.* 1: 17-36.

Haard, N.F. 1997. Enzymes from marine organisms. *Biotechnol.* 2: 78-85.

Haard, N.F. 1998. Specialty enzymes from marine organisms. *Food Technol.* 52: 64-67.

Haard, N.F. and Simpson, B.K. 2000. Seafood Enzymes. Marcel Dekker Inc., New York, NY, p. 696.

Halfyard, L.C.; Parrish, C.C.; Watkins, J. and Jauncey, K. 2000. Fatty acid and amino acid profiles of eggs from the common wolffish, Anahichas lupus. Aquaculture Canada 2000 – Aquacul. Assoc. Canada Spec. Publ. No. 4 60-63.

Hancock, R.E.W. and Lehrer, R. 1998. Cationic peptides: a new source of antibiotics. *Tibtech.* 16: 82-88.

Hancock, R.E.W. 2000. Cationic antimicrobial peptides: towards clinical applications. *Exp. Opin. Invest. Drugs* 9: 1723-1729.

Jia, X.; Patrzykat, A.; Devlin, R.H.; Ackerman, P.A.; Iwama, G.K. and Hancock, R.E.W. 2000. Antimicrobial peptides protect Coho salmon from *Vibrio anguillarum* infections. *Appl. Environ. Microbiol.* 66: 1928-1932.

Laemmli, U.K. 1970. Cleavage of structural proteins during the assemblage of the head of the bacteriophage T4. *Nature* 227: 680-685.

Lemieux, H.; Blier, P. and Dutil, J.-D. 1999. Do digestive enzymes set a physiological limit on growth rate and food conversion efficiency in the Atlantic cod (*Gadus morhua*). *Fish Physiol. Biochem.* 20: 293-303.

Le François, N.R.; Dutil, J.D.; Blier, P.; Lord, K. and Chabot, D. 2001a. Tolerance and growth of juvenile Common wolffish (*Anarhichas lupus*) under low salinity and hypoxic conditions: preliminary results. Aquaculture Canada 2000 – Aquacul. Assoc. Canada Spec. Publ. No. 4: 57-59.

Le François, N.R.; Lemieux, H.; Blier, P. and Falk-Petersen, I.B. 2001b. Effect of three different feeding regimes on growth, metabolism and digestive enzyme activities, in the early life of Atlantic wolffish (*Anarhichas lupus*). Aquaculture Canada 2000 – Aquacul. Assoc. Canada Spec. Publ. No. 4: 53-56.

Le François, N.R.; Lemieux, H. and Blier, P. 2002. Biological and technical evaluation of the potential of marine and anadromous fish species for cold-water mariculture. *Aquac. Res.* 33: 1-14.

Moe, C.M. and Place, A.R. 1999. Distribution of chitinase in gastrointestinal tissues of eight fish species. In *Fish feeding ecology and digestion: GUTSHOP'98*. D. Mackinlay and D. Houlihan (eds.). AFS, Madison, WI.

Moksness, E. and Pavlov, D.A. 1996. Management by life cycle of wolffish, *Anarhichas lupus* L., a new species for cold-water aquaculture: a technical paper. *Aquac. Res.* 27: 865-883.

Pham, L.; Dahiya, R. and Rubinsky, B. 1999. An in vivo study of antifreeze protein adjuvant cryosurgery. *Cryobiol.* 38: 169-175.

Richards, R.C.; O'Neil, D.B.; Thibault, P. and Ewart, D.K.V. 2001. Histone H1: an antimicrobial protein of Atlantic salmon from Vibrio anguillarum infections. *Biochem. Biophys. Res. Commun.* 284: 549-555.

Rubinsky, B.; Arav, A. and Fletcher, G.L. 1991. Hypothermic protection – a fundamental property of antifreeze proteins. *Biochem. Biophys. Res. Commun.* 180: 566-571.

Scott, W.B. and Scott, M.G. 1988. Atlantic fishes of Canada. University of Toronto Press, Toronto, ON, p. 731.

Scott, G.K.; Hayes, P.H.; Fletcher, G.L. and Davies, P.L. 1988. Wolffish antifreeze protein genes are primarily organised as tandem repeats that each contain two genes in inverted orientation. *Mol. Cell. Biol.* 8: 3670-3675.

Simpson, B.K.; Sequeira-Munoz, A. and Haard, N.F. 1999. Marine enzymes. In *Encyclopedia of Food Science and Technology, 2nd ed.* F.J. Francis (ed.), Vol. 3, John Wiley & Sons Inc., New York, NY, pp. 1525-1534.

Solomon, R.G. and Appels, R. 1999. Stable, high-level expression of a type I antifreeze protein in *Escherichia coli*. *Protein Expres. Purif.* 16: 53-62.

Tong, L.; Lin, Q.; Wong, W.K.R.; Ali, A.; Lim, D.; Sung, W.L.; Hew, C.L. and Yang, D.S.C. 2000. Extracellular expression, purification and characterization of a Winter flounder antifreeze polypeptide from *Escherichia coli*. *Protein Expres. Purif.* 18: 175-181.

Yang, L.; Weiss, T.M.; Lehrer, R.I. and Huang, H.W. 2000. Crystallization of antimicrobial pores in membranes: magainin and protegrin. *Biophys. J.* 79: 2002-2009.

CHAPTER 6

COMPARISON OF DEHEADED GUTTED AND UNGUTTED MACKEREL (*SCOMBER SCOMBRUS*) PRESERVED AT -30°C UNDER OPTIMAL CONDITIONS

Nathalie Fillion and Marie-Élise Carbonneau

Centre technologique des produits aquatiques,
Direction de l'innovation et des technologies
96, montée De Sandy Beach, bureau 205, Gaspé
Québec, G4X 2V6, Canada

The effect of prolonged storage on the quality of deheaded gutted mackerel was compared with that of ungutted mackerel. Chemical and sensory analyses were performed once a month over a 12-month period on mackerel stored at –30°C. The gutted mackerel and ungutted mackerel were processed under optimal handling and preservation conditions using in-plant blast air freezer, and storage in a freezer at –30 °C. After storage for 12 months, no difference was observed between the gutted and ungutted fish. The sensory evaluation scores for the cooked fillet showed moderate deterioration over the storage period but all samples remained acceptable. The overall data on free fatty acids (FFA) indicated a slight increase of lipid deterioration. The thiobarbituric acid (TBA) value showed important variation during storage for both deheaded gutted and ungutted fish. However, TBA values were generally higher for deheaded gutted mackerel. Extractable protein nitrogen (EPN) remained stable throughout the storage period. Thus, deheaded gutted and ungutted mackerel were still acceptable for consumption after 12 months of storage.

INTRODUCTION

In Québec, mackerel remains underutilized. Its processing is primarily limited to fresh raw material and takes place over a short period only. Recently, however, processors have become interested in extending their production periods. In some European countries, deheaded gutted and ungutted mackerel have been stored at temperatures as low as –30 °C while awaiting processing. Although literature mentions a number of advantages to gutting fish (less softening of the tissues), this process can make the flesh more susceptible to oxidation by exposing it to the air, resulting in changes in color and the production of undesirable flavors and odors (Huss, 1988). To enable processors to further enhance mackerel and extend their production periods, a study was conducted at –30 °C to compare the shelf-life of deheaded gutted and ungutted mackerel.

MATERIALS AND METHODS

Mackerel handling and processing

A 250-pound batch of mackerel arrived in a slush-ice-filled insulated tanks at the *Les Pêcheries Gaspésiennes Inc.* plant in Rivière-au-Renard in September 1998. This mackerel had been line-fished that same morning. The mackerel were washed in salt water and then the batch was divided into two lots; 125 pounds were deheaded and gutted while the other 125 pounds were processed ungutted.

The two lots of fish were individually frozen in a blast air freezer at –30 °C for about 2½ hours. They were then glazed by placing the mackerel in holed trays. The trays were immersed in water for 3 to 4 s and returned to the freezer for a few minutes. This step was repeated two more times.

The boxes of gutted and ungutted mackerel were stored at –20°C before shipping to the assay centre in Gaspé next day. The mackerel in each box was packaged in plastic bags in lots of 6 fish per box. Upon arrival, the boxes were stored in a freezing room at –30 °C throughout the entire duration of the experiment.

Sensory evaluation

A descriptive analysis of the mackerel's odor, flavor and texture attributes was performed by 6 trained panelists. The panelists were staff members at the Centre technologique des produits aquatiques in Gaspé who had experience in the sensory analysis of fish. Training sessions were conducted during which the questionnaire was discussed and reference samples were presented to the panelists. The questionnaire used (Table 1) was a modification of the Torry score sheet for cooked herring (Regeinstein and Regeinstein, 1991; Robichaud and Cormier, 1996; Simeonidou et al., 1997). The thawed filets of gutted and ungutted mackerel were baked in a steam cooker (Hobart 100G) at a pressure of 15 psi for 3 min 20 s and then served warm to the panelists for sensory evaluation. Each panelist received portions of each of the 6 mackerel filets that had been taken from different locations on the filets. The odor, flavor and texture of cooked fish samples were scored using a 1-to-5 point scale. The references used to assess the cooked mackerel included fresh and oxidised mackerel oil, fresh and aged seaweed, a 5% trimethylamine solution and a 0.15% dilute ammonia solution. The sensory evaluation was conducted in an isolated room according to the sensory assessment methods (Meilgaard et al., 1991).

Chemical analysis

Moisture and ash were determined by the gravimetric method of Woyewoda et al. (1986). Lipids were estimated using the Bligh et Dyer (1959) method. Proteins were determination during mineralisation (Kane, 1987) in a copper/titanium catalyst using the Kjeldahl assay (FisherTab™ TT-35 Kjeldahl tablets produced by Fisher Scientific). An automated device (Kjeltec 1035 by Tecator) was used to assay for the nitrogen (Woyewoda et al., 1986; Tecator, 1981). In converting nitrogen to protein, a factor of 6.25 was used. Extractable protein nitrogen (EPN) was determined by extraction in a 5% lithium chloride solution at a pH of 7.2 according to the method proposed by Kelleher and Hultin (1991). The extracted nitrogen was assayed according to the protein assay method described earlier. The free fatty acid (FFA) content was estimated by volumetric analysis (Woyewoda et al., 1986). The thiobarbituric acid value (TBA) was determined by the direct measurement of a trichloroacetic acid extract according to Vyncke (1970).

Statistical analysis

The data was subjected to statistical analysis using the U-test (Mann-Whitney) for sensory evaluation and ANOVA for chemical analysis. All statistical evaluations were done at a 95% confidence level.

RESULTS AND DISCUSSION

Proximate composition

The chemical composition of mackerel samples is presented in Table 2. The fat content of samples varied greatly, from 7.32 to 22.02% (average 13.65%) with a SD of 4.06%. These observations are in accordance with data reported by Gregoire *et al.* (1994) and Leu *et al.* (1981). These results reflect the natural variability occurring normally in mackerel populations, attributed to seasonal cyclic changes and also to a strong individual rancidity of pelagic fish.

Table 1. Sensory score sheet for cooked mackerel

	Odor	
Scale	**Descriptors**	**References presented**
5	Odor fresh, characteristic of the species.	. Fresh seaweed
4	Less fresh seaweed odor with a trace of oily odor	. Aged seaweed . Fresh mackerel oil
3	Marked oily odor, occasionally slightly oxidised; slight sweaty odor or salt fish odor	. Oxidised mackerel oil . Trimethylamine solution
2	Strong oxidised oil odor; salt fish odor, sweaty odor.	. Oxidised mackerel oil . Trimethylamine solution . Salted cod in water
1	Resembles #2 but odor is more pronounced, definitely unpleasant, ammonia, malt, sulfur or sour odors present.	. Dilute ammonia solution . Dilute KS solution

Samples : ____ : ____   ____ : ____

Table 1. continued...

	Taste	
Scale	**Descriptors**	**References presented**
5	Fresh taste, characteristic of the species, seaweedy and sweet taste.	. Fresh seaweed
4	Less sweet, weaker seaweed taste with traces of oily taste.	. Aged seaweed . Fresh mackerel oil
3	Marked oily taste, occasionally slightly oxidized, taste of old seaweed.	. Oxidized mackerel oil
2	Strong oxidized oil taste, older seaweed taste.	. Oxidized mackerel oil
1	Definite unpleasant oxidized oil taste, rancid taste, sweaty taste.	

Samples: ____ : ____   ____ : ____

	Texture
Scale	**Descriptors**
5	Firm, succulent and coherent, flaky
4	Succulent and coherent, less flaky
3	Moderately-succulent, become dry and tough after chewing, fibrous.
2	Dry and tough, fibrous, a little hard to chew.
1	Strongly dry and tough, hard to chew.

Samples : ____ : ____   ____ : ____

Table 2. Proximate composition of frozen mackerel flesh[a]

% w/w	Mean	± S.D.	Range
Moisture	65.67	3.39	58.38-71.08
Fat	13.65	4.06	7.32-22.02
Crude protein	18.89	0.84	17.25-19.88
Ash	1.19	0.08	1.08-1.34

[a] Number of replicates=12

Sensory evaluation

The odor, taste and texture of both deheaded gutted and ungutted mackerel was decreased over the frozen storage period (Fig. 1). Assessment of the cooked fillets indicated a moderate deterioration over the storage period. All scores were above the acceptable level (2.5) throughout the frozen storage period for both deheaded gutted and ungutted mackerel. The sensory evaluation showed no significant difference ($P > 0.05$) in scores for either sample forms. Both forms remained acceptable at the end of the storage period.

Chemical analysis

As for the FFA and EPN, no significant differences ($P > 0.05$) were observed between the two treatments. The FFA showed a slight increase due to lipid hydrolysis over the frozen storage period. Moreover, it was expected that very little lipid hydrolysis would result under these conditions since enzymatic activity was greatly reduced. The EPN remained stable and indicated no protein denaturation over the storage period. The same observation was made for protein denaturation, in addition to being stored under excellent conditions, it is recognized that the proteins of fatty fish are more resistant to denaturation than those of lean fish (Martin et al., 1982).

Although a significant difference in TBA ($P < 0.05$) was observed between the deheaded gutted and ungutted mackerel, no obvious trend was discernible (Fig. 1). The TBA of the deheaded gutted fish was much higher than that of ungutted mackerel. The deheading and gutting process appears to have slightly enhanced lipid oxidation of mackerel. However, the values obtained for ungutted mackerel earned an "excellent" grade according to the rancidity indices used to assess quality of mackerel (RIM) (TBA < 6.0 mg MA eq./kg, that is, 27.2 µmol MA eq. / kg) as proposed by Ke et al. (1975). For deheaded gutted mackerel, 60% of the time the TBA was generally below the RIM. These

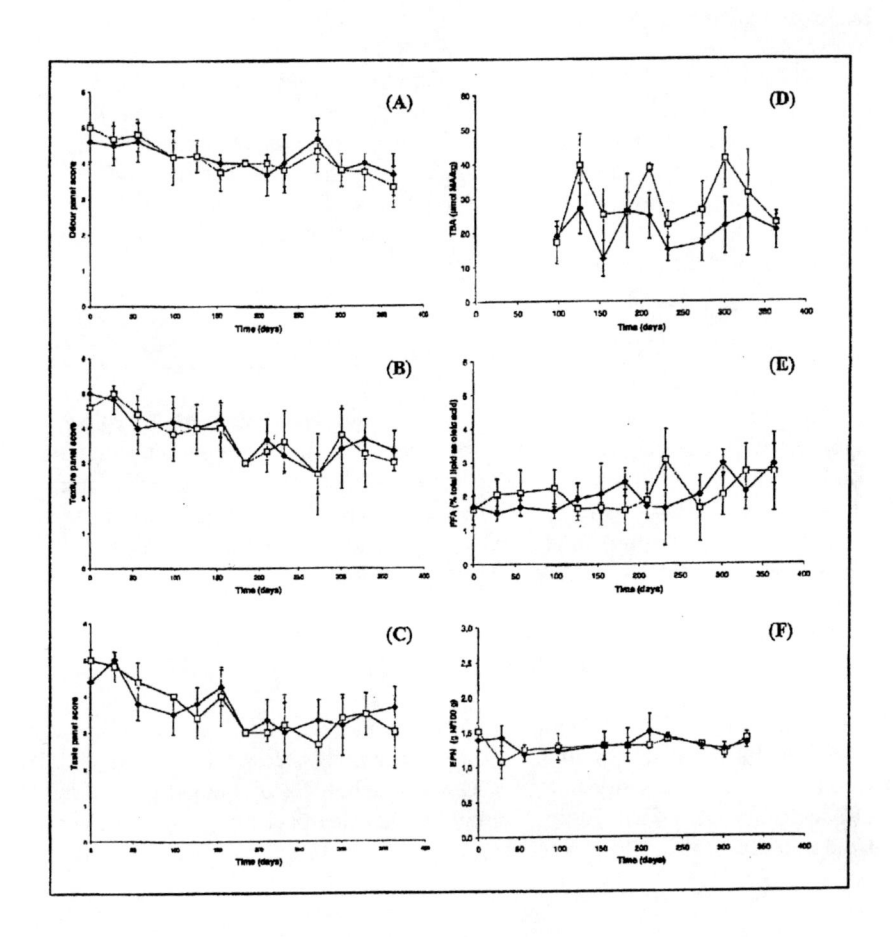

Figure 1. Sensory evaluation and chemical analysis of deheaded gutted (—□—) and ungutted (—♦—) mackerel over 12 months of frozen storage. Symbols are: **A,** Odor score; **B,** Texture score; **C,** Taste score; **D,** Thiobarbituric acid value (TBA); **E,** Free fatty acid (FFA) ; and **F,** Extractable protein nitrogen (EPN).

samples also received "excellent" grading according to this scale after being frozen for 264 days.

Deheaded gutted vs ungutted mackerel

When the fish was stored at −30°C, enzymatic reactions occurred at a much slow pace (Burt *et al.*, 1992). It is well recognized that self-digestion by enzymes is highly active in mackerel during periods of intensive feeding (Soudan, 1965). In Québec, these fish are sometimes caught during this intensive feeding period. Although literature presents a number of disadvantages to gutting, doing so before freezing could result in lower enzymatic activity before and during the storage period.

CONCLUSIONS

The sensory evaluation, and the FFA and EPN analysis revealed no significant differences ($P > 0.05$) between deheaded gutted and ungutted mackerel in contrast to the TBA ($P < 0.05$). The TBA seems to show that gutting of mackerel slightly increases lipid oxidation. However, sensory analysis indicated that both forms of mackerel were still acceptable for consumption after 12 months of storage at -30 °C under optimal conditions.

ACKNOWLEDGEMENTS

We would like to thank *"Pêcheries Gaspésiennes Inc."* for their collaboration; Jean Paradis for his help in preparing the samples; Josée Blais and Noëlla Coulombe for the sensory analysis; Diane Ouellet, Serge Latendresse, Jeanne D'Arc Rioux and Marie-Josée Moulin for the chemical analysis; and Francis Coulombe and Fabrice Pernet for their assistance with the statistical analysis.

REFERENCES

Bligh, E.G. and Dyer, W.J. 1959. A rapid method of total lipid extraction and purification. *Can. J. Biochem. Physiol.* 37: 911-917.

Burt, J.R.; Hardy, R. and Whittle, K.J. 1992. Pelagic fish The Resource and its Exploitation. The University Press. Cambridge, UK, p. 352.

Gregoire, F.; Dionne, H. and Levesque, C. 1994. Contenu en gras chez le marquereau bleu (*Scomber scombrus* L.) en 1991 et 1992. Rapport canadien a industrie sur les sciences hallieutiques, No. 220, Fisheries and Oceans Canada, 70 p.

Huss, H.H. 1988. Le poisson frais: qualité et altérations de la qualité. FAO. Fisheries Series, No.29, 132 p.

Kane, P.F. 1987. Comparison of HgO and $CuSO_4$ / TiO_2 as catalysts in manual Kjeldahl digestion for determination of crude protein in animal feed: collaborative study. *J. Assoc. Off. Anal. Chem.* 70: 907-911.

Ke, P.J.; Ackman, R.G. and Nash, D.M. 1975. Proposed rancidity indexes for assessing the storage life of frozen mackerel. Fisheries and Marine Service. New series circular No. 52. Halifax, NS, 5 p.

Kelleher, S.D. and Hultin, H.O. 1991. Lithium chloride as a preferred extractant of fish muscle proteins. *J. Food Sci.* 56: 315-317.

Leu, S-S.; Jhaveri, S.N.; Karakoltsidis, P.A. and Constantinides, S.M. 1981. Altantic mackerel (*Scomber scombrus* L.): seasonal variation in proximate composition and distribution of nutrients. *J. Food Sci.* 46: 1635-1638.

Martin, R.E.; Flick, G.J.; Hebard, C.E. and Ward, D.R. 1982. Chemistry & Biochemistry of Marine Food Products. AVI publishing company, Westport, CT, 474 p.

Meilgaard, M.; Civille, C.V. and Carr, B.T. 1991. Sensory evaluation techniques, 2e édition. CRC Press, Boca Raton, FL, 354 p.

Regenstein, J.M. and Regenstein, C.E. 1991. Introduction to fish technology. Van Nostrand Reinhold. New York, NY, 269 p.

Robichaud, R. and Cormier, A. 1996. Mackerel sensory qualities assessed by trained versus consumer panels. Food Research Centre, Univ. de Moncton, Moncton, NB, 13 p.

Simeonidou, S.; Govaris, A. and Vareltzis, K. 1997. Effect of frozen storage on the quality of whole fish and fillets of horse mackerel (*Trachurus trachurus*) andmediterranean hake (*Melluccius mediterraneus*). *Z. Lebensm Unters Forsch.* 204: 405-410.

Soudan, F.; Anquez, M. and Benezit, A. 1965. La conservation par le froid des poissons, crustacés et mollusques. J.-B. Ballière et fils. Paris, France, 514 p.

Tecator. 1981. Determination of Kjeldahl nitrogen content with Kjeltec Auto Systems, I, II, III and IV. Application note. Tecator, AN 30/81, 2 p.

Vyncke, W. 1970. Direct determination of the thiobarbituric acid value in trichloroacetic acid extracts of fish as a measure of oxidative rancidity, *Fette Seifen Amstrmittel* 72: 1084-1087.

Woyewoda, A.D.; Shaw, S.J.; Ke, P.J. and Burns B.G. 1986. Recommended Laboratory Methods for Assessment of Fish Quality. *Can. Tech. Rep. Fish. Aqua. Sci.* No. 1448, 143 p.

CHAPTER 7

QIM – A TOOL FOR DETERMINATION OF FISH FRESHNESS

Grethe Hyldig and Jette Nielsen

Ministry of Food, Agriculture and Fisheries, Danish Institute for Fisheries Research, Department of Seafood Research, DTU, Søltofts Plads Building 221, DK-2800, Lyngby, Denmark

In order to ensure the safety of food it is important to keep the quality of fish at a high level in each link of the whole complex chain from catch to consumer. The Quality Index Method (QIM) is expected to soon become the leading reference method for assessment of fresh fish within the European community. Sensory perception is most important for assessing freshness and quality in the fish sector and inspection services. Sensory methods performed properly are rapid and accurate tools for providing unique information about the food. European fisheries research institutes have developed QIM as a new tool by which sensory assessment is performed in a systematic and reliable way and used as a truly objective method of analysis rather than subjectively. Three European fisheries institutes, namely the Danish Institute of Fisheries Research, RIVO Netherlands Institute for Fisheries Research and Icelandic Fisheries Laboratories have decided to establish a strategic alliance called QIM - Eurofish. The mission of the alliance is to promote and implement the use of QIM as a versatile quality tool within fisheries distribution and production chains in Europe.

INTRODUCTION

Food quality is by no means clearly defined and a great deal of confusion can be created when communicating research and development with marketing sectors. Botta (1995) cites 15 different definitions of quality. These range from general statements to consumer definitions. Each definition can be used in a specific situation, but none of them are comprehensive. Quality cannot be defined in a simple manner, as the definition changes with the particular context where it is applied, is dependent on the multitude of species and the influence of biological (season, spawning period) and technological (handling, temperature, time) parameters. Quality must be defined in each link of the chain from the point of catch to the consumer and it is necessary to develop quantitative methods suitable for all links.

The quality concept is frequently described using terms related to nutritional, microbiological and physiochemical characteristics alone, but none of these terms serve as adequate indices of quality – sensory perception and consumer acceptability must be included. Sensory testing can be both objective and subjective. The objective tests include discriminative (triangle test, forced choice) and descriptive (profiling, structured scaling) tests. Both groups of test are analytical measurements of the intrinsic (e.g. species, fat content, smell, taste, appearance) quality of the product, whereas affective (subjective test) methods are used for consumer testing and to measure the attitude and emotional response of the consumer towards the product including both intrinsic and extrinsic (e.g. price, convenience, origin, handling) quality parameters (Ophuis and Van Trijp, 1995). Sensory assessment of fish and fish products has for years played a natural part of the fishery chain. There is, however, a need for new methods of sensory analysis that can be used as integrated objective markers and systems for bridging between research, development, industry, marketing and consumer. A structured scaling method, QIM is suggested for evaluation of fresh fish in production management, in seafood inspection and other parts of the chain.

Quality index method (QIM)

QIM is a rapid scaling method, which establishes the exact data reflecting different quality levels of fish in a simple and well-documented way. Since late 80's, the staff of the seafood laboratories in the Nordic and European countries have worked with development of QIM (Luten and Martinsdottir, 1997; Hyldig and Nielsen, 1997) based on a method suggested by Bremner *et al.* (1986). QIM is developed for raw fish (cod, saithe, red fish, sardines, haddock, pollock, plaice, dab, sole, brill, turbot, herring, Atlantic mackerel, horse mackerel, salmon and trout). The aim is to establish a linear correlation between the sensory quality expressed as a demerit score and storage life. At the Danish Institute for Fisheries Research the method has been tailored to the seafood industry needs to fit the production of a variety of products (Larsen *et al.*, 1992; Hyldig and Nielsen, 1997; Warm *et al.*, 1998; Jonsdottir *et al.*, 1999; Warm *et al.*, 2000). Manuals can support QIM in an excellent way in quality assurance and production. The manuals must contain the total plan for evaluation, explanation of the evaluation terms, and color photos illustrating the different levels of quality of fish.

For fish the progress in sensory quality can be divided into four phases: 1) the fish is very fresh and has a sweet, seaweedy and delicate taste, 2) there is a loss of the characteristic odor and taste, the flesh becomes neutral but has no off-flavors, the texture is still pleasant, 3) there are signs of spoilage and a range of volatile, unpleasant smelling substances is produced depending on the fish species and type of spoilage (aerobic, anaerobic), and 4) during the later stages sickening sweet, cabbage-like, ammoniacal, sulphurous and rancid smells develop, the texture becomes either soft and watery or tough and dry and the fish can be characterized as spoiled and putrid. The shelf life can be determined by profiling of a cooked sample using descriptors based on the four phases. Given the maximum storage in ice as the day where the fish is rejected, the ideal demerit curve can be drawn and used for evaluation of QIM.

QIM is based on significant sensory parameters for whole fish using many weighted parameters and a score system from 0 to 3 demerit points. The scores for all the characteristics are added to give an overall sensory score, the so-called quality index. QIM gives scores of zero for very fresh fish and an increasingly larger total result as the fish deteriorates.

The selection of parameters for QIM is determined as a combination of the best descriptors for the spoiling fish, which also fulfil the aim that it is possible to predict the remaining shelf life. The remaining shelf life (in days on ice) can be calculated on the basis of this line and knowledge about the corresponding quality index at the time of rejection. There is a linear correlation between the sensory quality expressed as the quality score and storage time in ice, which makes it possible to predict the remaining shelf life in ice. Figure 1 shows a batch of fish reaching a sum of 10 points - corresponding to 12 days in ice - and having a remaining shelf life of 9 days in ice.

The first development of QIM (Bremner, 1986) was on the basis of cooperation between industry and research. An obvious consideration has been how this relates to the consumers' perception. This question can now be partly answered with the modification of QIM to a consumer version. A simplified QIM (Table 2) for consumer testing (C-QIM) has recently been suggested by Warm (2000).

C-QIM is developed with the use of an external panel testing their own vocabulary in comparison with QIM terms of experts. The limits of the C-QIM are such that only the intrinsic quality parameters are considered and the method must be combined with subjective consumer tests.

Figure 2 describes which links form part of the chain from the point of catch to consumer. It is very important to ensure that fish and fish products with a high sensory quality reach the consumer. Therefore, it is necessary to use a tool measuring the sensory quality when it transferred from one link to

Table 1. Quality index scheme for whole farmed salmon (*Salmo salar*) containing a ranking of the description for each parameter and the given scores in succession from 0 to 3.

Quality parameters		Description	Points
Skin:	**Color/ Appearance**	Pearl-shiny all over the skin	0
		The head is still pearl-shiny, but the rest less, perhaps yellow	1
	Mucus	Clear and not clotted	0
		Milky and clotted	1
		Yellow and clotted	2
	Odor[1]	Fresh seaweeds, cucumber	0
		Neutral to metal, dry grass, corn	1
		Sour	2
		Rotten	3
Eyes:	**Pupils**	Clear and black, metal shiny	0
		Dark grey	1
		Mat, grey	2
	Form	Flat	0
		Little sunken	1
		Sunken	2
Abdomen:	**Blood in abdomen**	Blood light red/not present	0
		Blood more brown	1
	Odor	Neutral	0
		Corn	1
		Sour	2
		Rotten/Rotten kale	3
Gills[2]:	**Color/ appearance**	Red/dark brown	0
		Light red/brown	1
		Grey-brown, grey, green	2
	Mucus	Transparent	0
		Yellow, clotted	1
		Brown	2
	Odor	Fresh, seaweed	0
		Metal	1
		Sour	2
		Rotten	3
Texture:	**Elasticity**	Finger mark disappears immediately	0
		Finger leaves mark over 3 seconds	1
Quality Index (0 - 22) SUM:			

1 Turn the salmon and smell the skin on the other side
2 Examine the side that has not been cut through

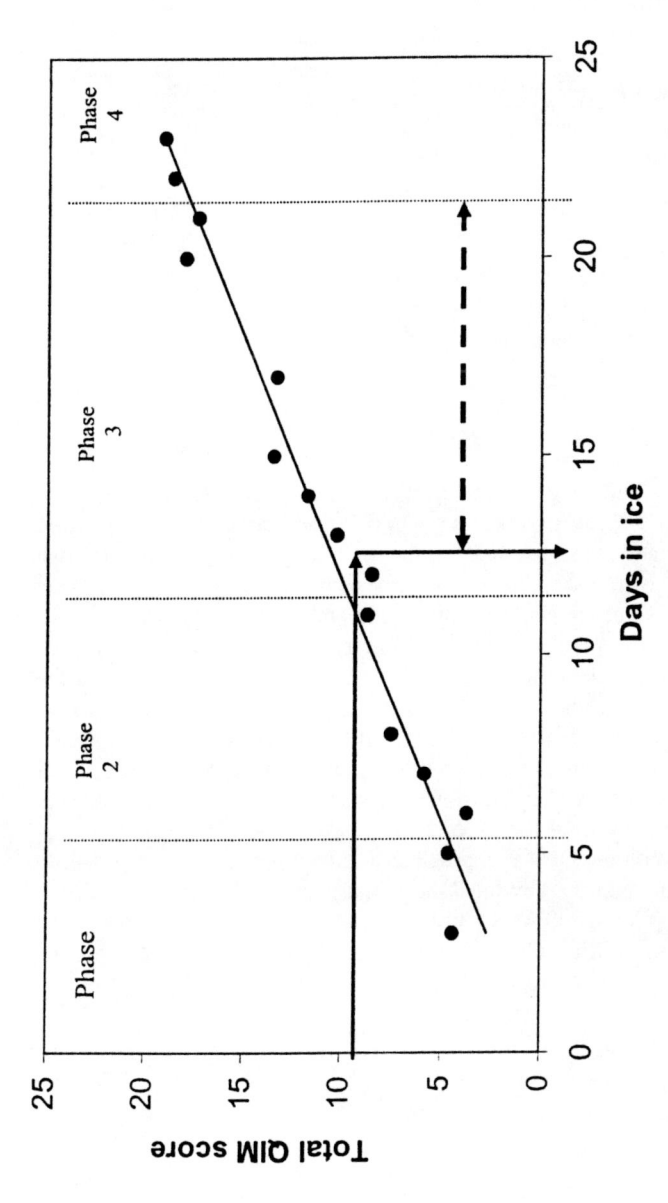

Figure 1. QIM – linear calibration curve (quality index versus days in ice) for salmon ($Y = 0.8299x + 0.1794$; $R^2 = 0.9694$). When a batch of fish reaches a sum of e.g. 10 points it corresponds to 12 days in ice and the remaining shelf life is 9 days in ice.

Table 2. Consumer QIM for assessment of raw whole non-specific fish species

Parameter	Instruction	Description	Point
Appearance	Brightness of skin	Bright	0
		Reduced brightness	1
		Dull	2
Odor	Belly	Sea, Seaweed	0
		Neutral	1
		Off odor	2
Texture	Press with thumb and forefinger along the back of the fish	Firm	0
		Firm/soft	1
		Soft	2
Sum of demerit point			

another in the chain. The QIM can, however, be a significant tool in many parts of the chain because QIM in different modifications contain important key properties that trigger acceptance and that can help to bring a good quality fish to the consumer. It is important to use the same terms/language between buyers and sellers of fish, fish processing, fish inspection and the consumer.

The EU scheme for quality assessment (Anonymous, 1996) are now used in the inspection service and in the fishing industry. There are three levels in the EU-scheme: E (Extra), A and B where E is the highest quality and below B is the level where fish is rejected for human consumption. The primary processors have a demand for more than these three levels. The EU scheme is not designed to be used in Quality Assurance (QA) in the processing industry, where the users attach numbers to the grades and carry out arithmetic on these numbers (Howgate, 2000). With the QIM it is possible to give more detailed information of the sensory quality and thereby fulfil the primary processors demand. For the processor it is important that QIM is a very fast sensory method, and can be used for prediction of shelf life. The influences of transportation/storage on the sensory quality and the remaining shelf life can also be measured by QIM.

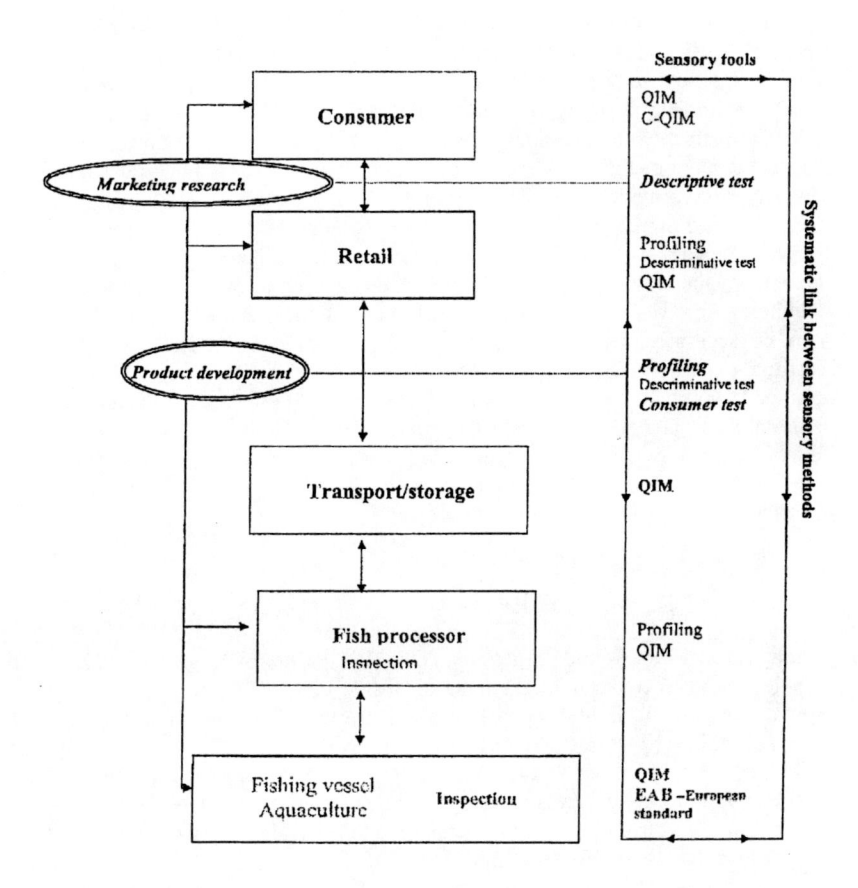

Figure 2. The chain from catch to consumer. Utilization of sensory analysis is shown and the need for a link between the sensory methods is indicated.

The sensory quality of the product the consumer buys depends on the quality of the raw material at the retail. Therefore, the QIM will be very useful for the retailer, but C-QIM can also be important in the shops. Introduction of the C-QIM may help the consumer to get a good sensory quality product and learn more about the quality and its variation. QIM is also suggested as a reference tool for verifying quality labelling of fish.

CONCLUSIONS

It is foreseen that the QIM will be useful to give feedback to fishermen concerning the quality of their catch, which may initiate better handling on board. A so-called 'catch-index' containing QIM points may contribute to quality assurance in the whole chain. Fish processing plants would have a better tool to measure freshness and thereby the possibility to control the freshness stage of their raw material. QIM-evaluation of raw material kept on ice will provide accurate and precise information concerning the freshness and prediction of freshness of fillets to be later inspected by the buyers. Sensory evaluation might also contribute to a 'processing index'. Developments and possibilities provided by the modern computer technology has made it possible to develop an Internet version of QIM that can be accessed (see www.dfu.min.dk/qim). Three European fisheries institutes, namely the Danish Institute of Fisheries Research, RIVO-Netherlands Institute for Fisheries Research and Icelandic Fisheries Laboratories have decided to establish a strategic alliance called QIM Eurofish (http://www.qim.eurofish.com/). The mission is to promote and implement the use of QIM as a versatile quality tool within fisheries distribution or production chains in Europe.

REFERENCES

Anonymous. 1996. Council regulation (EC) No 2406/96 of November 1996 laying down common marketing standard for certain fishery products.

Botta, J.R. 1995. Evaluation of seafood freshness quality. Food Sci. Technol., VHC Publishers Inc. New York, NY.

Bremner, H.A.; Olley, J. and Vail, A.M. A. 1986. Estimating time-temperature effects by a rapid systematic sensory method. In: *Seafood Quality Determination*. D.E. Kramer and J. Liston (eds.), Elsevier, Amsterdam, New York, NY, pp. 413-435

Hyldig, G. and Nielsen, J. 1997. A Rapid Sensory Method for Quality Management. In *Methods to determine the freshness of fish in research and industry*. Institut International du Froid, Paris, pp. 297-305.

Howgate, P. 2000. Personal communication. phowgate@rsc.co.uk

Larsen, E.P.; Heldbo, J.; Jespersen, C.M. and Nielsen, J. 1992. Development of a standard for quality assessment on fish for human consumption. In *Quality Assurance in the Fish Industry*. H.H. Huss, M. Jacobsen and J. Liston (eds.). Elsevier, Amsterdam, The Netherland, pp. 351-358.

Luten, J.B. and Martinsdottir, E. 1998. QIM: a European tool for fish freshness evaluation in the fishery chain. In *Methods to determine the freshness of fish in research and industry*. Institut International du Froid, Paris, France, pp. 287-296.

Jónsdóttir, S.M.; Hyldig, G.; Nielsen, J.; Bleechmore, T. and Silberg, S. 1999. Rapid PC based sensory methods. *Infofish Int.* 2: 54-56.

Ophuis, P.A.M.O. and Van Trijp, H.C.M. 1995. Perceived quality: a market driven and consumer oriented approach. *Food Qual. Pref.* 6: 177-183.

Warm K. 2000. Sensory quality criteria for new and traditional fish species of relevance to consumer needs. Ph.D. thesis, Danish Institute for Fisheries Research and The Royal Veterinary and Agricultural University, Copenhagen, Denmark.

Warm, K.; Nielsen, J.; Hyldig, G. and Martens, M. 2000. Sensory quality of five fish species. *J. Food Quality* 23: 583-602.

York, R.K. and Sereda, L.M. 1993. Sensory assessment of quality in fish and seafood. In *Seafood: Chemistry, Processing Technology and Quality*. F. Shahidi and J.R. Botta (eds.) Blackie Academic and Professional, New York, NY, pp. 234-262.

CHAPTER 8

HIGH PRESSURE PROCESSING OF SEAFOODS

A. LeBail, M. de-Lamballerie-Anton, M. Hayert and D. Chevalier

GEPEA-ENITIAA, la géraudière BP 82225 F-44322 Nantes cedex 3, France

The effect of high pressure (several hundreds of MPa) in biomaterial has been investigated since the last century. High pressure applied to food material has been reinvestigated during the last decades. High pressure causes a volume change of the matter; this change in volume results in modification of the structure of the biomaterial (i.e. proteins) or in a modification of the phase change temperature. From an applied point of view, high pressure can inactivate microorganisms, affects the activity of enzymes, and permit rapid thawing or freezing. The impact of high pressure on seafood, including the effect on proteins, on enzymatic reactions and on micro-organisms is present. The use of high pressure applied to seafood thawing and freezing is also detailed. Present applications and future of high pressure processing applied to seafood is dicsussed.

INTRODUCTION

High pressure processing (HPP) of biological material was pointed out by Bridgman (1914) who showed that a pressure of 700 MPa was able to coagulate albumin. Scientists reactivated research on HPP from the end of the 1980's. This HPP process consists of applying an isostatic pressure to a food (packaged usually) placed in a high-pressure vessel. The compressibility of water yields a slight temperature increase during pressurization and decreases during depressurization (around 4K/100 MPa). Three main applications of HPP are sanitation/stabilization (enzymatic activity or inactivation of microorganisms), texturization (protein denaturation) and phase change. Industrial applications are dealing with pressures of up to 400-500 MPa. The pressurization fluid is water or an aqueous solution (water-glycol for sub-zero applications).

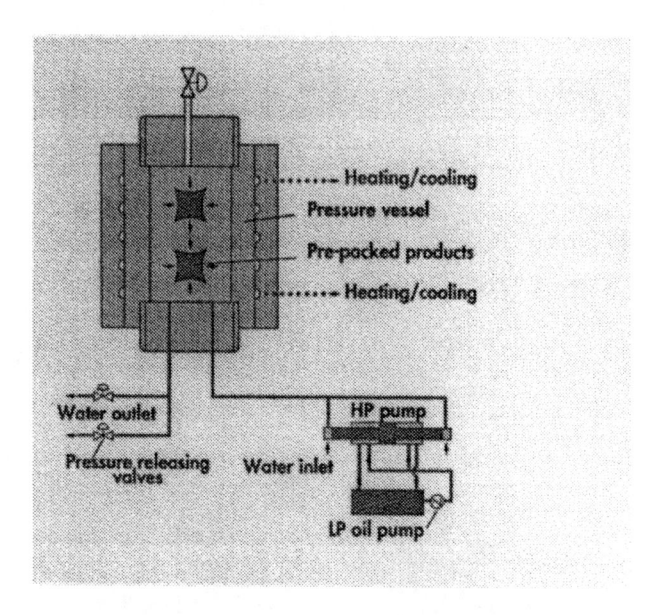

Figure 1. View of a High Pressure unit (ACB-Pressure System, France)

SEAFOOD PROTEINS, COLOR AND HPP

The sensitivity of proteins to high pressure is a function of the matrix in which the protein is treated. Differential scanning calorimetry (DSC) can be used to monitor the denaturation of proteins. Iso and co-workers (1994a,b) have shown that the total enthalpy change was lower for pressure-treated than for control samples. These results (from 7 fish species) showed that HPP and thermal denaturation could be comparable but are not strictly equivalent. Protein is a generic name which comprises sarcoplasmic (water-soluble), stroma (insoluble whatever the ionic strength at mild temperature) and myofibrillar (soluble at high ionic strength – 0.5 M NaCl). *Sarcoplasmic proteins*, which represent 30% of muscular proteins, are mainly glycolytic and ageing enzymes as well as heme proteins. The effect of HPP on these proteins is correlated with the effect of high pressure on enzymes and on muscle color. Okazie and Nahanum (1992) have shown that the texturization of fish sarcoplasmic protein was possible above 50 mg/ml. These authors studied the effect of selected parameters and concluded that HPP can be used for texturization of fish sarcoplasmic proteins. It was also shown that fish species can be identified by

SDS polyacrylamide gel of sarcoplasmic proteins even after HPP (Etienne *et al.*, 2001). *Stroma proteins* which represent 10 % of muscular proteins are made up of collagen and elastin. The contribution of collagen to the cohesiveness of the flesh doesn't seem to be affected by HPP (Suzuki *et al.*, 1993). Moreover, HPP reduces the activity of collagenases (Dufour *et al.*, 1996). *Myofibrillar proteins* are the main fraction of muscle proteins (60%) and can be classified in three groups (contractile proteins-myosin-actin, regulatory proteins- tropomyosin-troponins- α-actinin, β-actinin and Cytoskeletal proteins-nebulin-titin-desmin). HPP has a significant effect on myofibrils, but the effect depends on the matrix in which the product is treated (Suzuki *et al.*, 1991). HPP causes myofibril fragmentation, solubilization and depolymerization of myofibrillar proteins (Ko *et al.*, 1990; Jung *et al.*, 2000). Myofibrillar proteins are thus responsible for the texture and gelation properties. The effect of HPP (100-800 MPa) on the texture of cod muscle, tilapia filets (Kuo and Wen, 1998) carp (Yoshioka and Yamamoto, 1998) and turbot (Chevalier *et al.*, 2001) has been investigated. HPP was found effective for pressure between 150 and 300 MPa, differing from the texture of non-pressurized fish by a chewier and gummier texture. Fish paste (i.e. surimi) texturization is possible with pressure treatment at 200-400 MPa for Alaska pollack (Shoji *et al.*, 1990). Strong gels can be obtained with HPP in comparison to those obtained by thermal treatment. Among process parameters, the salt contents, temperature and pressure holding were most important (Serennes *et al.*, 1996; Kawasaki *et al.*, 1996; Perez *et al.*, 1997). It was shown that high pressure induced gels from blue whiting muscle exhibited a higher breaking force [200 MPa, 3°C, 10 min], but a lower water holding capacity in comparison with a heat induced gel [0 MPa, 37°C-30 min, 90°C-50 min] (Perez *et al.*, 1997) HPP of squid meat paste permitted improvement of its low thermal gelation ability (Nagashima *et al.*, 1993). The modifications induced by HPP on turbot proteins was increasing with increased pressure between 100 and 200 MPa (treatment at 4°C) as shown by DSC (Chevalier *et al.*, 2001). A low temperature treatment (−5°C) induced less protein denaturation than at a higher temperature in carp fillets (Sequeira-Munoz, 2001). Oysters can be shucked by HPP (Voisin, 2001; Mermelstein, 2000). A higher digestibility is expected from high pressure induced gels due to a lower extent of protein denaturation for high pressure gels. Pressure treatment is less effective than heat treatment in the inactivation of endogenous proteases (Sareevoravitkul, 2000), thus indicating that these gels might be less stable during storage.

Color of the flesh of fish is affected by pressures between 150 and 200 MPa, but the color change is always lower than the one obtained by heat treatment (Yoshioka and Yamamoto, 1998). The color modification has been related to protein and lipid denaturation and oxidation, respectively. The L and b parameters tend to increase in the L-a-b reference system (Carlez *et al.*, 1992). The metmyoglobine proportion increased from 37.7 % (atmospheric pressure) to 73.1% (pressure between 150 and 300 MPa) (Murakami *et al.*, 1997), perhaps related to the color change (Carlez *et al.*, 1992). Indeed below 150 MPa, no modification of L, a and b values was observed. The increase of the L value (lightness) is related to the denaturation of globine. The a value (redness) is usually not affected and is related to the oxidation of myoglobine. The b-value (yellowness) is usually affected, but this seems to depend on the species (Shoji *et al.*, 1990; Sareevoravitkul *et al.*, 1996; Lopez *et al.*, 2000; Ashie and Simpson, 1996; Ohshima *et al.*, 1993). The oxidation of lipids seems to be enhanced by HPP (Ohshima *et al.*, 1993) and this oxidation may also modify the color of fish. Color changes due to HPP in seafoods are effective for pressures starting from 150 MPa. This is a function of species, temperature and parameters modifying lipid oxidation (temperature, ionic content, and water activity, among others). Thus, it appears that the effect of HPP and heat on proteins is different. HPP is associated with a moderate thermal treatment (with or without pressure) and may thus be of interest for the gelation of fish paste. HPP offers the possibility of realizing gels with more flexibility on the formulation (i.e. lower salt content).

ENZYMES, LIPIDS AND HPP

Proteolytic enzymes are involved in the deterioration of seafoods. HPP may cause (i) inactivation, (ii) activation or (iii) no effect depending on their activation volume (Cheftel, 1992). The degradation of ATP and its compounds in carp muscle (200, 350 and 500 MPa) was studied (Shoji and Saeki, 1989). The level of IMP (inosine-5'-monophosphate; compound of the ATP degradation responsible of umami taste in seafood) stopped decreasing when fish was treated at 350 and 500 MPa. This phenomenon was attributed to the inactivation of enzymes involved in the degradation of ATP. Accumulation of free fatty acids under HPP of cod muscle during storage (Ohshima *et al.*, 1993). The inhibition of enzyme responsible for phospholipid degradation was observed for pressures higher than 405 MPa and a pressure holding time longer than 15 min. Several enzymes were extracted from two bluefish and sheephead and subjected to a pressure between 100 and 300 MPa with holding times of 5 to 30 min at room temperature (Ashie and Simpson, 1996). It was concluded that seafood enzymes were more sensitive than their warm-adapted bovine counterparts and inactivation was higher in the flesh matrix than for a crude enzyme extract. The protease activity decreased for pressures higher than 200

MPa for cod muscle (Angsupanich and Ledward, 1998). The alkaline protease activity and the acid protease activity were still effective after pressure treatment at 500 and 400 MPa, respectively. The neutral protease was most affected.

The change of Ca^{++} ATPase activity is an indicator of protein change. Results available in the literature showed very different activity changes and no final conclusion can be reached for behavior of Ca^{++} ATPase enzymatic activity due to high pressure treatment. Sequeira-Munoz (2001) worked on carp and observed and increase of Ca ATP activity (140 MPa, 12 min., -14°C) Ashie et al. (1999) didn't observed any change (Tilapia fish, 200 MPa, 20 min, 5°C) and Arai (1977) observed a decrease of Ca-ATPase activity on carp myofibril.

Fish lipids are different from those of land animals in that they exhibit a higher level of polyunsaturated fatty acids, which are susceptible to autoxidation. The corresponding oxidative degradation associated with free fatty acids formation (by glycerolipids hydrolysis) affects the quality of seafoods (Ohshima et al., 1993) (color, flavor, and texture). The oxidative reaction was enhanced by HPP especially in the case of species rich in lipids such as sardines (Takama et al., 1979). An increase of the oxidative deterioration after treatment at 100, 200 and 300 MPa for 10 min was shown (Takama et al., 1979); Tanaha et al. (1991) did not see any influence of HPP on lipids themselves (extracted from the flesh, 500 MPa /60 min, room temperature) whereas the opposite was observed for a model system made of sardine lipids and defatfed meat (50, 100 and 180 MPa for 30 and 60 min at room temperature). Thus, sardine lipids *per se* were partially oxidized during storage at 4°C whereas the oxidation was enhanced in the model system. This result was attributed by some authors to the denaturation of heme proteins by pressure, which facilitated their exposure to lipids (Wada and Ogawa, 1996). Lipid oxidation in sardine meat treated with high pressure was attributed to metal ions. Cod flesh exposed to 202, 404 and 608 MPa for 15 and 30 min showed an increase in peroxide value (indicator of the degree of oxidation) of the extracted oils (Ohshima et al., 1992).

The effect of myoglobin and water holding capacity on lipid oxidation in red meats of bonito and sardine was evaluated (100 and 200 MPa for 30 min at room temperature) (Wada and Ogawa, 1996). A decrease of the lipase activity and the denaturation of myoglobin by HPP (above 200 MPa) were attributed to the hydrophobic bonding between lipid and protein molecules. The change in water holding capacity was correlated with lipid oxidation confirming an association between these two phenomena. It seems that HPP acts on the conformation between protein and lipid and that the protection of lipids by water molecules which normally surround the lipids is affected; leading to an increase of the exchange surface between lipid and oxygen. Little effect of HPP on lipid oxidation below 400 MPa (20 min at room temperature) was observed (Angsupanich and Ledward, 1998), whereas a significant effect was noted for higher pressures. Similarly oxidation did not depend on the presence of oxygen during pressure treatment but rather related to the release of free metal ions (i.e. Fe, Cu) from complexes at around 400 MPa (Tanaka et al., 1991).

Temperature seems to be an important parameter. A decrease of lipid oxidation with the processing time was observed (carp fillets at 80 to 200 MPa for 8 to 22 min at 0-3°C) which is opposite to the general trend of available results in the literature. This was attributed to the low treatment temperature. An increase of free fatty acids formation with pressure level and processing was also observed. The effect of HPP on lipid oxidation seems to be negative in that some modification on lipids or lipids-enzyme system tends to increase lipid oxidation. Nevertheless, some contradictions remain in the literature, especially between species rich in lipids (significant increase of lipid oxidation) and species poor in lipid (not obvious) and further investigations are needed in this domain.

MICROBIAL QUALITY AND HPP OF SEAFOODS

Seafoods preservation can be achieved by freezing, canning, smoking or pickling. Hazards associated with pathogenic organisms are an important problem for seafoods. HPP is already used to sanitize oysters (*vibrio*). The processing time is of a few minutes at a pressure of 400 MPa. Inactivation of micro-organisms is probably the major application of this process for food purposes. The kinetic of inactivation has been studied by numerous authors who defined D and Z values similar to those established for heat treatment (logarithmic inactivation rate), except for the unit of Z which was determined at a given temperature and its unit is atmospheres (Butz et al., 1996; Ludwig et al., 1992; Mussa et al., 1999). The efficiency of HPP depends on the order (bacteria, yeast) and form (vegetative microorganisms) but also on the physiological stage. Cells in an exponential stage are more sensitive to pressure than in the stationary stage (Ochiai and Nakagawa, 1991; Hoover et al., 1989; Iwahashi et al., 1994). In a general trend, the gram-positive micro-organisms are more resistant to pressure than their gram negative counterparts. Another general trend concerns the temperature. HPP associated to temperature lower or higher than room temperature (i.e. 10 to 20°C below or above) yields the highest inactivation. Nevertheless, there is a lack of data in the subzero domain due to the interaction of phase change of water. Inactivation concerns mainly vegetative form; indeed, the pressure level required for spore inactivation is far too high to consider HPP for this purpose (above 800 MPa) (Sale et al., 1970; Okazaki et al., 1994). A reorganization of the cytosquelettum has been observed at 100–150 MPa (Hamada *et al.*, 1992; Sato *et al.*, 1996), but these modifications appear to disappear after a lag time (Kobori *et al.*, 1995). Membrane system seems to be affected especially at low temperatures and high pressures (Shimada *et al.*, 1993; Osumi *et al.*, 1992).

Sanitation of oysters is nowadays the first use in term of volume of HPP at an industrial scale for seafoods. Some companies such as Motivatit Sea Foods Inc. (Houma, LA) use HPP (batch process) to reduce harmful bacteria, mainly *vibrios*, to non-detectable levels to extend the shelf life. High pressure thawing seems to be accompanied by a significant inactivation with respect to the pressure level (below 200 MPa). Limited results are available in the literature for this application (Murakami *et al.*, 1997; LeBail *et al.*, 2000). HPP can significantly increase the shelf life of most seafoods (Ohshima, *et al.*, 1993; Miyao *et al.*, 1993; Lopez *et al.*, 2000). The temperature, pressure level and holding time are directly related to the rate of inactivation; below 100 MPa, no effect was observed. The pressurization phase and the depressurization phase seem to play a significant role on the overall inactivation even though there is no clear explanation about this in the literature (Ludwig *et al.*, 1992; Styles *et al.*, 1991; Oxen and Knorr, 1993; Heinz and Knorr, 1996; Smelt, 1998). First order kinetic is generally assumed. The effect of the compression/decompression kinetic has been studied by Makita (1992) who observed a correlation between the number of cycles and the degree of inactivation. Working at low temperatures (either positive or negative) enhances the effect of HPP on inactivation. The use of subzero temperatures combined with HPP (HP-BT Process of High Pressure – Low Temperature) appears as a promising field of investigation (Takahashi, 1992; Hashizume *et al.*, 1995). Other parameters such as pH and water activity have to be considered in the same way as processing under atmospheric pressure. It has also been shown that low pH (<5) increases the inactivation during HPP even though some contradictions have been observed for many micro-organisms depending on the matrix in which the HPP was realized (Kajiyama *et al.*, 1993; Mackey *et al.*, 1995). High concentration of NaCl, sugar and polyol may protect micro-organisms from the effects of HPP (Fujii *et al.*, 1996; Ogawa *et al.*, 1992).

In short, it appears that the treatment of seafoods which are often consumed raw (i.e. suchi, sachimi and sea urchin eggs) could be a promising field of application for HPP. Nevertheless, the color change observed from 150 to 200 MPa (increase of the lightness) remains a limitation if the product has to be eaten crude and if a "natural" color is expected by the consumer.

PHASE CHANGE: APPLICATION TO FREEZING AND THAWING

Water under pressure has been studied by Bridgman (1912) who established its phase diagram for pressures of up to 14,000 atmospheres (see Figure 2). The depression of the melting temperature of water is observed up to 210 MPa. Above this pressure the melting point increases.

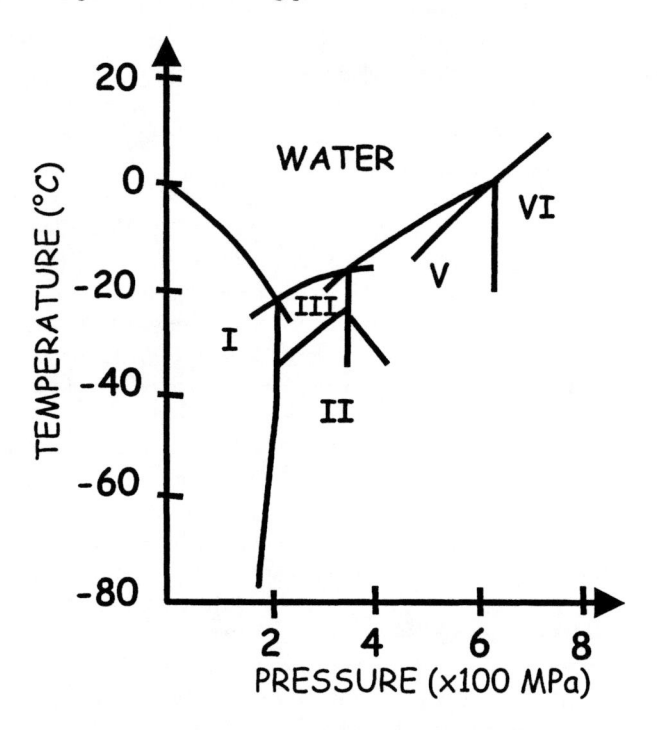

Figure 2. Phase diagram of water and ice under pressure (Bridgman, 1912). I, II, III, IV and V correspond to different ice crystal structure

Freezing of food under pressure can be realized through different pathways (Kalichevsky *et al.*, 1995). Pressure shift freezing (PSF) is probably the most interesting in that it permits increasing the ice-nucleation rate. It consists of refrigerating a food under pressure below the initial freezing temperature of the food. A rapid pressure release permits achieving a high supercooling and by the way numerous fine ice crystals. Supercooling can be as high as 10°K (Chevalier *et al.*, 200b). The maximum mass ratio of ice during adiabatic depressurization of water has been calculated (Sanz *et al.*, 1997) and is 0.36 kg ice/kg of water (pressure release from 210 MPa at -20°C). Freezing has to be completed in a conventional process.

A recent review on pressure assisted freezing and thawing has appeared (LeBail *et al.*, 2001). Fuchigami applied PSF to soybean curd, carrots, Chinese cabbage, eggplant and fruits, respectively (Fuchigami and Teramoto *et al.*, 1997; Fuchigami *et al.*, 1997; Fuchigami *et al.*, 1997; Fuchigami *et al.*, 1998; Fuchigami *et al.*, 1998; Otero *et al.*, 1998; Otero *et al.*, 2000). PSF has been applied to Norway lobster using SEM a refined microstructure (Chevalier *et al.*, 2000). PSF yielded a harder texture which was attributed to a synergistic effect of high pressure and freezing. More recently, PSF was applied to Turbot *Scophthalmus maximus*) and Carp (*Cyprinus carpio*) (Sequeira-Munoz, 2001; Chevalier *et al.*, 2000). All previous authors have confirmed the improvement of the microstructure (size of the ice crystals) with pressure. Chevalier *et al.*, (2000) observed a ten factor in the surface of the ice crystals (conventional freezing vs. PSF) and showed that the size of the ice crystals didn't change during storage (65 days). The pre rigor freezing yielded smaller ice crystals than post rigor conditions (Sequeira-Munoz, 2001). The total drip (thawing drip + cooking drip) could also be reduced by 50 % between PSF and air blast freezing (Sequeira-Munoz, 2001). The reduction observed for Turbot was not so important (10 to 30% reduction) (Chevalier *et al.*, 2000). PSF appears as a promising technology even though it might induce a significant additional cost in comparison to conventional processes. No study has been done on the effect of PSF on the microbial quality of food. The benefit in term of microstructure is quite obvious and this process will allow the production of a very high quality frozen food.

High pressure thawing can be considered as the opposite process. It consists of thawing of food under pressure taking advantage of the depression of its melting temperature. This phenomenon permits increasing the heat flux to the melting zone. Three major advantages are pointed out in the literature, namely (i) a reduction of the thawing time, (ii) inactivation of micro-organisms and (iii) reduction of he drip volume (LeBail *et al.*, 2001). The thawing time reduction is proportional to the increase of the temperature gap ratio between the melting temperature and the vessel temperature at atmospheric and under pressure (LeBail *et al.*, Chourot *et al.*, 1997). Available studies of high pressure thawing applied to food concerns beef (Deuchi and Hayashi, 1992), tuna and surimi (Takai *et al.*, 1991, tuna (Murakami *et al.*, 1992), and whiting (Chevalier *et al.*, 1999).

The reduction of the thawing time is the point of agreement by all authors. The reduction of drip volume was observed by Murakami *et al.* (1992) whereas this aspect was studied more deeply by Chavalier *et al.* (1999) who showed that a minimum drip volume (thawing and cooking) was obtained at 150 MPa with whiting. Moreover, it seemed that the drip volume reduction was related to the duration of the pressurisation. A lower drip volume was observed for longer pressurisation. This result was explained by a mass transfer between the drip and the tissue under pressure. The texturisation of the flesh was observed by Chevalier *et al.* (1999) on whiting filets and correlated to the extent of protein denaturation (observed by SDS electrophoresis). High pressure thawing of selected seafoods was proposed by Rouille *et al.* 2001 who also observed the positive effects of pressurisation duration on drip for aiguillat fish and scallops. Microbial inactivation was firstly observed (no growth) by Murakami *et al.* (1992) on tuna. More recently LeBail *et al.* (2000) obtained between 2 and 4 log-cycle reductions (indigenous microorganisms) in salmon during thawing at 200 MPa/20°C and 200 MPa/5°C, respectively. Similar results (significant reduction) were obtained during high pressure thawing of salmon flesh inoculated with *Listeria innocua* and *Schewanella putrefaciens* (10^6 CFU/g initial inoculation). These results tend to confirm the strong interest of high pressure thawing in term of sanitation. This process enjoys several major advantages. The reduction of the thawing time by a 2 to 5 factor was the most obvious one. Further investigations need to be done to evaluate the reduction of the drip volume and the reduction of the microbial load.

INDUSTRIAL APPLICATIONS AND CONCLUSIONS

Three applications of HPP are presented in Figure 3. Texturation can be used to process fish or seafood paste in order to texturize a preparation. Sanitation can potentially inactivate vegetative micro-organisms of can modify the enzymatic activity. HP thawing of HP freezing permits to achieve high phase change rate with benefit in terms of microstructure or texture. These potential applications are presented below and then discussed. Recent applications and outlook on existing or potential applications of HPP in the seafood industry are also proposed.

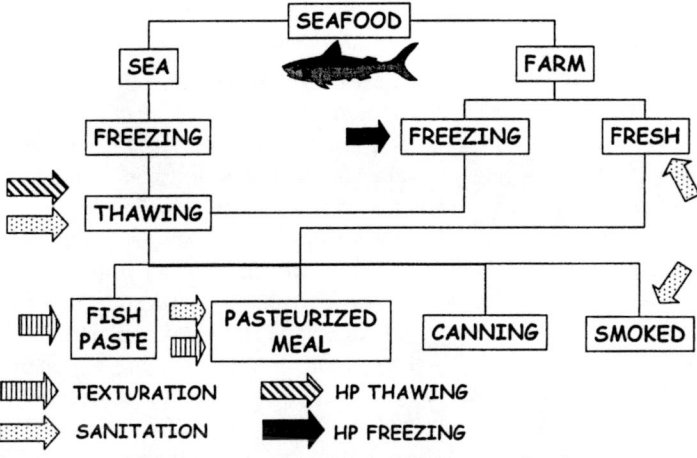

Figure 3. Applications of HPP to seafood.

Marinated kipper and plain kipper processed by high pressure are sold in Japan since 2000. The treatment is realised at low temperature in the frozen state, probably to sanitise the surface of the fish. Other applications are not related to seafoods, but will certainly help in promoting the application of HPP to seafoods (Sanitation of fruit products such as juices, jam, avocado paste, HP cooking of rice, sanitation of Ham–Espuna Spain, etc.). The pressure level of the industrial equipment is between 400 and 500 MPa. HPP is usually associated with a control of the temperature of the vessel and with a refrigerated storage. High pressure low temperature (HPLT) inactivation of micro-organisms appears as a promising field of application for which very few studies are available. Indeed, the researcher has to deal with pressure, temperature, and phase change (melting, freezing of water) which make the investigation more complicated. The use of HPP to sanitise and to shuck oysters is the major industrial application of this technology at the present time (3 companies are operating this process in the USA), as presented by Lopez *et al.* (2000) and Hermelstein. A patent has recently been published on the use of HPP applied to shellfish (Voisin, 2001). HPP could be applied for the treatment of wastewater from marine products (Kajiyama *et al.*, 1992). HPP was applied (up to 400 MPa) accompanied by a specific treatment such as filtration, and centrifugation coagulation; HPP allowed a better retention of proteins when filtration was used.

The cost of the equipment remains to be one the limitation of this process even though the cost of the HPP is not unaffordable (evaluated between 0.1 and 1 US $/kg but in most of the case is below 0.5 US $ /kg). It seems that, the numerous applications recently developed, the food industry is changing its mind and that the cost of HPP equipment may reach an acceptable level. High pressure thawing will probably be the next application in food industry because it solves most of the processing challenges (time, microbial quality and drip). Food texturation (gel forming, etc) might also be a future application in term of gel quality and to the possibility of combining inactivation and texturation with moderate temperatures. More generally, the results available in the literature show that HPP has to be combined with a good control of the temperature during treatment but also after treatment (storage refrigerated). More research is needed on the formulation and process parameters. The impact of HPP on the digestibility could provide enormous advantages as proteins are not denatured in the same way by HP as by thermal treatment. Research is needed to demonstrate a potential interest in this domain and development of regulations (i.e. the European Community NOVEL FOOD regulation, 1997) which require the industry to demonstrate that a new process yields safe and healthy products to the consumer.

HPP represents a new challenge to the food industry; after 10 years of intense investigation, numerous industrial applications have recently appeared (mainly in 2000) whereas existing applications are still growing (sanitation of fruit products mainly). Fundamental research is still needed and will facilitate further development of this innovative technology. Toxicology and studies on the biochemical modification of macromolecule involved by HPP will be of the first order of importance in the near future in order to address the requirements of the new regulations and to guarantee safe and healthy products for consumers. Seafood, a fragile tissue, seems to be a good candidate for high pressure processing.

ACKNOWLEDGEMENT

The authors would like to thank the Région Pays de Loire, The Pole Agronomique Ouest (PAO) and the French Ministry of Agriculture for their financial support.

REFERENCES

Angsupanich, K. and Ledward, D.A. 1998. High pressure treatment effects on cod (Gadus morhua) muscle. *Food Chem.* 63: 39-50.

Ashie, I.N.A.; Lanier, T.C. and MacDonald, G.A. 1999. Pressure-induced denaturation of muscle proteins and its prevention by sugars and polyols. *J. Food Sci.* 64: 818-822.

Ashie, I.N.A. and Simpson, B.K. 1996. Application of high hydrostatic pressure to control enzyme related fresh seafood texture deterioration. *Food Res. Int.* 29: 569-575.

Bridgman, P.W. 1914. The coagulation of albumin by pressure. *J. Biol. Chem.* 19: 511.

Bridgman, P.W. 1912b. Water in the liquid and five solid forms under pressure. *Proc. Am. Acad. Arts Sci.* 47: 411-558.

Butz, P.; Funtenberger, S.; Haberditzl, T. and Tauscher, B. 1996. High pressure inactivation of Byssochlamys nivea ascospores and other heat resistant moulds. *Lebensm. Wiss. und Technol.* 29: 404-410.

Carlez, A.; Cheftel, J.C.; Rosec, J.P.; Richard, N.; Saldana, J.L. and Balny, C. 1992. Effects of high pressure and bacteriostatic agents on the destruction of Citrobacter freundii in minced beef muscle. *High Pressure Biotechnol.* 224: 365-368.

Cheftel, J.C. 1992. Effect of high hydrostatic pressure on food constituents: an overview, in High Pressure and Biotechnology, Colloque INSERM, R.H. C. Balny, K. Heremans and P. Masson (eds.). John Libbey Eurotext, London, UK, pp. 195-209.

Chevalier, D.; LeBail, A. and Ghoul, M. 2001. Effects of high pressure treatment (100-200MPa) at low temperature on turbot (*Scophthalmus maximus*) muscle. *Food Res. Int.* 34: 425-430.

Chevalier, D.; Bail, A.; Chourot, J.M. and Chantreau, P. 1999. High pressure thawing of fish (whiting): influence of the process parameters on drip losses. *Lebensm. Wiss. und Technol.* 32: 25-31.

Chevalier, D.; Sequeira-Munoz, A.; LeBail, A.; Simpson, B.K. and Ghoul, M. 2000. Effect of freezing conditions and storage on ice crystal and drip volume in turbot (Scophthalmus maximus). Evaluation of pressure shift freezing vs. air-blast freezing. *Innov. Food Sci. Emerg. Technol.* 1: 193-202.

Chevalier, D.; LeBail, A. and Ghoul, M. 2000b. Freezing and ice crystals formed in cylindrical model food. Part II. Comparison between freezing at atmospheric pressure and pressure shift freezing. *J. Food Eng.* 46: 287-293.

Chevalier, D.; Sentissi, M.; Havet, M. and Bail, A. 2000. Comparison of airblast and pressure shift freezing on Norway lobster quality. *J. Food Sci.* 65: 329-333.

Chourot, J.M.; Boillereaux, L.; Havet, M. and Bail, A. 1997. Numerical modeling of high pressure thawing: application to water thawing. *J. Food Eng.* 34: 63-75.

Deuchi, T. and Hayashi, R. 1992. High pressure treatements at subzero temperature: application to preservation, rapid freezing and rapid thawing of foods. In: *High Pressure and Biotechnology.* INSERM, C.B., et al. (eds.) John Libbey Eurotext Ltd., London, UK, pp. 353-355.

Dufour, E.; Dalgalarrondo, M.; Hervé, G.; Goutefongea, R. and Haertlé, T. 1996. Proteolysis of type III collagen by collagenase and cathepsin B under high hydrostatic pressure. *Meat Sci.* 42: 261-269.

Etienne, M.; Jerome, M.; Fleurence, J.; Rehbein, H.; Kuendiger, R.; Mendes, R.; Costa, H. and Martinez, I. 2001. Species identification of formed fishery products and high pressure-treated fish by electrophoresis: a collaborative study. *Food Chem.* 72: 105-112.

EuropeanCommunity, Réglement CE n° 258/97 du Parlement Européen et du Conseil du 27 janvier 1997 relatif aux nouveaux ingrédients alimentaires. In: *Journal Officiel des Communautés Européennes.* 27 January 1997, pp. L 43-7.

Fuchigami, M.; Kato, N. and Teramoto, A. 1998. High-pressure-freezing effects on textural quality of Chinese cabbage. *J. Food Sci.* 63: 122-125.

Fuchigami, M. and Teramoto, A. 1997. Structural and textural changes in kinu-tofu due to high-pressure-freezing. *J. Food Sci.* 62: 828-832.

Fuchigami, M.; Teramoto, A. and Ogawa, N. 1998. Structural and textural quality of kinu-tofu frozen-then-thawed at high-pressure. *J. Food Sci.* 63: 1054-1057.

Fuchigami, M.; Kato, N. and Teramoto, A. 1997. High-pressure-freezing effects on textural quality of carrots. *J. Food Sci.* 62: 804-808.

Fuchigami, M.; Miyazaki, K.; Kato, N. and Teramoto, A. 1997. Histological changes in high-pressure-frozen carrots. *J. Food Sci.* 62: 809-812.

Fujii, S.; Obuchi, K.; Iwahashi, H.; Fujii, T. and Komatsu, Y. 1996. Saccharides that protect yeast against hydrostatic pressure stress correlated to the mean number of equatorial OH groups. *Biosci. Biotech. Biochem.* 60: 476-478.

Hamada, K.; Nakatomi, Y. and Shimada, S. 1992. Direct induction of tetraploids or homozygous diploids in the industrial yeast, Saccharomyces cerevisiœ by hydrostatic pressure. *Current Genetic* 22: 371-376.

Hashizume, C.; Kimura, K. and Hayashi, R. 1995. Kinetic analysis of yeast inactivation by high pressure treatment at low temperatures. *Biosci. Biotech. Biochem.* 59: 1455-1458.

Heinz, V. and Knorr, D. 1996. High pressure inactivation kinetics of Bacillus subtilis cells by a three-state-model considering distributed resistance mechanisms. *Food Biotechnol.* 10: 149-161.

Hermelstein, N.H. 2000. Seafood processing. *Food Technol.* 54: 66-68.

Hoover, D.G.; Metrick, C.; Papineau, A.M.; Farkas, D.F. and Knorr, D. 1989. Biological effects of high hydrostatic pressure on food microorganisms. *Food Technol.* 43: 99-107.

Iso, S.; Mizuno, H.; Ogawa, H.; Mochizuki, Y.; Mihori, T. and Iso, N. 1994. Physico-chemical properties of pressurized carp meat. *Fisheries Sci.* 60: 89-91.

Iso, S.; Mizuno, H.; Ogawa, H.; Mochizuki, Y. and Iso, N. 1994. Differential scanning calorimetry of pressurized fish meat. *Fisheries Science*, 60: 127-128.

Iwahashi, H.; Obuchi, K.; Fujii, S.; Kaul, S.C. and Komatsu, Y. 1994. Hydrostatic pressure is like high temperature and oxidative stress in the damage it causes to yeast - Saccharomyces cerevisiae. *High Pressure Biosci.* 4: 136-143.

Jung, S.; Lamballerie, A.M.D. and Ghoul, M. 2000. Modifications of ultrastructure and myofibrillar proteins of post-rigor beef treated by high pressure. *Lebensm. Wiss. und Technol.* 33: 313-319.

Kajiyama, N.; Akizumi, K.; Abei, K.; Nagata, M.; Egashira, T. and Miyake, Y. 1993. Sterilization of Escherichia coli by high pressure. *J. Jpn. Soc. Food Sci.* 40: 406-413.

Kajiyama, N.; Imada, I.; Abei, K.; Egashira, T. and Miyake, Y. 1992. The treatment of waste water of marine products by high pressure. *Nippon Shokuhin Kogyo Gakkaishi* 39: 641-646.

Kalichevsky, M.T.; Knorr, D. and Lillford, P.J. 1995. Potential food applications of high-pressure effects on ice-water transitions. *Trends Food Sci. Technol.* 6: 253-259.

Kawasaki, K.; Funatsu, Y. and Ito, Y. 1996. Texturization with frozen surimi and sardine lipid by high pressure treatment. *J. Jpn. Soc. Food Sci. Technol.* [Nippon Shokuhin Kogyo Gakkaishi] 43: 146-156.

Ko, W.C.; Tanaka, M.; Nagashima, Y.; Taguchi, T. and Amano, K. 1990. Effect of high pressure treatment on the thermal gelation of sardine and Alaska pollack meat and myosin pastes. *J. Jpn. Soc. Food Sci. Tech.* [Nippon Shokuhin Kogyo Gakkaishi] 37: 637-642.

Kobori, H.; Sato, M.; Tameike, A.; Hamada, K.; Shimada, S. and Osumi, M. 1995. Ultrastructural effects of pressure stress to the nucleus in Saccharomyces cerevisiae: a study by immunoelectron microscopy using frozen thin sections. *FEMS Microbiol. Lett.* 132: 253-258.

Kuo, C.H. and Wen, C.K. 1998. Changes in processing quality of tilapia meat stored under high hydrostatic pressure. *Food Sci. Taiwan* 25: 428-436.

LeBail, A.; Mussa, D.; Rouillé, J.; Ramaswamy, H.S.; Chapleau, N.; Anton, M.; Hayert, M.; Boillereaux, L. and Chevalier, D. 2000. High Pressure Thawing. Application to selected sea-foods. In *High Pressure Bioscience, Biotechnology*. Elsevier, Kyoto, Japan.

LeBail, A.; Chevalier, D.; Mussa, D. and Ghoul, M. 2002. High pressure freezing and thawing of foods; a review. *Int. J. Refrigeration.* 25: 504-513.

Lopez, C.M.E.; Perez, M.M.; Montero, P. and Borderias, A.J. 2000. Oyster preservation by high-pressure treatment. *J. Food Protect.* 63: 196-201.

Lopez, C.M.E.; Perez, M.M.; Borderias, J.A. and Montero, P. 2000. Extension of the shelf life of prawns (Penaeus japonicus) by vacuum packaging and high-pressure treatment. *J. Food Protec.* 63: 1381-1388.

Ludwig, H.; Bieler, C.; Hallbauer, K. and Scigalla, W. 1992. Inactivation of microorganisms by hydrostatic pressure. In *High Pressure and Biotechnology*, C. Balny (ed.) INSERM/John Libbey Eurotext Ltd, Montrouge, pp. 25-32.

Mackey, B.M.; Forestière, K. and Isaacs, N. 1995. Factors affecting the resistance of Listeria monocytogenes to high hydrostatic pressure. *Food Biotechnol.* 9: 1-11.

Makita, T. 1992. Application of high-pressure and thermophysical properties of water to biotechnology. *Fluid Phase Equilibr.* 76: 87-95.

Mermelstein, N.H. 2000. Seafood processing. *Food Technol.* 54: 66-68.

Miyao, S.; Shindoh, T.; Miyamori, K. and Arita, T. 1993. [Effects of high pressure processing on the growth of bacteria derived from surimi (fish paste).]. *J. Jpn. Soc. Food Sci. Technol.* [Nippon Shokuhin Kogyo Gakkaishi] 40: 478-484.

Murakami, T.; Kimura, I.; Miyakawa, H.; Sugimotot, M. and Sataka, M. 1997. High pressure thawing of frozen fish. In *High Pressure Bioscience*. K.R.S. Hayashi; S. Shimada; A. Suzuki, (eds.). San-Ei Suppan Co., Kyoto, Japan, pp. 304-311.

Murakami, T.; Kimura, I.; Yamagashi, T. and Sujimoto, M. 1992. Thawing of frozen fish by hydrostatic pressure, High Pressure and biotechnology. In *Colloque*, C.B. (ed.). INSERM/John Libbey Eurotext Ltd, London, pp. 329-331.

Mussa, D.M.; Ramaswamy, H.S. and Smith, J.P. 1999. High-pressure destruction kinetics of Listeria monocytogenes on pork. *J. Food Protect.* 62: 40-45.

Nagashima, Y.; Ebina, H.; Tanaka, M. and Taguchi, T. 1993. Effect of high hydrostatic pressure on the thermal gelation of squid mantle meat. *Food Res. Int.* 26: 119-123.

Ochiai, S. and Nakagawa, Y. 1991. [Sterilization by high-pressure treatment. I. Sterilization effect against yeasts]. *J. Antibact. Antifung. Agents*, Japan, 19: 269-273.

Ogawa, H.; Fukuhisa, K. and Fukumoto, H. 1992. Effect of hydrostatic pressure on sterilization and preservation of citrus juice. *High Pressure Biotechnol.* 224: 269-278.

Ohshima, T.; Nakagawa, T. and Koizumi, C. 1992. Effect of high hydrostatic pressure on the enzymatic degradation of phospholipids in fish muscle during storage. In *Seafood Science and Technology*, E.G. Bligh, (ed.) Fishing News Books, Oxford, UK, pp. 64-75.

Ohshima, T.; Ushio, H. and Koizumi, C. 1993. High-pressure processing of fish and fish products. *Trends Food Sci. Technol.* 4: 370-375.

Ohshima, T.; Ushio, H. and Koizumi, C. 1993. High pressure processing of fish and fish products. *Trends Food Sci. Technol.* 4: 370-375.

Okazaki, E. and Nakamura, K. 1992. Factors influencing texturization of sarcoplasmic protein of fish by high pressure treatment. Nippon Suisan Gakkaishi 58: 2197-2206.

Okazaki, T.; Yoneda, T. and Suzuki, K. 1994. Combined effects of temperature and pressure on sterilization of Bacillus subtilis spores. *J. Jpn. Soc. Food Sci. Technol.* [Nippon Shokuhin Kogyo Gakkaishi] 41: 536-541.

Osumi, M.; Yamada, N.; Sato, M.; Kobori, H.; Shimada, S. and Hayashi, R. 1992. Pressure effects on yeast cell ultrastructure: change in the ultrastructure and cytoskeleton on the dimorphic yeast, Candida tropicalis. in Colloque INSERM - High pressure and biotechnology. John Libbey Eurotext Ltd.

Otero, L.; Solas, M.T.; Sanz, P.D.; Elvira, C.D. and Carrasco, J.A. 1998. Contrasting effects of high-pressure-assisted freezing and conventional air-freezing on eggplant tissue microstructure. Zeitschrift fuer Lebensmittel Untersuchung und Forschung. *Food Res. Technol.* 206: 338-342.

Otero, L.; Martino, M.; Zaritzky, N.; Solas, M. and Sanz, P.D. 2000. Preservation of microstructure in peach and mango during high-pressure-shift freezing. *J. Food Sci.* 65: 466-470.

Oxen, P. and Knorr, D. 1993. Baroprotective effects of high solute concentrations against inactivation of Rhodotorula rubra. *Food Sci. Technol. Int.* 26: 220-223.

Perez, M.M.; Lourenco, H.; Montero, P. and Borderias, A.J. 1997. Rheological and biochemical characteristics of high-pressure- and heat-induced gels from blue whiting (Micromesistius poutassou) muscle proteins. *J. Agric. Food Chem.* 45: 44-49.

Rouillé, J.; LeBail, A.; Ramaswamy, H.S. and Leclerc, L. 2002. High pressure thawing of fish and shelfish. *J. Food Eng.* 53: 83-88.

Sale, A.J.H.; Gould, G.W. and Hamilton, W.A. 1970. Inactivation of bacterial spores by hydrostatic pressure. *J. Gen. Microbiol.* 60: 323-334.

Sanz, P.D.; Otero, L.; deElvira, C. and Carrasco, J.A. 1997. Freezing processes in high pressure domains. *Int. J. Refriger.* 20: 301-307.

Sareevoravitkul, R.; Simpson, B.K. and Ramaswamy, H.S. 1996. Comparative properties of bluefish (Pomatomus saltatrix) gels formulated by high hydrostatic pressure and heat. *J. Aquat. Food Prod. Technol.* 5: 65-79.

Sato, M.; Kobori, H.; Ishijima, S.A.; Hai Feng, Z.; Hamada, K.; Shimada, S. and Osumi, M. 1996. Schizosaccharomyces pombe is more sensitive to pressure stress than Saccharomyces cerevisiae. *Cell Struct. Funct.* 21: 167-174.

Sequeira-Munoz, A. Use of high pressure for improving the quality and shelf life of frozen fish, in Dept. of Food Science and Agricultural Chemistry. 2001, McGill University: Montreal, Canada, June 2001.

Serennes, F.; Chopin, C.; Mastail, M. and Vallet, J.L. 1996. Influence of high pressure on texturization of coalfish (Pollachius virens) pulp. *Sciences des Aliments* 16: 307-316.

Shimada, S.; Andou, M.; Naito, N.; Yamada, N.; Osumi, M. and Hayashi, R. 1993. Effects of hydrostatic pressure on the ultrastructure and leakage of internal substances in the yeast Saccharomyces cerevisiae. *Appl. Microbiol. Biotechnol.* 40: 123-131.

Shoji, T.; Saeki, H.; Wakameda, A.; Nakamura, M. and Nonaka, M. 1990. [Gelation of salted paste of Alaska pollack by high hydrostatic pressure and changes in myofibrillar proteins.]. *Bull. Jap. Soc. Sci. Fish.* [Nihon Suisan Gakkaishi] 56: 2069-2076.

Shoji, T. and Saeki, H. 1989. Use of high pressure in food. R. Hayashi, Editor. San-ei Publications: Kyoto, Japan, pp. 75-87.

Smelt, J.P.P.M. 1998. Recent advances in the microbiology of high pressure processing. *Trends Food Sci. Technol.* 9: 152-158.

Styles, M.F.; Hoover, D.G. and Farkas, D.F. 1991. Response of Listeria monocytogenes and Vibrio parahaemolyticus to high hydrostatic pressure. *J. Food Sci.* 56: 1404-1407.

Suzuki, A.; Suzuki, N.; Ikeuchi, Y. and Saito, M. 1991. Effects of high pressure treatment on the ultrastructure and solubilization of isolated myofibrils. *Agric. Biol. Chem.* 55: 2467-2473.

Suzuki, A.; Watanabe, M.; Ikeuchi, Y.; Saito, M. and Takahashi, K. 1993. Effects of high-pressure treatment on the ultrastructure and thermal behaviour of beef intramuscular collagen. *Meat Sci.* 35: 17-25.

Takahashi, K. 1992. Sterilization of microorganisms by hydrostatic pressure at low temperature. In: *High Pressure and Biotechnology.* C. Balny, *et al.* (eds). INSERM/John Libbey Eurotext Ltd, Montrouge, France, pp. 303-307.

Takai, R.; Kozhima, T. and Suzuki, T. 1991. Low temperature thawing by using high Pressure. in 17éme congrés international du froid. Montréal (Québec), CANADA.

Takama, K.; Hatano, M. and Zama, K. 1979. Developments in the protein utilization of abundantly caught fatty fish, V. Effects of pressure treatments on eliminated lipids and recovered proteins in fish. Bull. Faculty of Fisheries, Hokkaido University, Japan, 1: 74-83.

Tanaka, M.; Zhuo, X.Y.; Nagashima, Y. and Taguchi, T. 1991. Effect of high pressure on the lipid oxidation in sardine meat. *Bull. Jap. Soc. Sci. Fisheries* [Nihon Suisan Gakkaishi] 57: 957-963.

Voisin, E. 2001. Process of elimination of bacteria in shellfish, of shucking shellfish and an apparatus therefor. U.S. Patent 6,217,435.

Wada, S. and Ogawa, Y. 1996. High pressure effects on fish lipid degradation: myoglobin change and water holding capacity. In *High Pressure Bioscience and Biotechnology.* R.H.A.C. Balny (ed.) Elsevier Science B.V., London, UK, pp. 351-356.

Yoshioka, K. and Yamamoto, T. 1998. Changes of ultrastructure and the physical properties of carp muscle by high pressurization. *Fisheries Sci.* 64: 89-94.

CHAPTER 9

INTERACTION OF RAW MATERIAL CHARACTERISTICS AND SMOKING PROCESS ON SMOKED SALMON QUALITY

Mireille Cardinal[1], Camille Knockaert[1], Ole Torrissen[2], Sjofn Sigurgisladottir[3], Turid Mørkøre[4], Magny Thomassen[4], and Jean Luc Vallet[1]

[1]Laboratoire Génie Alimentaire, IFREMER, Nantes Cedex 3, France
[2]Inst. Marine Research, Austevoll Aquaculture Research Station, Storebo, Norway
[3]Matra, Technological Institute of Iceland, Keldnaholt, Iceland
[4]AKVAFORSK, Institute of Aquaculture Research Ltd, Aas, Norway

The relation between smoking parameters and characteristics of salmon raw material was investigated with respect to yield, color, flesh content of phenol and salt, and sensory properties. Three sets of farmed salmon, one from Iceland and two from Norway, with a fat content of 9.4, 20 and 16.8%, respectively were analyzed. Seven treatments were applied on fresh or frozen raw material combining dry or brine salting with cold smoking at 20 or 30°C. Electrostatic smoking was tested on dry-salted salmon fillets. The results show a lower yield after filleting of small size fish. Although freezing had little effect on yield, total loss was slightly greater, especially for fish with low fat content. Sensory differences were also apparent. The brine salting reduced the loss. Fish with higher fat content showed a better yield after processing. With brine salting, salt uptake was higher for smaller, leaner fish. The phenol content of flesh depended on the technique and/or smoking temperature used, regardless of the fish studied. However, for a smoking temperature of 30°C, the flesh of smaller, leaner fish showed a higher phenol level. Smoking conditions and preliminary treatment such as freezing produced similar differences in sensory characteristics, regardless of the fish studied, although smaller, leaner individuals appeared to be more sensitive to these processes.

© 2004 ScienceTech Publishing

INTRODUCTION

Important changes in food consumption and distribution systems during the second half of the 20th century have been associated with extensive industrialization of production structures. In this context, smoked salmon, which was initially regarded as a luxury product, has gradually become an ordinary item.

In France, the raw material is mainly of aquaculture origin and generally from the Scandinavian countries, but the rearing procedures can vary considerably from one farm to another. These variability factors are numerous and concern rearing conditions as well as slaughtering, packaging and transport (Torrisen, 1995). Thus, the producer must have complete knowledge of the raw material and the effects of the processes used, so that the characteristics of the finished product will be consistent with market demand and profitability requirements.

Several factors are involved in this notion of profitability and production control (Sigurgisladottir *et al.*, 1997). The most important ones are processing yields and a respect for the characteristics of the finished product. The latter, independent of health considerations, concerns chemical (salt, water, lipid and phenol contents), physical (texture and color) and sensory aspects.

For smoked salmon, yield and quality are related to the operations involved in fillet production, such as head removal, filleting and trimming (Rorå *et al.*, 1998), the choice of the processes used (salting, drying and smoking techniques), and the parameters relating to these processes, such as brine concentration, length of treatment, temperature or hygrometry (Chan *et al.*, 1975; Horner, 1993), all of which are closely dependent on the characteristics of the raw material.

The purpose of the present study was to analyze interactions between the characteristics of the raw material and certain processing parameters relative to the yields and quality of smoked salmon. Two salting techniques (brine or dry salt) were associated with two cold smoking temperatures (20 and 30°C). The deposition of smoke compounds by an electrostatic smoking method was also studied, as well as the effect of freezing the raw material. Our intention was to propose technical recommendations for better control of the finished product based on a relevant choice of the characteristics of the raw material with respect to processes available. This study is part of a larger undertaking on the "Interaction between raw material characteristics and smoking process on quality of smoked salmon".

MATERIALS AND METHODS

Materials

Three groups of 105 fish each, from different origins and with different rearing characteristics as described by Sigurgisladottir et al. (2000), were compared in this study. The first sample (Group A) was obtained from ocean ranched salmon harvested in July 1998 in Iceland. The average weight of these fish was 2.6 kg ± 300 g (individuals weighing 4 kg were expected, but proved impossible to obtain because of planning problems). In these conditions, no sorting according to weight was performed. The second sample consisted of farmed salmon harvested in October 1998 (Group B) with an average weight of 4 kg ± 150 g. The third group, also consisting of farmed salmon, was harvested in Norway in April 1999 (Group C) and had an average weight of 3.7 kg ± 500 g.

Methods

Sample preparation. Gutted fish were weighed and filleted. Each fillet was trimmed by hand and processed according to the preparation scheme in Figure 1. Each group of 105 fish was processed 7 days after slaughtering according to seven different treatments, including different salting and smoking techniques. To study the effect of freezing, 30 fish within each group were sent to the french partner, one month earlier and were frozen (1 hour) with cryogenic equipment (CO_2 at –60°C) and stored at –20°C. A thawing step in air was performed at 4°C for 24 h. Fifteen salmon were used for each procedure, and fillet weight was recorded at each step to calculate yields.

Salting. Two salting techniques were studied. The first was a dry salting technique in which fillets salted by hand with refined salt (Salins du Midi, France) were left for 6 h (4 h for ocean ranched salmon because of their small size) at 12°C. The fillets were then rapidly rinsed with water (15°C) on grids and stored at 2°C until smoking.

The brine salting technique used saturated brine (360 g/l) maintained at 12°C in which the fillets were placed [ratio 50/50 (w/v)]. After 6 h (4 h for ocean-ranched salmon), the fillets were removed, rapidly rinsed and stored overnight (12 h) at 2°C until smoking. Fillets were weighed before smoking.

Whole gutted salmon (*Salmo salar*)
A. Ocean ranched (average weight 2.6 kg), July 1998
B. Farmed salmon from northern Norway (4 kg), October 1998
C. Farmed salmon from western Norway (3.7 kg), April 1999

```
┌─────────────┐     ┌─────────────────┐
│  Frozen     │     │   Fresh Fish    │
│  Fish       │     │   (in ice)      │
└──────┬──────┘     └────────┬────────┘
       │                     │
       └──────────┬──────────┘
                  │
       ┌──────────────────────────────┐
       │ Manual filleting and Trimming │
       └──────────────┬───────────────┘
                      │
       ┌──────────────┴───┬────────────┐
       │   Dry salting    │ Brine salting│
       └──────────────┬───┴────────────┘
                      │
       ┌──────────────────────────────┐
       │ Rinsing and storage 12h at +2°C │
       └──────────────────────────────┘
```

20°C	30°C		20°C	30°C	T°C amb.		20°C	30°C

Traditional   **Traditional**   **Electrostatic**   **Traditional**

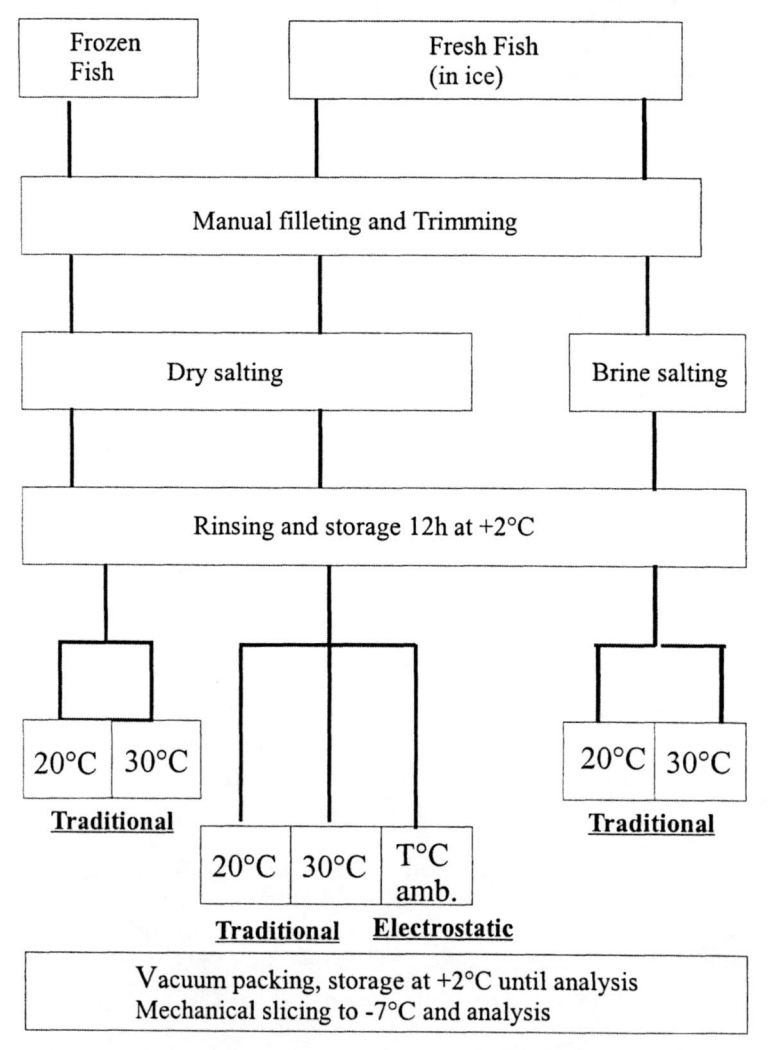

Vacuum packing, storage at +2°C until analysis
Mechanical slicing to -7°C and analysis

Figure 1. Process steps for production of whole gutted salmon.

Drying–Smoking. A traditional cold-smoking process was carried out at two different temperatures. After storage for 12 h at 2°C, the smoking process began with a drying step in the smoking oven for 30 min at 20°C, followed by a smoking step at 20°C ±1°C, with a relative hygrometry of 65 ± 3% and air speed of 2 ms^{-1} above the products and 18 ms^{-1} at the end of the air channel (HMI Thirode -PC90 Model). Smoking was performed for 2.5 h. The same procedure was carried out for fish smoked at 30°C, with a relative humidity of 50%. With electrostatic smoking (Collignan *et al.*, 1992, 1993; Bardin *et al.*, 1997), fillets were maintained at +2°C for 12 h and then placed in the smoking tunnel for 15 min.

Slicing and packaging. One day before sensory sessions and chemical analyses, fillets were frozen by the cryogenic method (Airgaz) for 20 min at –60°C to an inner temperature of -7°C and mechanically sliced (PNP, France). Skin was removed before slicing. Sliced fillets were immediately vacuum-packed (Boulanger, France). Samples were stored at +2°C until analysis.

Analytical methods

Yield measurement. Filleting and trimming loss = 100 * (gutted weight of fish - weight after filleting and trimming) / gutted weight of fish
Salting loss = 100 * (fillet weight after trimming – fillet weight after salting) / fillet weight after trimming
Smoking loss = 100 * (fillet weight after salting – fillet weight after smoking) / fillet weight after salting
Total loss after salting and smoking = 100 * (fillet weight after trimming – fillet weight after smoking) / fillet weight after smoking.

Chemical analyses. Dry matter content was analyzed by oven drying of 2 g of smoked salmon at 105°C until a constant weight was reached. Total fat content was determined by non-destructive computer-assisted X-ray tomography, as described by Rye (1991), and salt content was measured with Chloride Analyzer 926 (Corning, Halstead, England). Total phenols were quantified by the method described in the French standard for smoked salmon (NF V 45-065, 1995). All these analyses were performed on the front part of each smoked fillet.

Sensory evaluation. Descriptive and quantitative analysis (Stone *et al.*, 1974; Stone and Sidel, 1985) was performed to evaluate the sensory characteristics of smoked salmon. Twenty panelists (IFREMER staff) were trained on sensory descriptors chosen during a preliminary step. The descriptors chosen related to the appearance, odor, flavor and texture of smoked salmon slices — odor:

smoke intensity, wood fire, acrid smell of smoke, toasted bread, fresh salmon; appearance: pink color, orange color, homogeneity of color; darkness of the slice edge, translucent appearance, fatty aspect, white stripes; flavor: smoke intensity, wood fire, salty taste, fresh salmon; and texture: firmness, melting texture, fatty texture, pasty texture.

Four sessions were proposed to the panel for each raw material. Samples smoked at 20°C and those smoked with the electrostatic process were evaluated twice, after 8 and 9 days of storage at +2°C. Salmon smoked at 30°C were also scored twice, after 12 and 13 days. Samples dry-salted and smoked at 20°C were used as a common reference in these four sessions. The storage conditions at +2°C, resulted in samples that were considered to keep the same characteristics between 8 and 13 days of storage.

Sessions were performed in individual partitioned booths equipped with a computerised system (Fizz system, Biosystèmes, Dijon, France). These conditions were conducive to concentration and avoided communication between assessors and disturbance by external factors. Panelists rated the sensory attributes on a continuous scale presented on a computer screen, from low intensity (0) to high intensity (10). Products were assigned 3-digit numbers, randomized and served simultaneously. Twelve sessions were organized to test all the raw material and processes.

Color measurement. Color was determined by a Hunterlab miniscan XE using the D65 light source and a 10° observer. Reflectance from 400 to 700 nm was recorded as well as L^*, a^*, b^* (CIE, 1976). Color was measured in the anterior, median and posterior parts of the dorsal and ventral side of raw and smoked fillet. Results are shown as the mean value of six measurements per fish fillet.

Statistical analysis. The mean, standard deviation, analysis of variance and Duncan's multiple range test were performed using Statgraphics Plus software (Sigma Plus, Paris, France). The significant statistical level was set at $p < 0.05$. Multivariate data processing was performed with Uniwin Plus 3 software (Sigma Plus, Paris, France). To adjust for variations among assessors in the scoring range, sensory data were standardised using an isotropic scaling factor according to the procedure proposed by Kunert and Qannari (1999). Scores of the dry-salted product smoked at 20°C were used as a reference at each session, thereby allowing data from different sessions to be compared. A principal component analysis was performed on the means of these data.

RESULTS AND DISCUSSION

Influence of raw material and processing on yields

Raw material effect on loss after filleting. Flesh lipid content showed significant differences among the three raw materials: 9.4, 20.0 and 16.8% of wet weight, respectively, for ocean ranched salmon, farmed salmon slaughtered in October 1998 and farmed salmon slaughtered in April 1999. Analysis of variance for loss after filleting showed a significant difference between different raw materials. Ocean-ranched salmon from Iceland, which were smaller and leaner than the fish in the other groups, had the highest loss (Figure 2). According to Rora *et al.* (1998), this result can be regarded as an effect of weight more than fat content.

Effect of the salting technique on yields after processing. The results showed better yields with the brine-salting than the dry-salting technique because in brine salting case, salmon fillets are already soaked in a liquid solution, so that water diffusion is reduced and losses are lower (Figure 3). The effects are greater to the extent that the raw material is lean (i.e. with high water content), as observed in group A for the loss recorded after salting. These results are in accordance with those of Boury (1934) and Jason (1965), who demonstrated the importance of the size factor and the barrier effect of lipids.

Moreover, losses after dry salting were generally slightly higher with frozen material, especially when fish were small and lean. Difference in losses between fresh and frozen raw material was 0.1% in October, 0.3% in April 1999 and reached 0.6% in July 1998. The difference was only statistically significant in the ocean ranched group. However, losses with frozen material after dry salting were generally low compared to the total loss. This may seem contradictory with the results of many previous studies indicating that all treatments affecting cell integrity, such as freezing, increase solute intakes at the expense of water (Ponting, 1973; Dussap and Gros, 1980; Saurel *et al.*, 1994). However, in our study, the cryogenic method used for freezing, as well as the short storage period (one month), limited tissue destruction and allowed the raw material to be preserved and to retain its good functional properties, as reported by Sivertsvik (1994) for frozen salmon (*Salmo salar*).

Effect of the smoking technique or drying/smoking temperature on yields after processing. The speed of water diffusion in the flesh of lean fish is faster than in fat fish, which leads to more rapid drying and higher losses for wild salmon (group A) (Table 1). The drying-smoking step at 20°C leads to greater losses than at 30°C due to a better control of humidity. At 30°C, the water diffusion rate for fat fish was lower than the rate of surface evaporation, which induces case-hardening and thus a blockage of transfer and a reduction of loss occurs. This phenomenon has been described by Bimbenet (1978), who noted that too low a hygrometry associated with too high a temperature alters drying profiles. This case-hardening effect can be detrimental to product quality since it prevents effective drying and can reduce product life.

For smoking at 30°C, the results were the same whether the raw material was frozen or not. As the drying conditions were more favorable at 20°C and water more available in the flesh after the freezing step, the losses were greater.

For ocean ranched salmon, electrostatic smoking led to losses equivalent to those for smoking at 20 or 30°C. Given the very short smoking period and the absence of a drying step, this suggests that losses occurred during storage in the cold room (12 h at 2°C) before the smoking operation. During this storage period, it is likely that dry-salted fillets reached a level of hygrometry similar to that maintained in the smoking unit during processing. As losses in brine after salting were lower, processing at 20 and 30°C still had an effect on weight loss. In the case of fatty farm fish (groups B and C), in which water was less available, overnight storage at +2°C was inadequate to produce a notable drying effect. Thus, drying continued during the drying-smoking step.

Overall effect of salting and smoking conditions on yields after processing. Regardless of the processing technique used, losses were higher with ocean ranched salmon, particularly when frozen (Table 1). This was essentially attributable to the size effect in direct relation with chemical composition. Regarding the salting technique used, the brine salting shows the lowest losses specially when fish has a higher fat content.

Effect of smoking techniques and temperature on quality parameters

Salt content. Mean salt content in the groups of smoked fish ranged from 2.2 to 4 g/100 g of wet flesh (Table 2). Two main factors influenced salt uptake, the nature of the raw material (particularly size and composition) and the kind of salting technique used. It is the reason why we observed a salt content from 2.2 to 3.4% for fatter fish and from 3.2 to 4% for lean fish. Products salted in saturated brine generally had a higher salt content than dry-salted samples because of better contact between surfaces and the salting medium. Salt content in smoked products was slightly higher for frozen raw material than fresh material, especially for leaner fish. This resulted from the freezing step, which modified cell structure slightly, increasing salt diffusivity.

Table 1. Weight loss after smoking step and total loss (%)

Drying / Smoking process		Dry salting (Frozen material) 20°C	Dry salting (Frozen material) 30°C	Brine (Fresh material) 20°C	Brine (Fresh material) 30°C	Dry salting (Fresh material) 20°C	Dry salting (Fresh material) 30°C	E
Loss after smoking %	Group A	6.8^d	4.5^{abc}	5.2^c	5.1^{bc}	4.7^{abc}	4.3^a	4.4^{ab}
	Group B	5.4^e	3.4^b	4.1^{cd}	3.7^{bc}	4.4^d	3.75^{bc}	2.35^a
	Group C	4.9^d	3.2^b	4.4^c	3.35^b	4.9^d	3.25^b	2.7^a
Total Loss %	Group A	13.7^c	11.4^b	9.0^a	8.7^a	11.3^b	8.7^b	10.9^b
	Group B	9.1^d	6.7^b	5.4^a	5.1^a	7.9^c	7.3^{bc}	5.9^a
	Group C	8.9^d	7.25^c	5.85^b	5.0^a	8.9^d	6.9^{bc}	6.4^b

Table 2. Salt and phenol content

Drying / Smoking process		Dry salting (Frozen material) 20°C	Dry salting (Frozen material) 30°C	Brine (Fresh material) 20°C	Brine (Fresh material) 30°C	Dry salting (Fresh material) 20°C	Dry salting (Fresh material) 30°C	E
Salt content (g/100g wet flesh)	Group A	3.77 (0.38)	3.88 (0.39)	4.0 (0.77)	3.77 (0.77)	3.29 (0.38)	3.38 (0.37)	3.49 (0.4)
	Group B	2.74 (0.40)	2.8 (0.4)	2.79 (0.46)	2.7 (0.5)	2.43 (0.26)	2.27 (0.17)	2.55 (0.35)
	Group C	3.04 (0.36)	2.85 (0.23)	3.19 (0.35)	3.40 (0.42)	3.05 (0.28)	2.63 (0.32)	2.73 (0.42)
Phenol content (mg/100g wet flesh)	Group A	0.76 (0.24)	2.97 (0.51)	1.17 (0.37)	2.34 (0.47)	0.86 (0.23)	2.15 (0.48)	0.56 (0.15)
	Group B	1.06 (0.16)	2.03 (0.2)	0.94 (0.24)	1.44 (0.24)	0.95 (0.19)	1.70 (0.5)	0.33 (0.15)
	Group C	0.88 (0.17)	1.96 (0.48)	0.86 (0.32)	1.87 (0.38)	1.03 (0.22)	1.87 (0.35)	0.21 (0.09)

Loss after filleting and trimming(%)

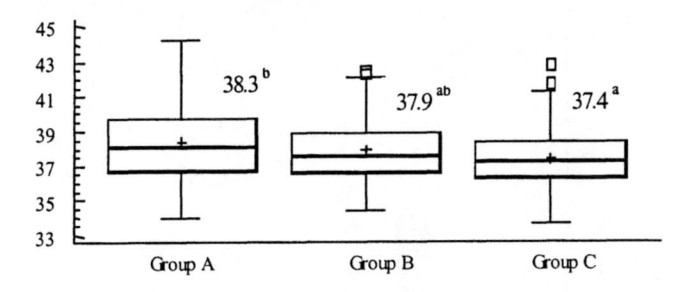

[Box plot showing three groups: Group A = 38.3^b, Group B = 37.9^{ab}, Group C = 37.4^a, with y-axis from 33 to 45]

Figure 2. Loss after filleting and trimming for three raw materials (%)

Phenol content. Phenol content allowed samples to be differentiated according to the smoking technique or smoking temperature (Table 2). The lowest levels (0.2 to 0.56 mg/100 g of flesh) were observed for the electrostatic process. With smoking at 20°C, mean phenol content in the groups ranged from 0.75 to 1.17 mg/100 g, whereas a 30°C temperature involved levels of up to 3 mg/100 g. Various reasons could account for these differences, including the type of smoke production and the type of smoke deposit. Despite the use of oak instead of beechwood with the electrostatic technique, processing time could not exceed 15 min in order to limit surface coloration. Therefore, phenol deposits were reduced. The 30°C temperature used with traditional smoking allows compounds with higher molecular weight, such as phenolic compounds, to remain in the vapor phase mainly involved in the smoking effect (Potthast, 1977, 1978; Foster, 1961a, 1961b; Girard, 1988). Therefore, samples can reach high phenol values.

A raw material effect was also noted for the 30°C smoking technique, i.e., the lower the fat content in muscle, the higher was the phenol content in smoked samples. This may have been due to water availability relative to fat content. The relative thinness of the fillet was another reason for higher phenol values in small fish.

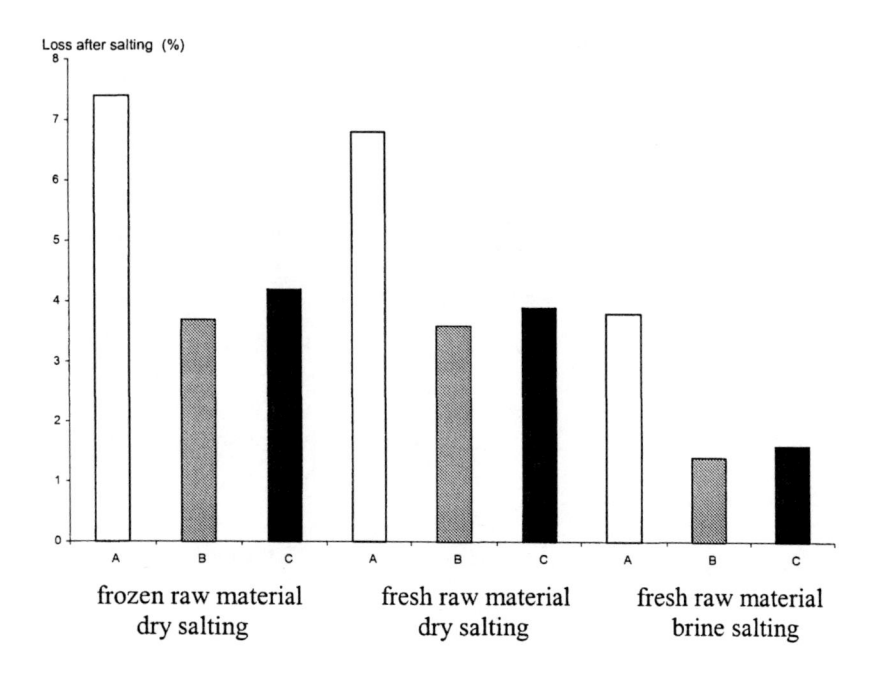

Figure 3. Loss after salting (%).

Color. Color measurements of raw and smoked fillet surface are given in Figure 4. Before processing, ocean ranched fish never fed pigments had lower a* and b* values than the two farmed fish groups. There was no significant difference in color parameters between the two reared fish groups. Smoking led to a reduction of a* values, as previously reported by Skrede and Storebakken (1986), Choubert *et al.* (1992) and Rorå *et al.* (1998), an increase of b* values regardless of the raw material, and a reduction of L* values (mainly for farmed fish) (results not shown).

Moreover the use of 30°C smoking temperature instead of 20°C increases b* values, i.e. products had a more intense yellowish tone. Regardless of the raw material, b* values for raw and smoked fillets were higher when fish had been frozen. No color difference between samples was observed relative to the salting technique.

Sensory properties. Figure 5 shows the main characteristics of each product in terms of principal component analysis performed on mean panel scores. Regardless of smoking temperature, a freezing step before processing affected the quality of smoked salmon, especially if fish were small and lean. Samples whose raw materials were frozen had saltier taste and less translucent slices. Their texture, which was close to that of the electrostatic smoked sample, was less firm and more pasty than that of other products. A texture slightly modified by freezing might actually increase the perception of salt more than the salt content level itself. The results of Sigurgisladottir *et al.* (2000) also show a significant effect of freezing/thawing on microstructure and texture.

The smoking temperature or technique allowed products to be differentiated according to the intensity of smoke odor and flavor. From the lowest to the highest intensity, the panel ranked samples smoked by the electrostatic method first, followed by those smoked at 20°C and finally those smoked at 30°C. These results are in agreement with those of Simon *et al.* (1966) and Girard (1988), who found that a higher temperature increases the deposit of smoke compounds. For products smoked by the electrostatic technique, the panel noted a special toasted bread odor in addition to low flavor intensity. These characteristics could be attributable to various factors, such as the kind of wood used, the smoke production method, pyrolysis temperature, smoke density, or smoking time (Cardinal *et al.*, 1997). Moreover, this sample appeared to be fatter, and the texture was described as fatty and melting in the mouth. Samples smoked at 20°C also showed these characteristics but to a lesser extent. These sensory properties were observed, regardless of the raw material. It is noteworthy that muscle fat content could have modified sensory perception of smoked salmon, as reported by Sheehan *et al.* (1996). For the same fat level, the kind of smoking procedure may also modify the perception of texture.

In summary, these results show that smoking conditions and preliminary treatment of fish (such as freezing) are major factors determining the sensory characteristics of smoked salmon, even though effects are reduced when fat content is increased in flesh.

CONCLUSIONS

This study confirms the results of previous work concerning the relation between fat content and yields or the final characteristics of smoked salmon. It also dealt with the nature of the interaction between the process used and the initial raw material characteristics. It would appear that losses after processing increase when fish have reduced lipid content (less than 10%) as compared to fatter fish, especially in the case of frozen fish. A fat fish is less

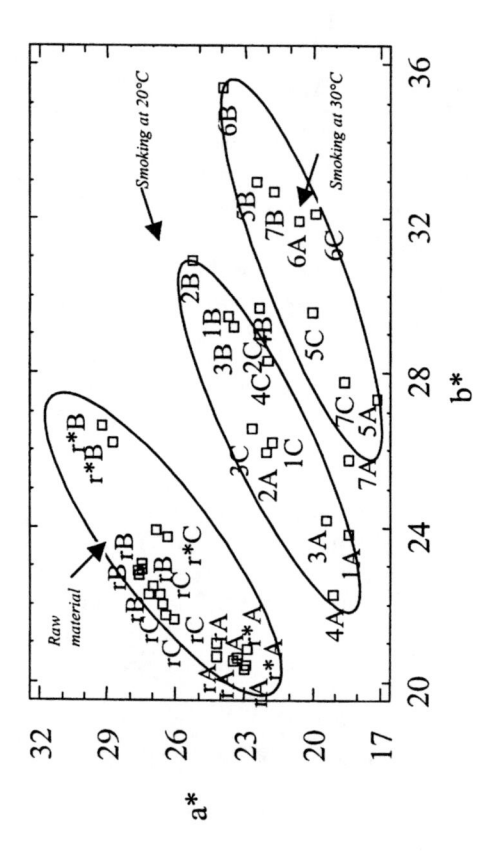

Figure 4. Color measurements. a* and b* values of raw and smoked salmon. Color system L*, a*, b*, CIE 1976 A: ocean ranched salmon (July 1998); B: farmed salmon (October 1998); C: farmed salmon (April 1999); r: fresh raw material, r*: frozen raw material smoking procedures: 1 = dry salting and smoking at 20°C, 2 = frozen raw material, dry salting and smoking at 20°C, 3 = brine salting and smoking at 20°C, 4 = dry salting and electrostatic smoking, 5 = dry salting and smoking at 30°C, 6 = frozen raw material, dry salting and smoking at 30°C, 7 = brine salting and smoking at 30°C. Each point is the mean of 6 readings per fillet for 15 fish.

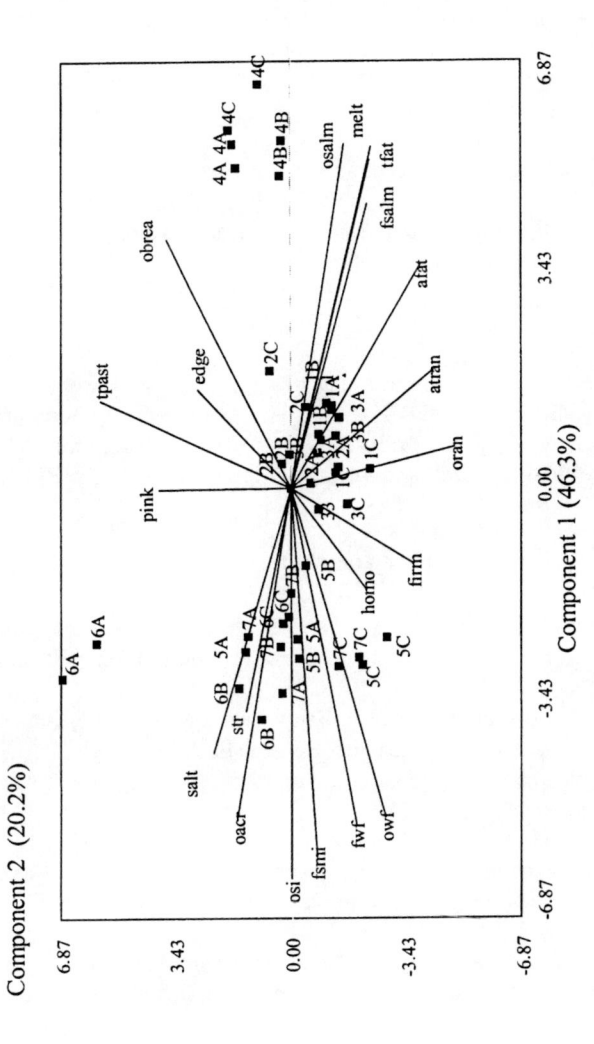

Figure 5. Projection of variables and samples in the 1-2 plane of principal component analysis on sensory descriptors. A: ocean ranched salmon (July 1998); B: farmed salmon from northern Norway (October 1998); C: farmed salmon from western Norway (April 1999). Smoking procedures: 1 = dry salting and smoking at 20°C, 2 = frozen raw material, dry salting and smoking at 20°C, 3 = brine salting and smoking at 20°C, 4 = dry salting and electrostatic smoking, 5 = dry salting and smoking at 30°C, 6 = frozen raw material, dry salting and smoking at 30°C, 7 = brine salting and smoking at 30°C.

sensitive to a freezing step, and the yield after processing is increased, regardless of the kind of smoking procedure used. Nevertheless, for this type of raw material, smoking temperature needs to be reduced to prevent too high a desiccation of the surface (which generally produces intense surface coloring but little diffusion of smoke compounds into the flesh). Regarding the step of salting, we can advice to use a brine salting to reduce the losses after processing.

For the three raw materials analyzed, the processes studied showed the same kind of effect on smoked salmon properties, but with greater differences between treatments in the case of small lean fish.

But to really conclude about a recommended process, it would be necessary to complete the study with an analysis of the quality criteria of smoked salmon after packaging, throughout the storage time in order to analyse the effect of fat content on color, fat release and preservation.

ACKNOWLEDGEMENTS

This research was performed with the financial support of the European Union in the context of FAIR project CT-95-1101. The authors are grateful for the assistance of the project co-ordinator, Dr. H. Hafsteinsson (Matra, Technological Institute of Iceland) and for the work of the panel of IFREMER.

REFERENCES

Bardin, J.C.; Desportes, G.; Knockaert, C. and Vallet, J.L. 1997. Improvement in devices for electrostatic smoking of meat products. French Patent No. 9708177, filing date: June 25.

Bimbenet, J.J. 1978. Le séchage dans les industries agricoles et alimentaires. Cahiers du Génie Industriel Alimentaire. Sepaic (Ed.), Paris.

Boury, M. 1934. Etudes sur le salage du poisson. *Revue des travaux de l'Institut des Pêches Maritimes* 26: 195-221.

Cardinal, M.; Berdagué, J.L.; Dinel, V.; Knockaert, C. and Vallet, J.L. 1997. Effet de différentes techniques de fumage sur la nature des composés volatils et les caractéristiques sensorielles de la chair de saumon. *Science des Aliments* 17: 679-696.

Chan, W.S.; Toledo, R.T. and Deng, J. 1975. Effect of smokehouse temperature, humidity and air flow on smoke penetration into fish muscle. *J. Food Sci.* 40: 240-243.

Choubert, G.; Blanc, J.M. and Courvalin, C. 1992. Muscle carotenoid content and color of farmed rainbow trout fed astaxanthin or canthaxanthin as affected by cooking and smoke-curing procedures. *Int. J. Food Sci. Technol.* 27: 277-284.

Collignan, A.; Knockaert, C.; Raoult-Wack, A.L. and Vallet, J.L. 1992. Procédé et dispositif de salage séchage et de fumage à froid de produits alimentaires carnés. Patent No. 92/08958.

Collignan, A.; Knockaert, C.; Raoult-Wack, A.L. and Vallet, J.L. 1993. Process for salting-drying and smoking cold meat products and a device for carrying this out. Patent No. 93430009.6, filing date: July 10.

Dussap, G. and Gros, J.B. 1980. Diffusion-sorption model for the penetration of salt in pork and beef muscle. *Food Process Engineering*, Applied Science Publishers, Amsterdam, The Netherland, pp. 407-411.

Foster, W.W. 1961a. Studies of the smoking process for food. 1-The importance of vapours. *J. Sci. Food Agric.* 5: 363-374.

Foster, W.W. 1961b. Studies of the smoking process for food. 2-The role of smoke particles. *J. Sci. Food Agric.* 9: 635-644.

Girard, J.P. 1988. La fumaison: Technologie de la viande et des produits carnés. *Ed Lavoisier* 6: 171-214

Horner, W.F.A. 1993. Preservation of fish by curing (drying, salting and smoking) *In Fish Processing Technology*. G.M. Hall (ed.) Black Academic & Professional, VCH Publishers, New York, NY, pp. 30-72.

Jason, A.C. 1965. Drying and dehydration Excerpt from "Fish as Food" G. Borgstrom (ed.) *Academic Press*, London, UK, pp. 111: 1-54.

Kunert, J. and Qannari, E.M. 1999. A simple alternative to generalized procrustes analysis: Application to sensory profiling data. *J. Sensory Studies* 14: 197-208.

NF V 45-065 1995. *Poisson transformé. Saumon fumé.* Association Française de Normalisation (AFNOR), Paris, France.

Ponting, J.D. 1973. Osmotic dehydration of fruits. Recent modifications and applications. *Proc. Biochem.* 8:18-20.

Potthast, K. 1977. Determination of phenols in smoked meat products. *Acta Alimentaria Polonica* 3: 189-193.

Potthast, K. 1978. Smoking methods and their effects on the content of 3,4-benzo(a)pyrene and other constituents of smoke in smoked meat products. *Die Fleischwirtschaft* 58: 371-375.

Rorå, A.M.B.; Kvale, A.; Morkore, T.; Rorvik, K.-A.; Steien, S.H. and Thomassen, M.S. 1998. Process yield, color and sensory quality of smoked Atlantic Salmon in relation to raw material characteristics. *Food Res. Int.* 31: 601-609.

Rye, M. 1991. Prediction of carcass composition in Atlantic salmon by computerized tomography. *Aquaculture* 99: 35-48.

Saurel, R.; Raoult-Wack, A.L.; Rios, G. and Guilbert, S. 1994. Mass transfer phenomena during osmotic dehydration of apple. Part 2: Frozen plant tissue. *Int. J. Food Sci. Technol.* 29: 543-550.

Sheehan, E.M.; O'Connor, T.P.; Sheehy, P.J.A.; Buckley, D.J. and FitzGerald, R. 1996. Effect of dietary fat intake on the quality of raw and smoked salmon. *Irish J. Agric. Food Res.* 35: 37-42.

Sigurgisladottir, S.; Torrissen, O.; Lie, O.; Thomassen, M. and Hafsteinsson, H. 1997. Salmon Quality: Methods to determine the quality parameters. *Rev. Fish. Sci.* 5: 223-252.

Sigurgisladottir, S.; Ingvarsdottir, H.; Torrissen, O.J.; Cardinal, M. and Hafsteinsson, H. 2000. Effects of freezing/thawing on the microstructure and the texture of smoked Atlantic salmon (*Salmo salar*). *Food Res. Int.* 33: 857-865.

Simon, S.; Rypinski, A.A. and Tauber, F.W. 1966. Water-filled cellulose casings as model absorbents for wood smoke. *Food Technol.* 20: 114-118.

Sivertsvik, M. 1994. Processing and Preservation Technology: The influence of different packaging materials during storage of whole frozen Atlantic salmon. Norconserv. Institute of Fish Processing and Preservation Technology.

Skrede, G. and Storebakken, T. 1986. Instrumental color analysis of farmed and wild atlantic salmon when raw, baked and smoked. *Aquaculture* 53: 279-286.

Stone, H.; Sidel, J.L.; Oliver, S.; Woolsey, A. and Singleton, R.C. 1974. Sensory Evaluation by Quantitative Descriptive Analysis. *Food Technol.* 28: 24-34.

Stone, H. and Sidel, J.L. 1985. Sensory evaluation practices. Academic Press Inc., New York, NY.

Torrissen, O. 1995. Norwegian salmon culture: 1 million tons in 2005. *Aquaculture Europe* 19: 6-11.

CHAPTER 10

MODELING HEAT AND MOISTURE TRANSFER DURING SHRIMP COOKING

Michael Ngadi

Agricultural and Biosystems Engineering Department
McGill University, Macdonald Campus
Ste-Anne-de-Bellevue, Quebec, Canada

A mathematical model of heat and moisture transfer was developed to predict temperature and moisture distributions in breaded butterfly jumbo shrimp. The model considered actual shape of the shrimp samples. It incorporated thermal properties of shrimp as a function of temperature and moisture. Predictions of the model were validated using experimental data obtained from batch laboratory deep-fat frying and from industrial pilot plant continuous fryer. Experimental temperature and moisture histories closely matched the predicted results. Mathematical model can be used as a tool for a priori evaluation of frying parameters and optimization of product safety and quality.

INTRODUCTION

Shrimp processing is an important component of the Canadian seafood industry. Canadian coldwater shrimp export was worth more than $303 million in 2000, an increase of 13% in value and an increase of 5% in volume over figures for 1999 and this has expanded modestly since. Shrimp is marketed in different forms including fresh, frozen (in-shell or shelled) and prepared/preserved forms. The major destinations of Canadian shrimp products are exported to the United States, Denmark and United Kingdom. Nearly a third of all shrimp exports from Canada are in the prepared or preserved form. This amounts to an increase of 34% in value and an increase of 46% in volume in year 2000 as compared to that in 1999 (Barnett, 2002). These facts are primarily due to increasing demand and consumption of shrimp.

Shrimps have a unique flavor and are used in a variety of food service applications. They can be used as "peel and eat" product, sprinkled on salads and incorporated in sandwiches, sauces, stuffing, and pizza, among others They are also increasingly marketed as stand alone fried breaded products. In order to preserve the expanding market quota, it is vital to maintain both the quality and safety of processed shrimp. Shrimp processing is usually by cooking, including boiling, grilling, frying and microwaving. Heat affects sensory and textural qualities of the product and causes yield loss due to moisture loss. For instance, Northern pink shrimp will toughen and lose its flavor if subjected to too much heat during processing (Barnett, 2002). Therefore accurate prediction of heat transfer and temperature distribution will be a vital tool for process control and optimization of quality during processing.

Chau and Snyder (1988) modelled temperature distribution in shrimp during thermal processing. The authors simplified the heat transfer problem by approximating the shape of shrimp with segments of right circular cones and neglecting curvature of the product. They assumed constant thermal properties for shrimp. The finite difference approach was used to solve the resulting heat transfer equations. Erdogdu et al. (1998) improved the mathematical modelling approach of Chau and Snyder (1988) by incorporating variable thermal property and dimensional shrinkage to predict temperature distribution in shrimp during thermal processing. However, circular cross-sectional area at different segments was also used and reported sufficient to describe heat transfer in tiger shrimp. Thermal properties namely thermal conductivity and specific heat previously reported for beef were used in the model. Erdogdu et al. (1999) used a mathematical model to predict yield loss during cooking of shrimp. In this model, shrimp was approximated by a series of short cylinders with diminishing radii. Yield loss during cooking depended on shrimp size, cooking temperature and time. Mallikarjunan et al. (1996) assumed a simple cylindrical shape to develop mathematical model for microwave cooking of cocktail shrimp. The model was coupled to a first order reaction kinetics to predict microbial inactivation. Mathematical models for predicting heat transfer in standard shaped objects are available in the literature. However, some shrimp products do not have regular standard shapes. Therefore, a model that more closely approximates the actual shape will be an improvement. Also considerable changes in structure and thermal properties occurred during heating for different species of shrimps. These difficulties make mathematical modelling of shrimp cooking challenging. Since moisture content of shrimp changes during cooking, accounting for the coupled heat and moisture transfer will also improve prediction of moisture and temperature distributions during cooking. Most food service shrimp products are breaded. Breading characteristics may influence heat and moisture transfer characteristics. The objectives of this work were to develop a heat and moisture transfer model in breaded shrimp during cooking as in deep fat frying and to determine the influence of breading pickup on heat and moisture transfer and finally to validate the mathematical model.

Model Development

Heat transfer in the core region of the product was considered to be due to temperature gradient given as:

$$\rho c_p \frac{\partial T}{\partial t} = \nabla \cdot (k \nabla T) \qquad (1)$$

where T is temperature (°C), t is time (s), ρ is density (kg/m^3), k is thermal conductivity (W/m.°C) and c_p is heat capacity (J/kg.°C). At the crust region (defined as the regions with temperatures equal or above vaporization temperature, 100°C), heat transfer is due to conduction resulting from temperature gradients as well as due to moisture vaporization given as:

$$\rho c_p \frac{\partial T}{\partial t} = \nabla \cdot \left[(k + L\rho \delta D_m) \nabla T\right] + \nabla \cdot (L\rho D_m \nabla C_m) \qquad (2)$$

where C_m is moisture concentration (kg/kg dry weight), δ is thermal gradient (1/°C) and D_m is moisture diffusivity (m^2/s).

Moisture transfer in the given as:

$$\frac{\partial C_m}{\partial t} = \nabla \cdot D_m (\nabla C_m + \delta \nabla T) \qquad (3)$$

The boundary conditions for heat and moisture transfers are:

$$k \frac{\partial T}{\partial n} = h(T_\infty - T_s) \qquad (4)$$

$$D_m \frac{\partial C_m}{\partial n} = h_m (C_{m\infty} - C_{ms}) \qquad (5)$$

where n is the direction normal to the surface of the product, h and h_m are heat and mass transfer coefficients, respectively. The governing equations 1 to 3 were solved using the finite element method. Appropriate product properties and parameters were obtained from literature (Ngadi *et al.*, 2000).

Experimental Setup and Model Validation

Breaded butterfly shrimp was used in this study. Shape and dimension of the shrimp product were determined using a digital video camera to take still images of samples as shown in Figure 1. Over 200 individual samples of shrimp were selected randomly and used for size and shape determination. The images were digitized to determine dimensions.

Two sets of experiments were conducted namely a batch laboratory scale frying at 140°C oil temperature and a continuous industrial scale par-frying at 200°C oil temperature. For batch frying, shrimp samples were manually predusted, subsequently placed in a batter solution and then coated with breading. Bread pick-up and thickness were verified before the product was fried. Temperature of product and oil during frying were monitored using T type thermocouples. For industrial frying experiments, pre-dust, batter and breading were applied on-line as products moved on a conveyor belt. Oil temperature measurement in the industrial scale continuous fryer was more challenging than the measurement in the batch fryer. A multi-point temperature recorder (DataPaq, Wilmington, MA) was used to monitor temperature profiles in the continuous fryer. The recorder was enclosed in an insulated case with a custom made jig that enabled it to mount on fryer belt and flow through the fryer with frying products. Moisture content of samples was determined using a freeze drying method.

RESULTS AND DISCUSSION

Figure 2 shows temperature profiles in butterfly shrimp during batch laboratory scale frying. Oil temperature was maintained at 140°C. The temperature was measured at a location close to the surface of the shrimp. Temperature of product initially increased rapidly but a slower increase in temperature was observed as product temperature approached 100°C, the vaporizing temperature. This is attributed to the latent heat during moisture vaporization at 100°C. Mass average moisture change in shrimp is shown in Figure 3. Moisture content of shrimp decreased from 3 to 1.5 g/g db. Moisture content of bread coating decreased from 1 to 0.2 g/g db. The bread portion of the product was directly in contact with high temperature frying oil and tended

Cross Sections:

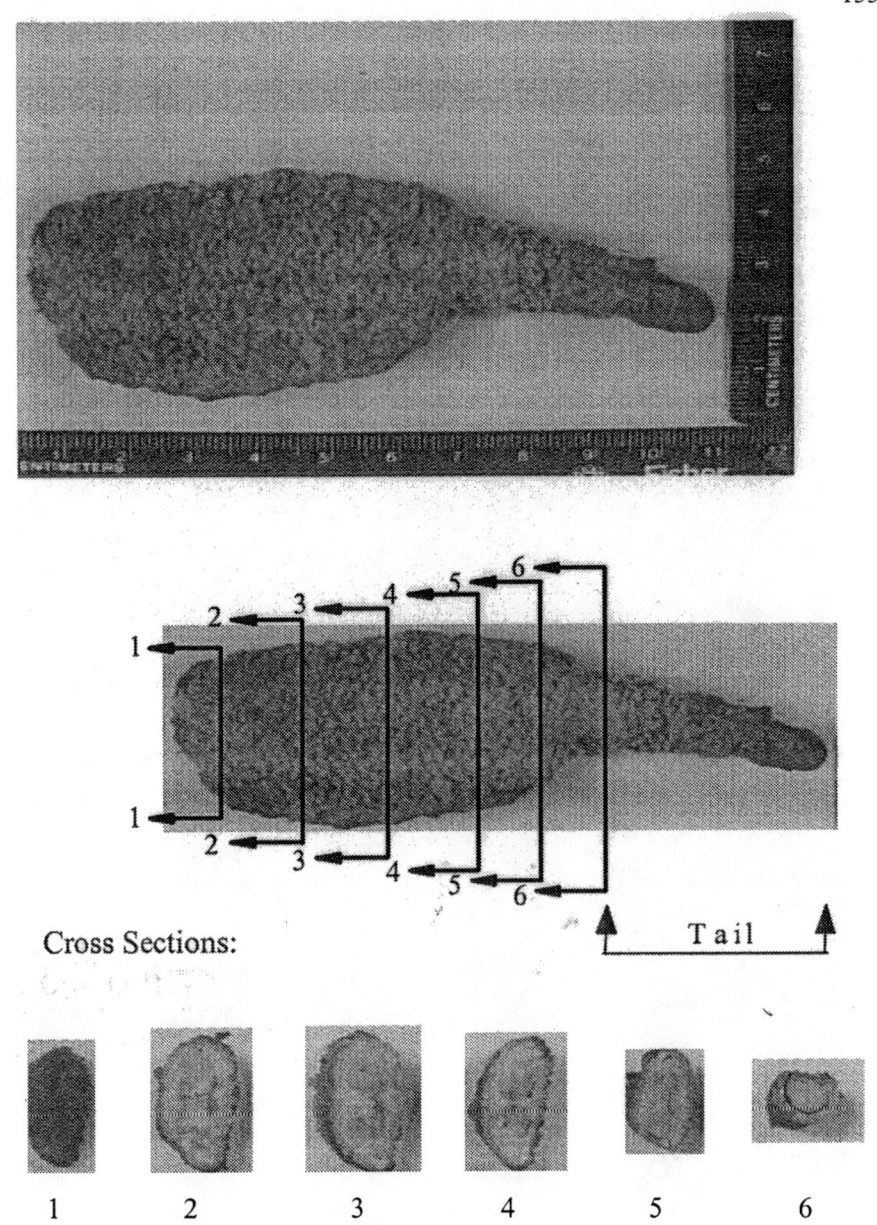

1　　　2　　　3　　　4　　　5　　　6

Figure 1. Shape and size determination for breaded shrimp

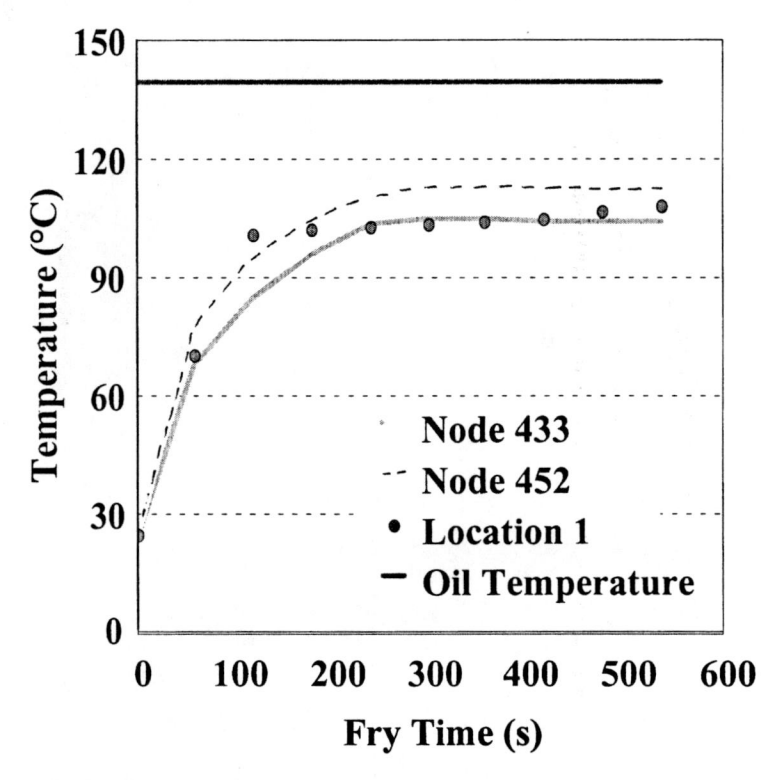

Figure 2. Predicted and measured temperature histories in butterfly shrimp during deep-fat frying at 140°C in a batch laboratory scale fryer.

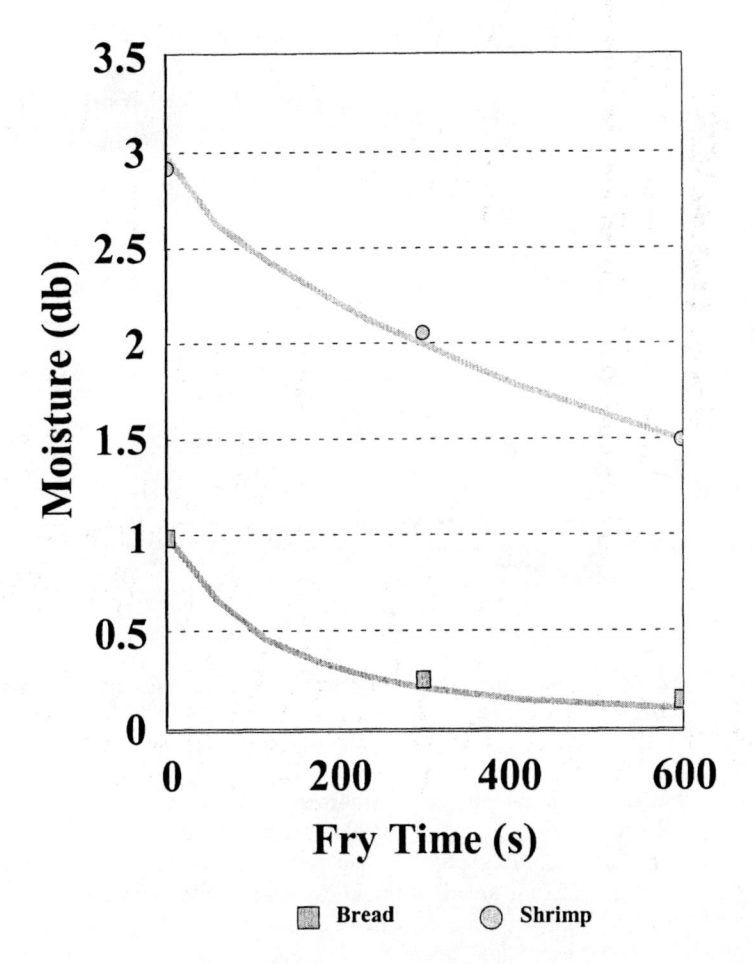

Figure 3. Measured and predicted mass averaged moisture histories for butterfly shrimp fried at 140°C in a batch laboratory fryer.

to lose moisture more easily than the inner shrimp portion. The mathematical model closely predicted temperature and moisture profiles in butterfly shrimp during batch deep-fat frying in a laboratory fryer. Figure 4 shows temperature profiles in the continuous industrial fryer. Lower temperatures were obtained in the shrimp because of the lower frying time. Predicted temperatures were

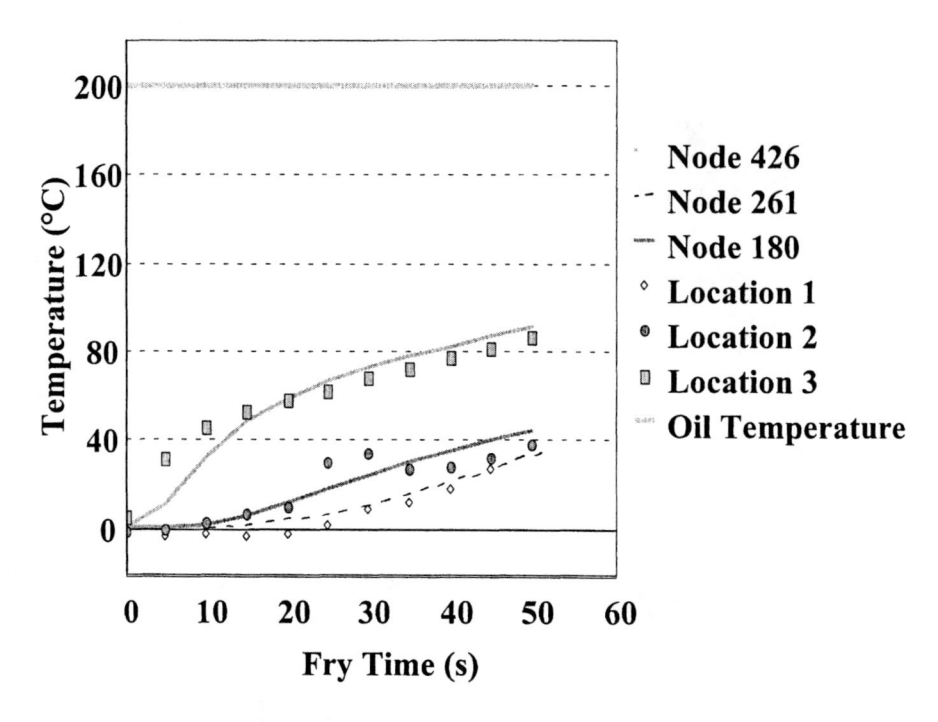

Figure 4. Measured and predicted temperature histories for breaded butterfly shrimp fried at 200°C in a continuous deep-fat fryer

compared with experimented temperatures at 3 different locations in the shrimp. Temperature profiles predicted by the mathematical model were close to the measured temperature profiles. Simulated temperature profiles in shrimp samples with different breading thickness (that is pick-up) are shown in Figure 5. Increasing breading thickness from 2 to 4 mm decreased temperature of the product at corresponding locations. This is understandable since product thickness influences heat transfer characteristics. The result will have significance in quality and safety of fried products. Effect of different frying parameters can be simulated and used to optimize the frying process.

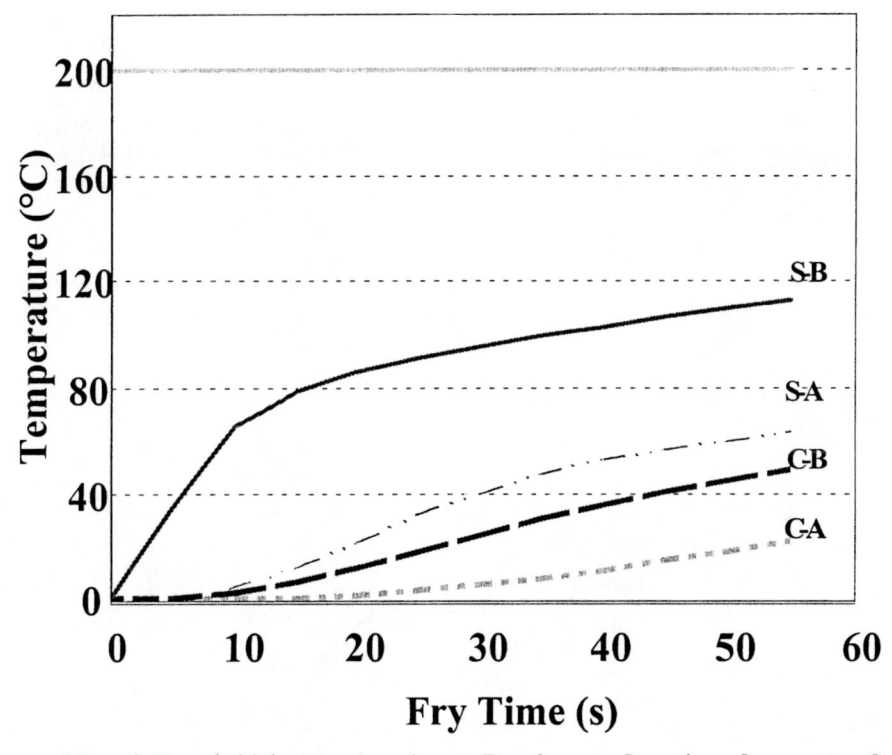

Legend: Bread thickness: A = 4 mm; B = 2 mm. Location: C = center; S = surface

Figure 5. Simulated temperature profiles in breaded shrimp with different breading thickness. Samples were fried at 200°C in a continuous deep-fat fryer.

REFERENCES

Barnett, J. 2002. Seafood Specialist, Food Bureau, Agriculture and Agri-Food Canada. Personal Communication.

Chau, K.V. and Snyder, G.V. 1988. Mathematical model for temperature distribution of thermally processed shrimp. *Trans. ASAE* 31: 608-612.

Erdogdu, F.; Balaban, M.O. and Chau, K.V. 1999. Mathematical model to predict yield loss of medium and large tiger shrimp (Penaeus Monodon) during cooking. *J. Food Process Eng.* 22: 383-394.

Erdogdu, F.; Balaban, M.O. and Chau, K.V. 1998. Modeling of heat conduction in elliptical cross section: II. Adaptation to thermal processing of shrimp. *J. Food Eng.* 38: 241-258.

Mallikarjunan, P.; Hung, Y-C. and Gundavarapu, S. 1996. Modeling microwave cooking of cocktail shrimp. *J. Food Process Eng.* 19: 97-111.

Ngadi, M.O.; Mallikarjunan, P.; Chinnan, M.S.; Radhakrishnan, S. and Hung, Y-C. 2000. Thermal properties of shrimps, French toasts and breading. *J. Food Process Eng.* 23: 73-87.

CHAPTER 11

EFFECT OF PRESSURE-SHIFT FREEZING OF CARP (*Cyprinus carpio*) VS AIR-BLAST FREEZING: A STORAGE STUDY

A. Sequeira-Munoz[1], D. Chevalier[2], B.K. Simpson,[1] A. Le Bail[2] and H.S. Ramaswamy

[1]Department of Food Science and Agricultural Chemistry, MacDonald Campus of McGill University, Ste. Anne de Bellevue, Quebec, H9X 3V9, Canada.
[2]ENITIAA-L.G.P.A., Nantes. Cedex 03, France

Intact carp (Cyprinus Carpio) fillets were packaged under vacuum and pressurized at 100, 140, 180 and 200 MPa for 15 and 20 min at 4°C. Changes in the lipid fraction, color and electrophoretic profiles of the fish proteins were studied to establish the best time and pressure conditions for pressure-shift freezing of carp fillets. Thiobarbituric acid (TBA) values, free fatty acid (FFA) content and color parameters (L*, a*, b*) increased as pressure and time treatments increased. After 15 min of treatment at any pressure level the intensity of the protein band with MW<36 kDa decreased. Based on these results carp fillets were frozen using either pressure-shift freezing (PSF) or air-blast freezing (ABF) and then stored at -20°C for 75 days. TBA values and FFA content were relatively lower in PSF samples than in ABF samples. The freezing procedure had no significant effect (p>0.05) on texture of carp fillets. PSF was more effective in reducing total drip loss on cooked samples and with smaller and more regular ice crystals compared to ABF samples.

INTRODUCTION

Frozen storage is a long-term preservation method that allows storage of fish under more controlled conditions. Ideally, there should be no difference between fresh fish and frozen fish after thawing. If appropriate conditions are used, fish in the frozen state can be stored for several months without appreciable changes in quality (Santos-Yap, 1995). The reality however is that freezing and frozen storage can adversely affect the quality of a variety of muscle foods such as beef, pork, fish and chicken. This phenomenon of quality deterioration during frozen storage is more pronounced in fish muscle than other muscle foods, with deterioration in texture, flavor, and color being the most serious problems.

During freezing of fish, the separation of water as pure ice crystals creates an environment that is conducive to protein denaturation and this phenomenon has been cited as a possible cause for deterioration in frozen fish. Love (1968) summarised the factors that influence the size and location of ice crystals formed during freezing of fish muscle tissue as follows: physiological status of the fish, the freezing rate, the storage time, and temperature fluctuations.

There is a lot of interest in the use of high pressure in food processing and the fish industry is one sector taking advantage of this technology (Farr, 1990). In the specific case of fish freezing, high pressure may become a new alternative method. Kalichevsky *et al.* (1995) and Le Bail *et al.* (1997) summarized the effects of high pressure on the solid-liquid phase diagram of water and its potential food applications. As pressure increases to 210 MPa, the freezing point of water decreases to -22°C. Increasing the pressure beyond 210 MPa results in an increase in the freezing point. The unfrozen region of water under high pressure presents a tremendous opportunity for food processing. The primary applications of pressure in relation to the water phase changes are the increased freezing rates obtained using pressure-assisted freezing, the increased thawing rates and also the possibility of non-frozen storage at subzero temperatures.

The use of high pressure during freezing of fish may offer one important advantage over traditional freezing techniques, such as air blast and plate freezing, i.e., the increase in freezing rate during the freezing process could result in a final product with better quality after thawing. The improvement of quality may be a direct result in the reduction of tissue or textural damage, and indirectly by the reduction of lipid deterioration on the final product. Fuchigami *et al.* (1997a,b) applied high pressure freezing to carrots and reported that pressures of 200, 340 and 400 MPa appeared to be effective in reducing histological damage and improving the texture of thawed carrots. Fuchigami and Teramoto (1997) observed that high pressure freezing between 200 and 400 MPa yielded products with better texture compared to conventional freezing. Similar results were obtained for Chinese cabbage subjected to high pressure freezing (Fuchigami *et al.*, 1998). The behavior of food products under pressure freezing is very complex, and this also is a consequence of the variation in food composition. For practical purposes it is not very accurate to extrapolate processing conditions for vegetable material to animal food. Even among muscle tissues, there are differences between beef, pork and fish tissues that call for a better understanding of the effects of high pressure for a specific commodity. Thus, our objective was to compare the biochemical properties of thawed carp fillets subjected to conventional air-blast freezing (ABF) versus pressure-shift freezing (PSF), and to correlate the observed biochemical properties to some objective measurements of texture.

MATERIALS AND METHODS

Biological Samples

Live carp (*Cyprinus carpio*) were slaughtered, cleaned and filleted. The fresh fish fillets were placed in moisture-impermeable polyethylene bags (La Bovida, France) and vacuum-packaged. Samples were stored in isothermal boxes in a cold room at 4°C and processed between 3-5 h after slaughter. For each pair of fillets obtained, one was subjected to pressure shift-freezing (PSF) and the other to air-blast freezing (ABF).

Freezing processes

Air-blast freezing (Servathin, France) was carried out at -20°C using an air speed of 4 m/s and the samples were designated as ABF. Pressure shift freezing was carried out in a 3L capacity high pressure vessel (ALSTOM, Nantes, France). Samples were placed in the high pressure vessel and the pressure was increased to 140 MPa at a rate of 100 MPa/min. When the temperature of the samples reached -14°C ($\approx$12 min), pressure was released at a rate of 10 MPa/sec. Samples were immersed in a stirred cooling bath of 50/50 (v/v) ethanol/water solution at -20°C for 7 min to complete freezing. Once frozen, the carp samples were stored in isothermal boxes in a cold room at -20°C for 2, 15, 30, 65 and 75 days. Thawed samples were analyzed for: thiobarbituric acid (TBA) value, free fatty acid (FFA) content, texture, drip losses, and ice crystal formation by microscopy.

Thiobarbituric acid (TBA) value

The TBA value was determined according to the procedure of Vyncke (1970).

Free fatty acids

The method of Kirk and Sawyer (1991) was used to determine the free fatty acid content.

Texture

The cooked samples were kept at 4°C for 12 h to allow them to cool down before measurement. Objective texture measurements were performed with a texture testing machine (Lloyd Instruments LR5K, UK) using a Kramer Shear Compression Cell (65 mm x 65 mm x 65 mm) with a 10 blade probe was used.

Microscopy Aanalysis

Microscopy analysis were carried out using the procedure described by Martino and Zaritzky (1986).

Drip Loss

Drip loss was determined using the procedure described by Sequeira-Munoz (2001).

Statistical Analysis

The effects of frozen storage time on parameters such as extraction of TBA number, release of fatty acids, texture, drip losses and size of ice crystals were analyzed by two-way analysis of variance according to the method of Sokal and Rolhf (1995).

RESULTS AND DISCUSSION

Lipid Oxidation and Free Fatty Acids Release

It is known that during frozen storage of fish, products of lipid hydrolysis and oxidation can accumulate and cause deterioration of the product (Mackie, 1993). Hydroperoxides are the first products of oxidation and their measurement is a useful indicator of the stage of oxidation (Khayat and Schwall, 1983; Shewfelt, 1981). Figure 1 shows the effect of PSF on the TBA numbers of intact carp fillets compared with those that were frozen using ABF. In general, TBA values of PSF fillets remained constant during frozen storage while those for the ABF fillets increased after the first month of storage. In all cases, the TBA values of the ABF fillets were significantly higher ($p<0.05$) than those of PSF fillets. This data suggests that PSF had a curtailing effect on the oxidation of lipids in carp fillets. Cheah and Ledward (1997) reported an increase in TBA values during a 6 day storage period of minced pork pressurised at 400 and 800 MPa for 20 min at 19°C. However, no significant increased rate in oxidation was observed in muscle samples treated at 300 MPa. In our study, the pressure level used was 190 MPa for only 12 min at -14°C unlike the condition used by Cheah and Ledward (1997). Ohshima *et al.* (1993) used pressure levels between 200 to 610 for 15 to 30 min and suggested that isolated extracted marine lipids were more stable against autoxidation than lipids present in intact muscle. On the contrary, Angsupanich and Ledward (1998) observed little changes in TBA values of cod muscle pressurised at 200 MPa for 20 min. However, in the same experiment TBA values increased when samples were treated at 400 MPa or higher for 20 min and continued to increase during the 7 days of storage. Temperature may be a contributing cofactor inducing autooxidation. It has been suggested that accelerated oxidation may be due to the release of iron from haemoglobin and myoglobin which promote autooxidation of lipid pressurised fish meat (Tanaka *et al.*, 1991; Ohshima *et al.*, 1992; Cheah and Ledward, 1996). In this study, the processing conditions (pressure level, time and temperature) were probably not severe enough to induce autoxidation.

Besides oxidation, lipids may undergo hydrolysis resulting in the accumulation of FFA. Myofibrillar proteins are considered to be the prime target of FFA and they become largely unextractable in their presence (Shenouda, 1980; Ohshima *et al.*, 1984). It is generally believed that the interaction between proteins and FFA occurs primarily through electrostatic, van der Waals, hydrogen and hydrophobic forces. These interactions may create more hydrophobic regions in place of polar or charged groups resulting in a decrease in protein solubility in aqueous buffer, or further intermolecular linkages extensive enough to decrease extractability (Shenouda, 1980). Figure 2 shows the effects of PSF and ABF on the FFA content of intact carp fillets, which indicates that the level of FFA increased significantly ($p<0.05$) throughout the storage period for carp fillets frozen using ABF. In the case of the carp fillets frozen by PSF, during the first 30 days of storage the concentration of FFA remained constant, then started increasing during the rest of the study. At the end of the storage study, there were no significant

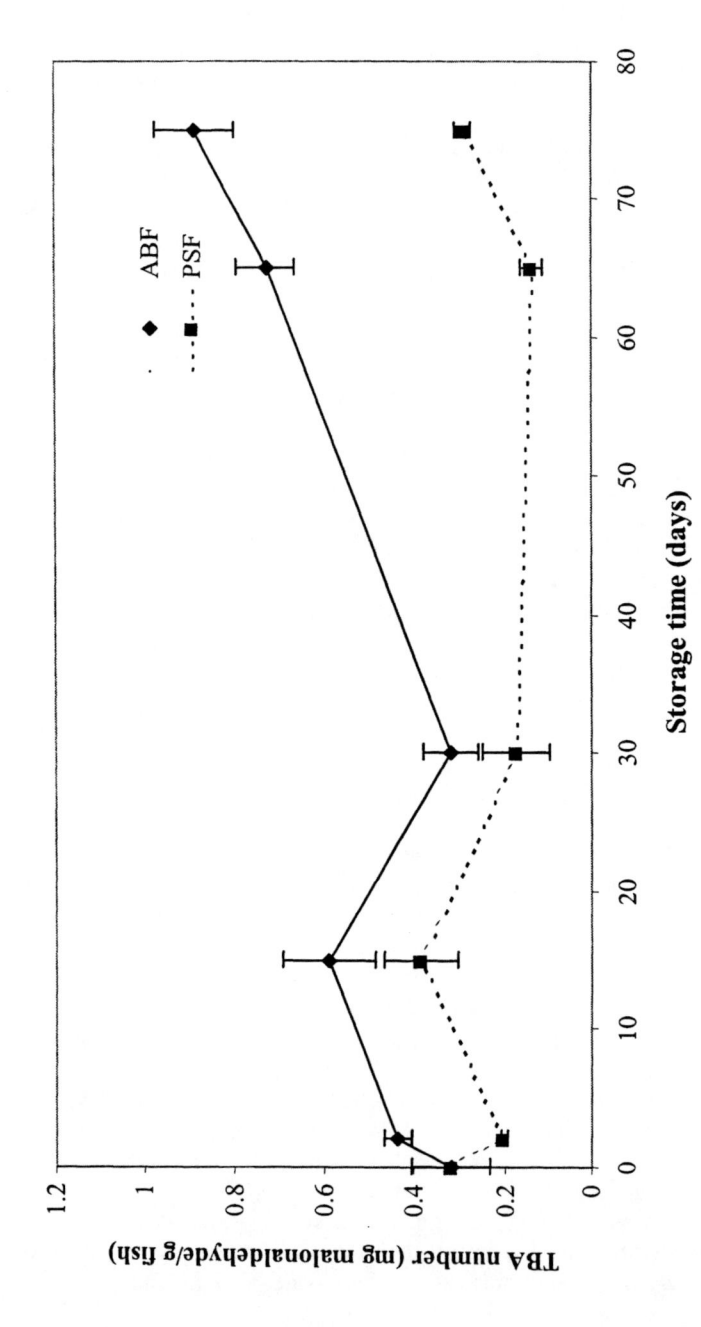

Figure 1. Effect of the freezing procedure on TBA number for carp fillets during frozen storage. Each marker represents the mean of 3 values.

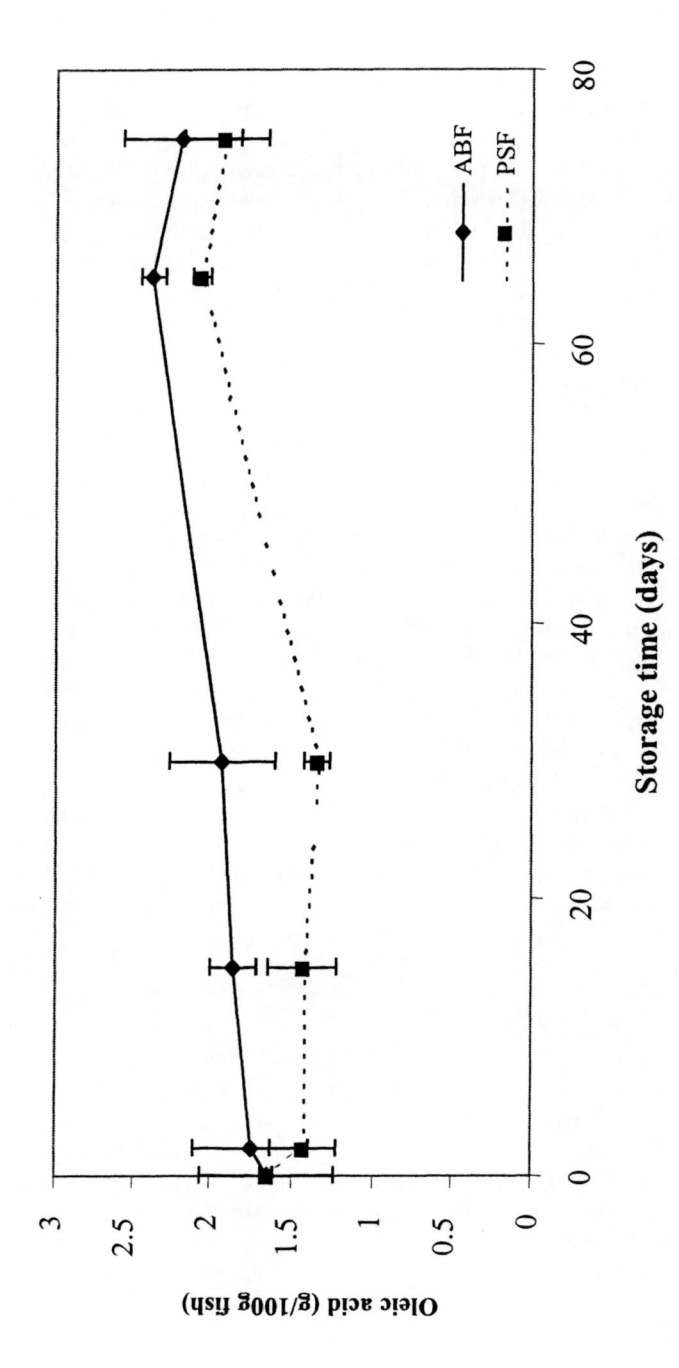

Figure 2. Effect of the freezing procedure on the level of free fatty acids on carp fillets during frozen storage. Each marker represents the mean of 3 values.

differences (p>0.05) between the FFA levels in the PSF samples versus the ABF samples. However it is important to note that between 15 to 65 days of storage, samples treated using PSF had a significantly lower FFA content than the samples frozen by ABF. It seems that PSF had a retardation effect on the release of FFA from lipid hydrolysis. Ohshima *et al.* (1992) showed that the enzymatic degradation of fish muscle phospholipids was effectively prevented by high hydrostatic pressure treatment. In this study, the pressure level used (140 MPa) may have had an initial inhibitory effect on the activity of phospholipases, but the enzymes may have gradually recovered their activity during the storage period.

Changes in Texture

Figure 3 shows the results for texture of cooked carp fillets, no significant differences were observed between ABF and PSF samples. Although it has been reported that toughness of fish flesh increases during frozen storage (Dias *et al.*, 1994; LeBlanc *et al.*, 1988), for carp fillets, toughness remained stable during frozen storage for both ABF and PSF treatments. It seems that carp fillets were less susceptible to quality deterioration during frozen storage than salt-water fish, probably because of the absence of the trimethylamine system responsible for the formation of formaldehyde, which promotes protein denaturation and deterioration of texture during frozen storage. Angsupunich and Ledward (1998) reported an increase in toughness after pressure treatment (400 and 600 MPa). However, at 200 MPa this effect was almost negligible. In this study, the fact that almost no differences were observed between the two treatments may be due to the low pressure level used and also to the fact the PSF reduces cooking losses in carp fillets (unpublished results). This increase in water content for cooked fish may counteract the effect of pressure denaturation resulting in little or no effect on hardness of the fish fillets.

Drip loss

Acceptability and consumption of frozen seafood is based in part on the overall flavor and textural quality of the cooked product. Loss of muscle fluid as drip may adversely influence the juiciness, tenderness and flavor of the cooked fish (Morrison, 1993). Otero *et al.* (1998) observed that high pressure assisted freezing was effective in reducing drip losses in eggplant. They suggested that massive nucleation obtained by high pressure-assisted freezing had less detrimental effect on cellular structure resulting in less drip losses. Similar observations were made by Koch *et al.* (1996) with potato cubes processed

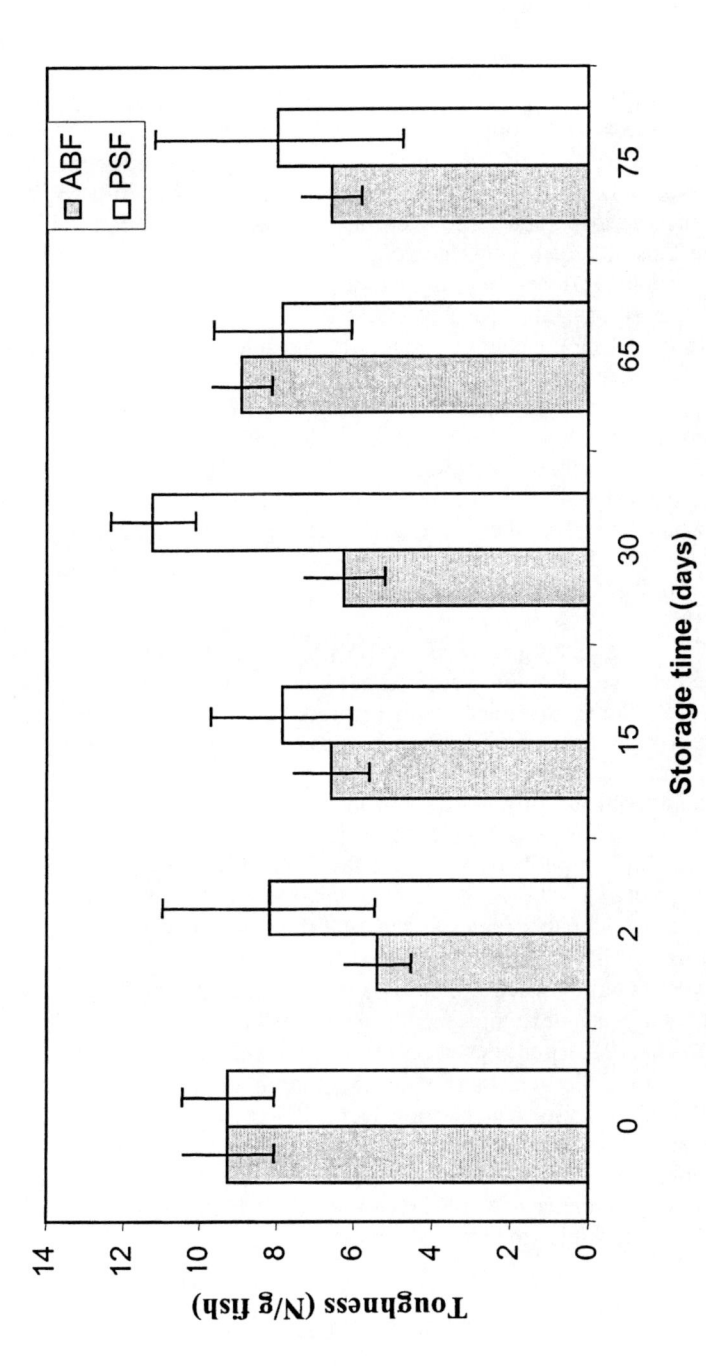

Figure 3. Effect of the freezing procedure on the toughness of carp fillets during frozen storage. Each bar represents the mean of 3 values.

using pressure-shift freezing. However, Ngapo et al. (1999) found no effect of freezing rate on drip losses in pork samples stored for 4 weeks.

In this study, carp fillets frozen by ABF showed significantly higher total drip losses ($p<0.05$) compared with the PSF samples (Figure 4). For carp fillets frozen by ABF, drip losses increased over time and became significantly higher ($p<0.05$) than those of fresh fish after 65 days of storage. Awad et al. (1969) observed that cooking losses increased during frozen storage of fresh water whitefish muscle. In contrast, total drip losses in PSF samples remained almost constant ($p<0.05$) during the storage period. These results are in agreement with those reported by MacFarlane (1973, 1974) that indicated reduced cooking loss in pre-rigor pressurized beef muscle. It was suggested that pressure treatment increase water holding capacity of meat as a result of a loosening of protein structure leading to increased hydration. An ionic mechanism was proposed and because of electrostriction effects between water molecules and exposed ions, the reaction would be expected to occur with a decrease in volume.

Histological analysis

Figure 5a shows micrographs of fresh carp muscle (control). The cross section of the untreated samples showed a uniform distribution of regular shape fibres. Figure 5b shows the micrograph of cross sections from ABF treated samples. Ice crystals formed during this freezing process had irregular shapes. The ABF process is considered a slow freezing process (0.9 cm/h) which may cause important shrinkage of the cells and formation of large extracellular ice crystals (Love, 1968; Fennema, 1973). These ice crystals induced deformations of the tissue resulting in pooling the muscle fibres as reported by Bevilacqua et al. (1979), Grujic et al. (1993) and Martino et al. (1998) for beef and pork.

Micrographs of the fish samples frozen by PSF process and stored 2 days at –20°C are presented in Figure 5c. Smaller and more uniform (rounded shape) ice crystals were obtained when PSF was used instead of ABF treatment. It is believed that a uniform and isotopic ice nucleation results in isotropic ice crystal size. Burke et al. (1975) indicated that freezing by pressure release from 150 MPa is theoretically able to achieve supercooling around –14°C, and thereby induce a 140 fold increase in nucleation rate. It was suggested that a large number of small and homogenous ice crystal nuclei formed during depressurization, grew during the second step of the PSF process (atmospheric pressure). The large number of small crystals prevents tissue deformations and shrinkage of cells yielding a good quality.

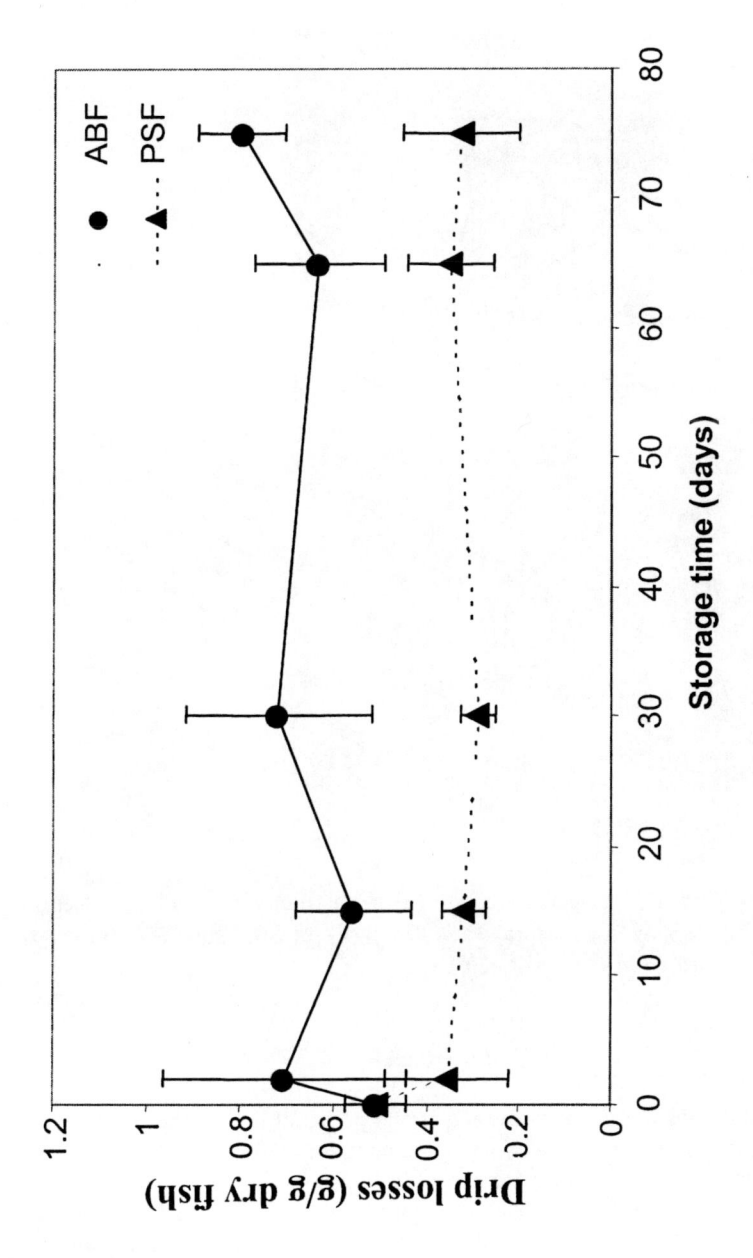

Figure 4. Effect of the freezing procedure on drip losses for carp fillets during frozen storage. Each marker represents the mean of 3 values.

150

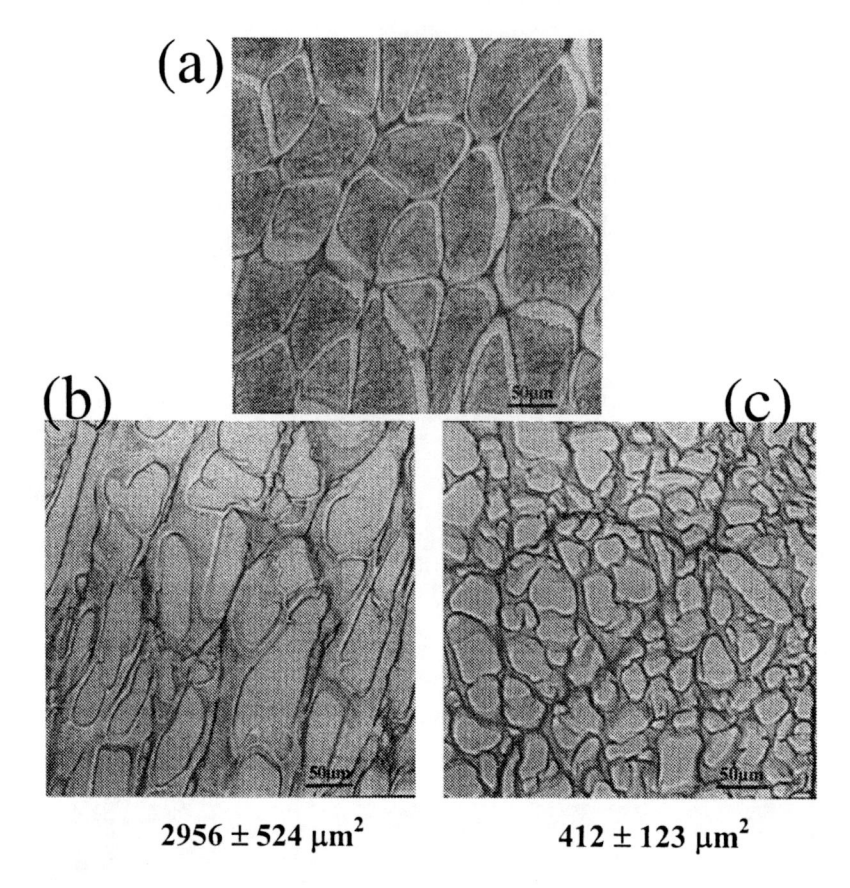

2956 ± 524 μm²   412 ± 123 μm²

Figure 5. Micrographs of carp fillet tissue. (a) fresh fillet, (b) air-blast frozen carp fillet after 2 days of frozen storage at –20°C; (c) pressure-shift frozen carp fillet after 2 days of frozen storage at –20°C.

CONCLUSIONS

This study shows the relative advantages and potential use of high-pressure technology to increase freezing rates while maintaining a good quality product. Using high-pressure technology reduced freezing time. However, more important was the reduction in lipid degradation such as oxidation and hydrolysis that may eventually determine the flavor of flesh fillets. Additionally, high pressure was effective in reducing the size of ice crystals within the fish samples

ACKNOWLEDGEMENTS

This work was funded by NSERC of Canada and the Centre de Cooperation Franco-Quebecois.

REFERENCES

Angsupanich, K. and Ledward, D.A. 1998. High pressure treatment effects on cod (*Gadus morhua*) muscle. *Food Chem.* 63: 39-50.
Awad, A.; Powrie, W.D. and Fennema, O. 1969. Deterioration of fresh-water whitefish muscle during frozen storage at −10 C. *J. Food Sci.* 34: 1-9.
Bevilacqua, A.; Zaritzky, N.E. and Calvelo, A. 1979. Histological measurements of ice in frozen beef. *J. Food Technol.* 14: 237-251.
Burke, M.J.; George, M.F. and Bryant, R.G. 1975. Water Relations of Foods. Food Science and Technology monographs. Academic Press, London, UK.
Cheah, P.B. and Ledward, D.A. 1996. High pressure effect on lipid oxidation in minced pork. *Meat Sci.* 43: 123-134.
Cheah, P.B. and Ledward, D.A. 1997. Catalytic mechanism of lipid oxidation following high pressure treatment in pork fat and meat. *J. Food Sci.* 62: 1135-1141.
Dias, J.; Nunes, M.L. and Mendes, R. 1994. Effect of frozen storage on the chemical and physical properties of black and silver scabbardfish. *J. Sci. Food Agric.* 66: 327-335.
Farr, D. 1990. High pressure technology in the food industry. *Trends Food Sci. Technol.* 1: 14-16.
Fennema, O.R. 1973. Nature of freezing process. In *Low Temperature Preservation of Food and Living Matter*. O.R. Fennema, W.D. Powrie and E.H. Marth (eds.) Marcel Dekker, New York, NY, pp. 126-145.
Fuchigami, M.; Kato, N. and Teramoto, A. 1998. High-pressure-freezing effects on textural quality of Chinese cabbage. *J. Food Sci.* 63: 122-125.
Fuchigami, M. and Teramoto, A. 1997. Structural and textural changes in kinu-tofu due to high-pressure-freezing. *J. Food Sci.* 62: 828-832, 837.
Fuchigami, M.; Kato, N. and Teramoto, A. 1997a. High-pressure-freezing effects on textural quality of carrots. *J. Food Sci.* 62: 804-808.
Fuchigami, M.; Miyazaki, K.; Kato, N. and Teramoto, A. 1997b. Histological changes in high-pressure-frozen carrots. *J. Food Sci.* 62: 809-812.
Grujic, R.; Petrocic, L.; Pikula, B. and Amidzic, L. 1993. Definition of the optimun freezing rate −1. Investigations of structure and ultrastructure of

beef *M. Longissimus dorsi* frozen at different freezing rates. *Meat Sci.* 33: 301-318.

Kalichevsky, M.T.; Knorr, D. and Lillford, P.J. 1995. Potential food applications of high-pressure effects on ice-water transitions. *Trends Food Sci. Technol.* 6: 253-258.

Khayat, A. and Schwall, D. 1983. Lipid oxidation in seafood. *Food Technol.* 37: 130-140.

Kirk, R.S. and Sawyer, R. 1991. Pearson's composition and analysis of foods. 9th Ed. Longman Scientific and Technical, London, UK, pp. 640-643.

Koch, H.; Seyderhelm, I.; Wille, P.; Kalichevsky, M.T. and Knorr, D. 1996. Pressure- shift freezing and its influence on texture, colour, microstructure and rehydration behaviour of potato cubes. *Nahrung* 40: 125-131.

Le Bail, A.; Chourot, J.M.; Barillot, P. and Lebas, J.M. 1997. Congélation-décongélation a haute pression. *Rev. Gen. Froid.* 972: 51-56.

LeBlanc, E.; LeBlanc, R. and Blum, I. 1988. Prediction of quality in frozen cod (*Gadus morhua*) fillets. *J. Food Sci.* 53: 328-340.

Love, M. 1968. Ice formation in frozen muscle, In *Low Temperature Biology of Foodstuffs*. J. Hawthorn and E.J. Rolfe (eds.) Pergamon, Oxford, UK, pp. 321-360.

MacFarlane, J.J. 1973. Pre-rigor pressurization of muscle: effects on pH, shear value and taste panel assessment. *J. Food Sci.* 38: 294-298.

MacFarlane, J.J. 1974. Pressure-induced solubilization of meat proteins in saline solution. *J. Food Sci.* 39: 542-547.

Mackie, I.M. 1993. The effects of freezing on flesh proteins. *Food Rev. Int.* 9: 575-610.

Martino, M. and Zaritzky, N. 1986. Fixing conditions in the freeze substitution technique for light microscopy observation of frozen beef tissue. *Food Microstructure* 35: 19-24.

Martino, M.N.; Otero, L.; Sanz, P.D. and Zaritzky, N.E. 1998. Size and location of ice crystals pork frozen by high-pressure assisted freezing as compared to classical methods. *Meat Sci.* 50: 303-313.

Morrison, C.R. 1993. Fish and shellfish. In *Frozen Food Technology*. C.P. Mallet (ed.) Blackie Academic & Professional. London, pp.196-236.

Ngapo, T.M.; Babare, I.H. and Mawson, R.F. 1999. The freezing and thawing rate effects on drip loss from samples of pork. *Meat Sci.* 53: 149-158.

Ohshima, T.; Nakagawa, T. and Koizumi, C. 1992. Effect of high hydrostatic pressure on the enzymatic degradation of phospholipids in fish muscle during storage. In *Seafood Science and Technology*. E.G Bligh (ed.) Fishing News Books, Oxford, UK, pp. 64-75.

Ohshima, T.; Ushio, H. and Koizumi, C. 1993. High pressure processing of fish and fish products. *Trends Food Sci. Technol.* 4: 370-375.

Ohshima, T.; Wada, S. and Koizumi, C. 1984. Effect of accumulated free fatty acid on reduction of salt-soluble protein of cod flesh during frozen storage. *Bull. Jpn. Soc. Sci. Fish.* 50: 1567-1572.

Otero, L.; Solas, M.T.; Sanz, P.D.; de Elvira, C. and Carrasco, J.A. 1998. Contrasting effects of high-pressure-assisted freezing and conventional air-freezing on eggplant tissue microstructure. *Zeitschrift-fuer-Lebensmittel-Untersuchung-und-Forschung* 206: 338-342.

Santos-Yap, E.M. 1995. Fish as Food. In *Freezing Effects on Food Quality*. L.E. Jeremiah (ed.). Marcel Dekker, Inc. New York, NY, pp. 109-133.

Sequeira-Munoz, A. 2001. Use of High Pressure for Improving the Quality and Shelf Life of Frozen Fish. Ph. D. Dissertation. McGill University, Montreal, Canada.

Shenouda, S.Y.K 1980. Theories of protein denaturation during frozen storage of fish flesh. *Adv. Food Res.* 26: 275-311.

Shewfelt, R. 1981. Fish muscle lipolysis - A review. *J. Food Biochem.* 5: 79-100.

Sokal, R.R. and Rolhf, F.J. 1995. *Biometry*. 3rd. W.H. Freeman & Co. New York, NY, pp. 328-367.

Tanaka, M.; Xueyi, Z.; Nagashima, Y. and Taguchi, T. 1991. Effect of high pressure on the lipid oxidation in sardine meat. *Nippon Suisan Gakkaishi* 57: 957-963.

Vyncke, W. 1970. Direct determination of the thiobarbituric acid value in trichloroacetic acid extracts of fish as a measure of oxidative rancidity. *Fette Seifen Anstrichmittel* 72: 12.

CHAPTER 12

BIOGENIC AMINES AND E-NOSE DEVICES: INDICATORS OF SEAFOOD QUALITY

Amaral Sequeira-Munoz, Wen-Xian Du, Francis R. Antoine, Maurice R. Marshall and Cheng-I Wei

Food Science and Human Nutrition Department, P.O. Box 110370, University of Florida, Gainesville, FL 32611-0370

Microbial assessment, biogenic amine determination and sensory analysis were performed on fresh salmon, mahi-mahi, and tuna stored at different temperatures for various time periods. Color, texture, appearance, and odor were evaluated by a trained sensory panel, while aroma/odor was evaluated using an AromaScan. The objectives were to determine and correlate the use of biogenic amines and their levels; and to evaluate an electronic nose device for predicting seafood quality. Biogenic amines may be used as quality and safety indicators for mahi-mahi and tuna. Human sensory analysis provides a useful means to monitor both changes in freshness and the onset of spoilage. In determination of seafood quality, good correlation exists between AromaScan and sensory analysis, and AromaScan and bacterial counts. Electronic nose devices can be used in conjunction with microbial counts and sensory panels to evaluate the degree of decomposition in mahi-mahi and tuna during storage.

INTRODUCTION

Seafoods are highly susceptible to spoilage and deterioration due to the growth of postmortem microbial populations (Dainty et al., 1983). Microbial activity on seafood products produces pronounced off-flavors and odors, a shorter shelf life, and heavy economic losses (Reddy et al., 1994). Biogenic amines are indicators of fish quality and spoilage, and are formed as a result of decarboxylation of free amino acids in fish, meat, and other protein rich food products. The decarboxylating enzymes are produced as a result of bacterial growth in the presence of the free amino acids. Histamine, one of the biogenic amines, is the causative agent of histamine poisoning, which results from eating certain economically important fish species (Arnold and Brown, 1978; Lerke *et al.*, 1978; Morrow *et al.*, 1991). Other biogenic amines called potentiators such as cadaverine, putrescine, among others, augment the effect of histamine during scombroid poisoning (Lyons *et al.*, 1983; Taylor *et al.*, 1984; Clifford *et al.*, 1989).

Scombroid poisoning is generally associated with decomposing scombroid fish such as tuna, mackerel and bonito, and non-scombroid fish such as mahi-mahi, blue fish and sardines. It is a chemical food-borne intoxication that results from the ingestion of foods containing relatively high levels of histamine (Arnold and Brown, 1978; Lerke *et al.*, 1976; Taylor *et al.*, 1984). Histamine intoxication is a worldwide problem which occurs in countries where fish, containing high levels of histamine, is consumed (Huss, 1994; Smart, 1992; Taylor and Sumner, 1987). Because the disease is generally mild, of short duration and self-limiting, it is usually not reported to health authorities. In, addition, it is sometimes misdiagnosed as "fish allergy" (Gellert *et al.*, 1992).

Taylor *et al.* (1984) and Clifford *et al.* (1989) concluded that histamine formation in marine products tends to be governed by their histidine content. They also stated that histamine formation is not proportional to the loss of histidine. Fletcher *et al.* (1995) concluded that histamine-producing bacteria found on fish are capable of producing histidine decarboxylase at ambient temperatures (20-25°C). At that temperature, bacterial activity results in the production of high levels of histamine. However, Baranowski *et al.* (1990) found that both mesophilic and psychrotrophic bacteria were responsible for biogenic amine formation in mahi-mahi and mackerel.

Amino acids, abundant in protein foods such as fish, also play a major role in the microbial spoilage of fish (Jay, 1992; Ingram and Dainty, 1971). They are the immediate precursors of biogenic amines such as histamine, cadaverine and putrescine, all of which are indicators of fish safety and quality (Sikorski *et al.*, 1990). As simple low molecular weight compounds, they provide essential nutrients for the rapid growth of microbes (Finne, 1992; Jay, 1992). Amino acids are broken down during bacterial spoilage yielding ammonia, a major total volatile basic nitrogen (TVB-N) component, and carbonyl compounds, both of which signal freshness deterioration and spoilage of fish tissue (Schmitt and Siebert, 1961; Liston, 1973).

In order to protect consumers against unwholesome fish, and therefore, possible scombroid poisoning, it is essential to establish controls not only for imported fish but also for the marketed catch from sports fishermen, fish processors, and retailers. Such control mechanisms, e.g. hazard analysis critical control point (HACCP) plans, as mandated by the government (FDA, 1995), require stipulation of specific objective parameters of quality (Kramer, 1966).

However, to establish standards and uniformity in measurement and assessment of quality, it is necessary to have objective methods for measuring specific quality attributes. Herein lies the importance of quality indicators such as levels of histamine, cadaverine, putrescine, and TVB-N compounds, which result from amino acid breakdown.

Organoleptic panels are generally the primary means used to evaluate fish quality, however, they are not always reliable as analytical tools, because of possible subjectivity and variations within the panel (Learson and Ronsivalli, 1969; Hollingworth and Throm, 1982; Bomio, 1998). Consequently, there is always the need for simpler and more reliable objective tests, which will objectively measure the quality of fish, however quality may be defined (Learson and Ronsivalli, 1966). It seems though that no practical universal method is available for measuring the overall quality of seafood (Jacober and Rand, 1982). The organoleptic panel, despite its limitations, remains to date the only current method available for measuring the overall quality of fish. However, measures must be taken to control and minimize factors responsible for variability in the organoleptic data (Kramer, 1966). Jacober and Rand (1982) pointed out that the inherent disadvantages of sensory evaluation range from standardization problems among laboratories to any personal prejudices a judge might have as to color, flavor, and odor.

Of all senses, smell has been the most difficult to define objectively. Only in the last several years has the ability to measure and characterize smell become possible. Advances in organic chemistry, electronics and computing have made possible the development of new digital aroma technology (electronic nose: AromaScan Inc. Hollis, NH) that parallels the human nose (Russel, 1995). AromaScan is an electronic nose unit that mimics the human nose. As aroma-odor vapors from test samples are drawn over an array of 32 organic polymer sensors, the volatiles are adsorbed and desorbed from the polymer surface. These temporary changes in electrical resistance due to interaction with the sensor can then be expressed as an aroma histogram. The combined resistance from the 32 sensors, which represent the continuous and real-time analysis of the overall aroma-odor properties for each sample, can be condensed into a single representative data point on a two-dimensional AromaMap. The axes representing Euclidean distances between each sample can be plotted and the differences among samples determined. Multiple samples appear as populations on the AromaMap and demonstrate the reproducibility of using AromaScan for quality determination (Russel, 1995). Very little sample preparation time is needed for analysis with AromaScan. Therefore, the analysis can be completed within a few minutes. An electronic nose, such as AromaScan, can be used in parallel with sensory panels or to complement the more costly and time-consuming chemical analysis for product quality determination (Aparicio *et al.*, 2000).

The primary advantage of using the electronic nose as a quality-assurance tool for the food industry is speed of analysis, in terms of data generation and interpretation. Rapid, meaningful data interpretation is possible with various chemometric (multivariate analysis) techniques. The artificial neural network of AromaScan is a data-processing of algorithms based loosely on the structure of the human brain. Quality control models can be developed by "training" the electronic nose, then routine samples can be tested against the model, providing a "goodness of fit" approximation or a sample accepted or rejected answer (Aparicio *et al.*, 2000: Borjesson *et al.*, 1996).

Given the importance of free amino acids in fish spoilage, their role as precursors of biogenic amines, and the continued need for objective parameters of fish quality, the objectives of this research were: 1) to determine and correlate the use of biogenic amines and their levels, and 2) to evaluate an electronic nose device for predicting seafood quality.

MATERIALS AND METHODS

Fish preparation and storage

Mahi–mahi. Fresh mahi-mahi (dolphin fish, *Coryphaena hippurus*) was obtained from commercial fisheries in the Gulf of Mexico. Three groups of fresh mahi-mahi fillets (6 pieces per group) were processed and stored at constant temperatures of 1.7, 7.2 and 12.8°C. Another group of fillets was stored at 1.7 °C for 1 day, than shifted to 12.8°C for 24 h before returning to 1.7°C (Temp. Ramp I). A fifth group was stored at 1.7° for 2 days, then shifted to 12.8°C for 24 h before returning to 1.7°C (Temp. Ramp II). Stored samples were examined at 0, 1, 3, 5 days for quality by sensory, aroma, bacteriological, and chemical evaluations.

Tuna. Fresh yellowfin tuna (*Thunus albacores*) loins of 15-20 kg each were purchased from a local seafood store in Gainesville, FL., and transported in ice to the Food Science and Human Nutrition Department, University of Florida. The outer layers of the tuna loin were then carefully removed with a sterile knife. The loins were cut to prepare 120 pieces of tuna steaks (12 x 10 x 2 cm) of about 250 g each. After each steak was placed in separate sterile, labeled Zipper bags (26 x 28 cm, Tenneco Packaging, Pittsford, N.Y.), the samples were divided randomly into four groups and stored at 0, 4, 10 and 22°C for 1, 3, 5 and 9 days.

At each interval, tuna steaks were removed from each temperature group for determination of bacterial counts and sensory quality using a 10-person taste panel and AromScan. Five pieces of tuna steaks were used as day 0 control to determine background bacterial loads and sensory quality, both by the taste panel and AromaScan.

Salmon. Fresh Atlantic salmon (*Salmo salar*, whole fish but eviscerated) was purchased from a local seafood store in Gainesville, Florida. Following washing under running tap water, the salmon was filleted to prepare skinless samples (15 x 12 cm) of about 280 g each. After each fillet sample was placed and labeled in separate sterile Whirl-Pak bags (Fisher Scientific, Orlando, FL), they were divided randomly into three groups and stored at -20, 4, $10°C$ for 3, 5, 7, 9, and 14 days. At each sampling interval, samples were taken for sensory analysis and AromaScan analysis. Six pieces of salmon were used as day 0 control to determine background bacterial loads and sensory quality both by taste panel and AromaScan.

Sensory Evaluation of Fish

On each day of testing, the fish samples from each temperature group were prepared for quality evaluation. Ten trained panelists wearing gloves were solicited to evaluate the fillets by touching, smelling, and visual observation of the samples, describing whether differences occurred between samples, and commenting on specific sensory attributes, such as color and odor. Panelists also used descriptive analysis to identify the characteristics that distinguish the samples and rate the intensity of important sensory attributes. The protocols and evaluations described in the National Marine Fisheries Service (NMFS) Fishery Products Inspection Manual (Section 1, Chapter 18, Part II, August 25, 1975) were followed. Quadruplicate samples were used for each storage temperature and the experiments were duplicated. The fillets were considered spoiled when they had a strong discoloration and off-odor, and were unfit for human consumption.

Sensory test, AromaScan mapping, and microbial tests were carried out on the same day of removal from the treatment conditions.

Microbial Analysis

Total Microbial Counts. For total microbial counts, about 20 g portions of fish fillet were cut from test samples at each sampling period and homogenized at high speed for 2 min in a sterile blender with nine volumes (1:9, w/v) of sterile Butterfield's buffer. The homogenates were serially diluted with sterile

Butterfield's buffer. Then, 0.1 mL aliquots of the diluents were surface plated on quadruplicate aerobic plate count (APC) agar (Difco, Detroit, MI) plates containing 1.5% NaCl. Pour plate method was also used for some homogenates as necessary. Bacterial colonies were counted after the plates were incubated at room temperature for 48 h.

AromaScan Analysis

A 10 g portion of each fish was weighed out and placed in analysis pouches. The pouches were evacuated then charged or inflated, with carbon filtered air. The bag humidity setting was 5 g/m^3 and the reference air humidity was 10 g/m^3. Each bag was individually incubated at 35°C for 10 min, so that the headspace was equilibrated. Prior to analysis the sensors were cleaned using reference air. The air was dried by passing it over dried silica gel. After each analysis the sensors were washed (1 min) with the headspace from the wash bottle filled with 2% isopropanol. The sensors were then allowed to react with reference air for 2.5 min before the next analysis. AromaScan sampling time was 120 s, and data was collected from the period 45 to 60 s of the total analysis time. AromaMap patterns were generated from all the samples analyzed.

Biogenic Amines

The gas chromatographic method of Rogers and Staruszkiewicz (1993) was modified for the analysis of biogenic amines. The procedures used 75% methanol in distilled water, instead of 100% methanol, for extraction of the amines, based on the work of Luten *et al.* (1992). A Perkin-Elmer 8500 gas chromatograph with flame ionization detector (FID) was used for amine analysis. Separations were achieved using a 15 m x 0.32 mm DB-1 column (J&W Scientific, Folsom, CA) with a film thickness of 3 µm, which was fitted with a fused silica 1 m x 0.32 mm untreated guard column (Supelco, Bellefonte, PA).

Histamine Analysis

Capillary Electrophoresis (CE). CE was performed on a BioFocus 2000 Capillary Electrophoresis System (Bio-Rad, Hercules, CA) using a coated capillary cartridge of 24 cm x 25 µm I.D. The cartridge temperature was maintained at 35°C. A constant-voltage at 10 kV was applied and detection of histamine was performed by monitoring the absorbance at 210 nm. Data were stored in a Pentium PC and processed using an integration BioFocus program (Bio-Rad).

Fluorometric Method. A modified AOAC fluorometric method for histamine analysis (AOAC Official Method 977.13, 1990) using 75% methanol as the extracting solvent was used for sample extraction (Rogers and Staruszkiewicz, 1997). Chromatographic polypropylene tubes, 200 x 7 mm (id), were each fitted with 45 cm of Teflon tubing and a flow control valve. The column was packed with an 8 cm bed of the Dowex 1-X8 resin, and the height of the tubes were adjusted to ensure that the flow rate was > 3 mL/min. The photofluorometer was fitted with an excitation wavelength of 360 nm narrow band pass (NB) filter, and a NB 440 emission wavelength filter. A 5 mL glass cuvette was used for all measurements.

Statistical Analysis

Statistical analysis was performed using pairwise correlation analysis to determine the relationships between the production of biogenic amines in each test species and the results of sensory evaluation, as well as the results of AromaScan and bacterial counts. Pairwise correlation analysis was also used for determining the correlation between different histamine analytical methods. Microsoft Excel for Windows 95 Version 7.0 (Microsoft Corporation, 1995) was used to calculate the mean and standard deviation of microbial, sensory, and amine data.

Multidimensional data were generated when a sample aroma was analyzed on AromaScan's unique 32 polymer sensor array. Sammon mapping technique was the statistical method used for AromaScan mapping. Multiple discriminant analysis was also used to process the AromaScan data. Significant differences between samples are shown by the spatial separation between the clusters.

RESULTS AND DISCUSSION

Mahi-mahi. Stored mahi-mahi fillets showed time-and temperature- related increases in bacterial loads. Those fillets stored at 7.2 and 12.8°C showed a dramatic increase in bacterial counts, reaching >9 $\log_{10}$ CFU/g in 3 days (Figure 1). Those fillets stored at 1.7°C did not reach this level until day 5. The changes in bacterial counts for the two temperatures ramp groups (Temp. Ramp I and II) were similar to the 7.2°C group. It was noted that, when the bacterial number reached 7 $\log_{10}$ or higher, the fish meat was considered spoiled and rejected for consumption. Mahi-mahi fillets stored at 12.8°C were rejected by day 1 (based on the microbiological standard). Fillets stored at all other temperatures were rejected by day 3.

The levels of putrescine, cadaverine, and histamine increased as decomposition progressed (Figure 2). Putrescine was not detected at initial decomposition but increased rapidly at advanced decomposition. Cadaverine was detected in small amounts initially but the quantity increased as decomposition progressed. Histamine was not detected in fresh muscle. Its formation did not occur until the stage of initial decomposition (7.2°C for 3 days). However, it increased dramatically after storage at higher temperature (12.8°C) for 5 days. The formation of putrescine and cadaverine in mahi-mahi stored at 1.7, 7.2°C and Temp. Ramp I was faster than that of histamine. The levels of putrescine and cadaverine at these three temperatures were also higher than that of histamine at each interval. In contrast, the level of histamine was higher than those of putrescine and cadaverine in mahi-mahi stored at 12.8°C on day 5 of storage. Good correlation was found between the sum of (Put+Cad+His) and bacterial count. The counts for mahi-mahi fillets were higher than 7 $\log_{10}$ CFU/g when (Put+Cad+His) was more than 20 ppm and this can be used as an indicator of mahi-mahi spoilage (Figure 3a). Most of the mahi-mahi fillets were rejected by the sensory panel when higher than 50 ppm (Put+Cad+His) were found (Figure 3b). Most of the fillets were classified as grade B or C products when more than 20 ppm (Put+Cad+His) was detected. This indicated that decomposition of fish had occurred and higher levels of biogenic amines were found. Sensory analysis was not a good indicator of the presence of high levels of biogenic amines in fish. (Figure 3b).

AromaScan could not differentiate odor differences of mahi-mahi fillets stored at 1.7°C by days (Figure 4). Sensory analysis of mahi-mahi indicated that fillets stored at 1.7°C were grade A products during the 5 days of storage. Mappings of fillets stored at 7.2°C started to separate from day 0 controls on day 5 when fillets were rated as grade C products; whereas, fillets stored at 12.8°C started to separate from day 0 controls on day 3, when fillets were rated as grade C products. Similar results were also found in the two temperature ramp groups. Mappings of these two groups separated from day 0 controls when they became grade C products. These results indicated that AromaScan can differentiate the odor difference between grade C and grade A products.

Clusters of mahi-mahi fillets stored at different temperatures with different microbial loads separated from each other in the canonical discriminant graphs (Figure 5). Clusters with close bacterial numbers were also closer to each other on the graphs. A high correlation was also found between the AromaScan and sensory analysis. The cluster of grade 3 fillets was farther away from those of grade 1 and 2 fillets in both temperatures. This result indicated that AromaScan is capable of detecting differences in odor profiles in mahi-mahi under different stages of decomposition, and that it can be used for quality and freshness evaluation. Figure 6 shows that AromaScan can be used to predict the content

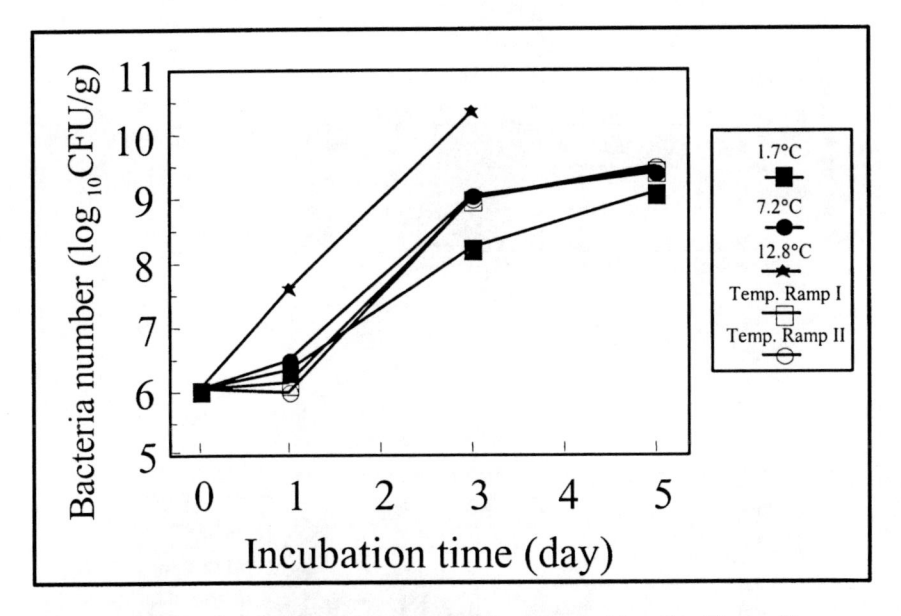

Figure 1. Time-related changes in bacterial load for mahi mahi. Temp. Ramp – 1.7°C for 24 (I) or 48 h (II), then 12.8°C for 24 h, return to 1.7°C

of biogenic amines in mahi-mahi stored under Temp. Ramp II condition. The total biogenic content reflected by (Put+Cad+His) could be used to indicate whether the fillets were fresh (<20 ppm), under initial decomposition (20-50 ppm), or under advanced decomposition (>50 ppm).

The results of this study demonstrate a correlation between AromaScan and sensory evaluation conducted by a panel. Comparison of the AromaScan results with microbiological measurements also shows a similar effect. Thus, loss of freshness in mahi-mahi can be objectively determined by AromaScan analysis.

Tuna. A comparison of the storage time on the mapping of tuna fillets stored at 0°C showed that the days of storage did not cause a separation of the AromaScan maping (Figure 7). This indicates that the quality of the tuna stored at 0°C did not change much during 9 days of storage. This low storage temperature also affected the release of volatiles for assessment by AromaScan. Storage time had a more profound affect on maping when tuna fillets were stored at temperatures higher than 0°C. As the storage temperature increased, samples of different storage time increasingly separated from each other, especially those at the higher storage temperature for a longer time. Thus,

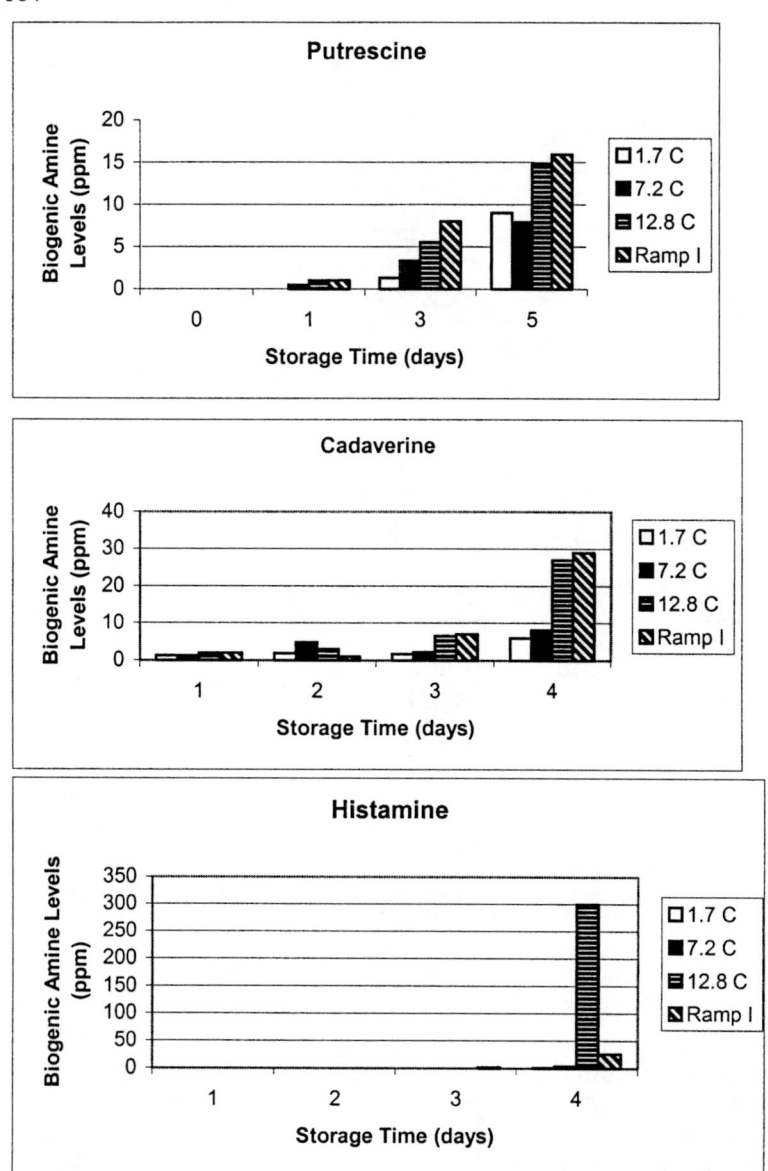

Figure 2. Time-related changes in biogenic amines for mahi mahi. Temp. Ramp – 1.7°C for 24 (I) or 48 h (II), then 12.8°C for 24 h, return to 1.7°C.

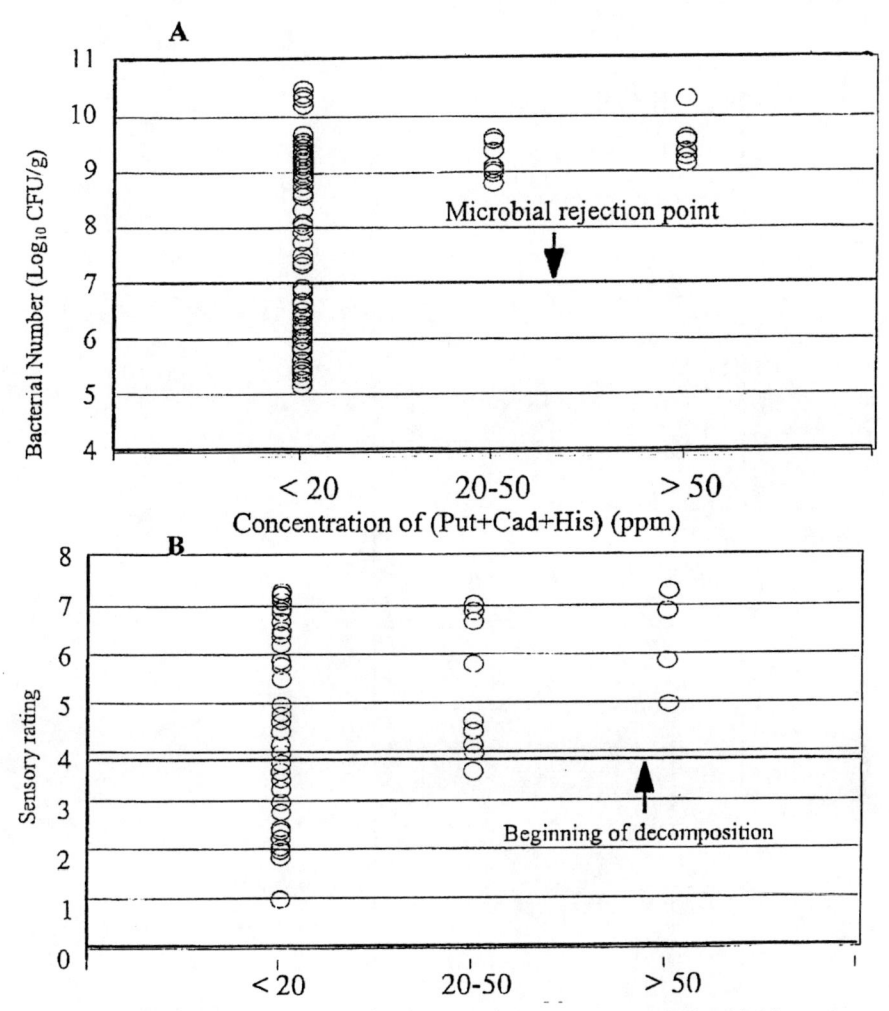

Figure 3. Correlation of bacterial load (A) and sensory scores (B) with biogenic amine levels.

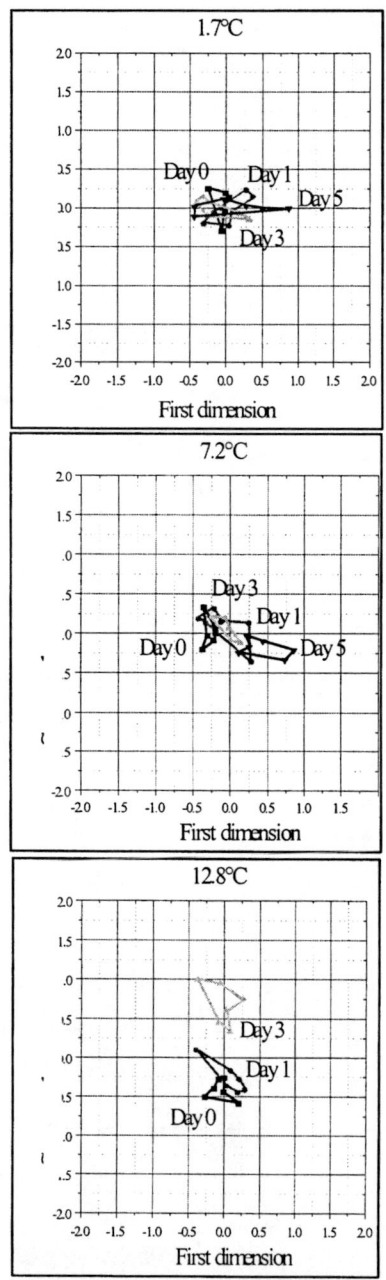

Figure 4. Aroma maps of mahi mahi stored at 1.7°C, 7.2 °C and 12.8 °C.

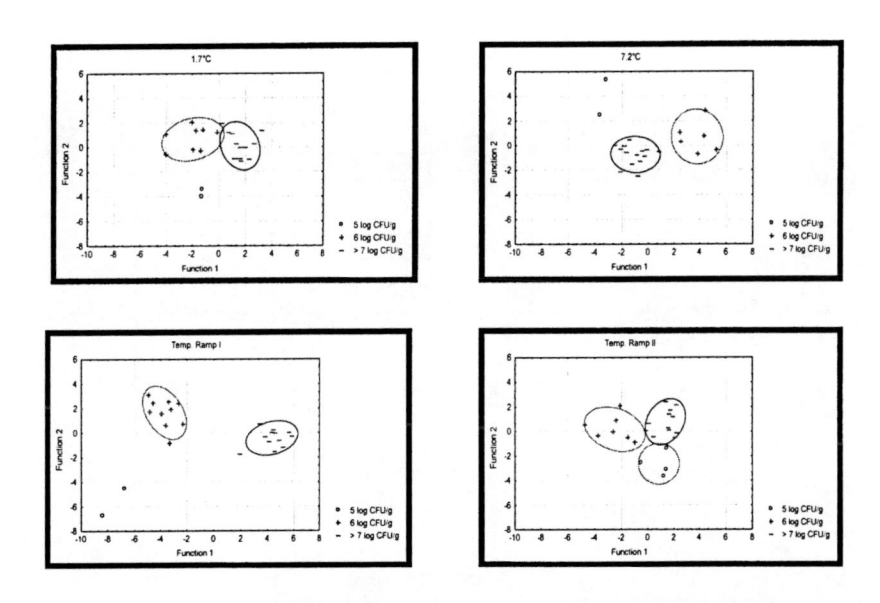

Figure 5. Correlation of AromaScan and bacterial load for mahi mahi using multiple discriminant analysis at various storage conditions.

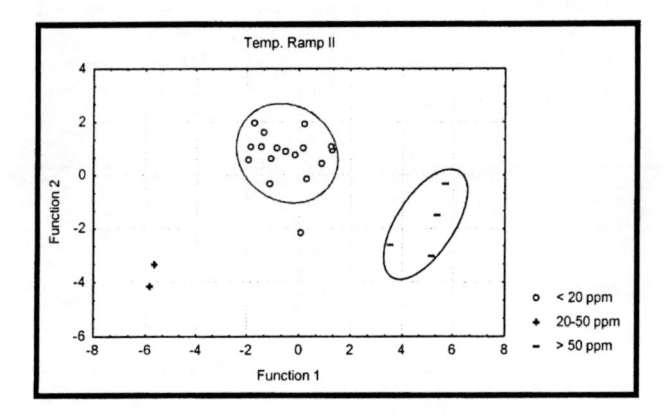

Figure 6. Correlation of AromaScan and biogenic amines for mahi mahi using multiple discriminant analysis at Ramp II storage condition.

168

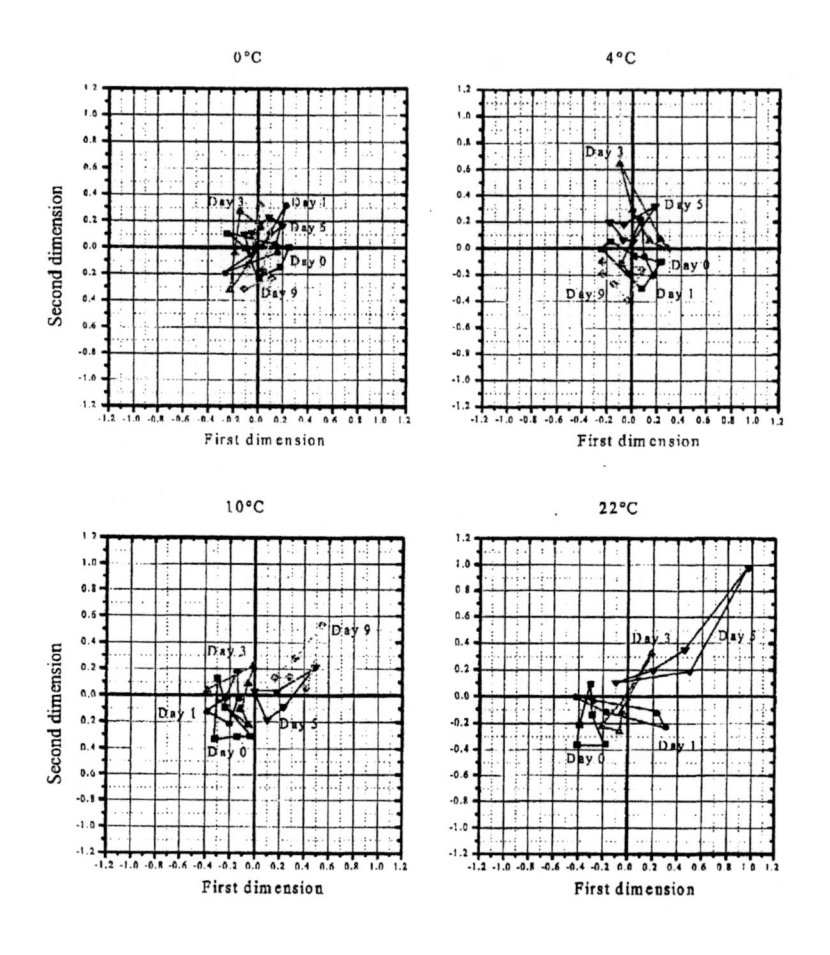

Figure 7. Aroma maps of Yellowfin Tuna stored at 0°C, 4 °C, 10 °C and 22 °C.

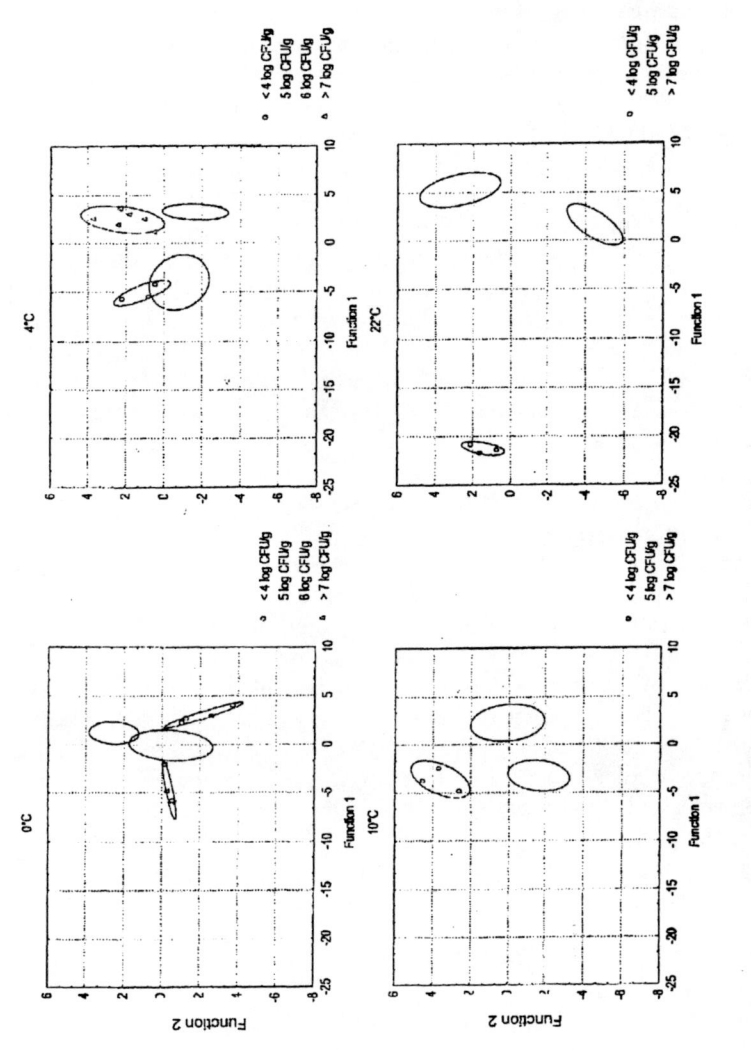

Figure 8. Correlation of AromaScan and bacterial load for Yellowfin tuna using multiple discriminant analysis at various storage conditions.

AromaScan can be used to differentiate the degree of decomposition in tuna during storage.

The change in quality as detected by AromaScan matched that of microbiological measurements. The clusters of tuna stored at different temperatures with different microbial loads separated from each other in the canonical discriminant graphs (Figure 8). They also showed a temperature-related separation in the plots; as the temperature increased, the clusters were separated further. The clusters of fillets with similar total bacterial counts were close to each other on the graphs. Thus, AromaScan results can be used as microbial-quality indicators of stored tuna fillets.

The AromaScan also predicted the grade of tuna fillets stored at 4, 10, and 22°C. A temperature-related relationship occurred with the cluster separation of sensory grade; the clusters of tuna fillets stored at 22° C were more separated than those at 4 or 10°C (data no show). Figure 9 showed that AromaScan can be used to predict the level of biogenic amines in tuna stored at 4°C. The clusters of fillets with different levels of histamine content were separated from each other in the graph. Fillets with the total biogenic amines (Put+Cad+His) less than 20 ppm can be separated from those having more than 20 ppm (Put+Cad+His) in the canonical discriminant graph. Thus, AromaScan can be used for quality and safety evaluation of tuna.

Salmon. Analysis by AromaScan of the sensory quality of salmon fillets stored at different temperatures (Figure 10) showed that mappings for the five groups were separated from that of the day 0 samples in a time-related fashion. Samples stored at 1.7°C for different days did not separate from each other. A time-related separation was found in samples stored at 7.2 and 12.8°C; their maps continued to separated further as storage time increased.

AromaScan can predict the grade of salmon fillets stored at the two temperature ramp groups. Compared to the clusters of salmon fillets stored at Temp. Ramp I, the cluster of grade B fillets stored at Temp. Ramp II was closer to that of grade A fillets. The cluster of grade C fillets was farther away from those of grade A and B samples in both temperature groups. This indicated that the grade C fillets had an odor profile far different from those of grade A and B fillets. Thus, AromaScan can be used for quality and freshness evaluation of salmon.

CONCLUSIONS

Storage temperature and microbial counts played important roles in causing spoilage of fish fillets. As the stored fillets showed time- and temperature-related increases in bacterial loads, they showed deterioration of quality.

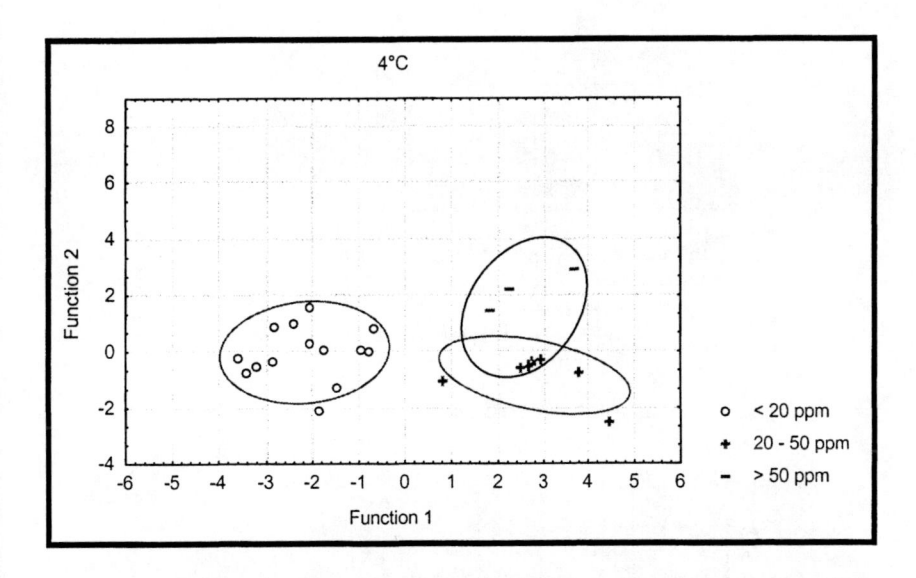

Figure 9. Correlation of AromaScan and biogenic amines for Yellowfin tuna using multiple discriminant analysis at 4 °C storage condition.

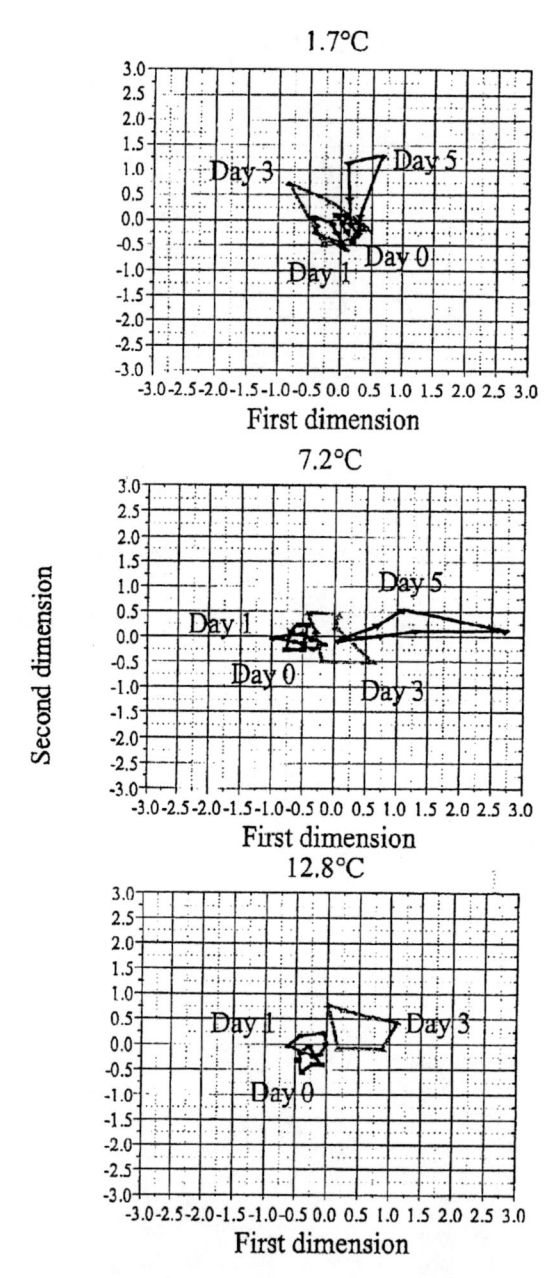

Figure 10. Aroma maps of salmon stored at 1.7°C, 7.2 °C and 12.8 °C.

Therefore, bacterial level is a useful and objective indicator of gross spoilage in fish fillets. Histamine showed little value in monitoring freshness. However, other biogenic amines such as cadaverine showed promise correlating well with TVB-N, putrescine and sensory analysis. The change of fish quality as detected by AromaScan matched that of microbiological measurements. AromaScan can be used to indicate the degree of decomposition during storage. Human sensory study provides a useful means to monitor both changes in freshness and onset of spoilage. AromaScan analysis may provide a useful means to monitor both changes in freshness and onset of spoilage. AromaScan analysis may provide a viable quantitative approach for determining fish freshness that could be used for quality control and inspection purposes.

REFERENCES

Aparicio, R.; Rocha, S.M.; Delgadillo, I. and Morales, M.T. 2000. Detection of rancid defect in virgin olive oil by the electronic nose. *J. Agric. Food Chem.* 48: 853-860.

Arnold, S.H. and Brown, W.D. 1978. Histamine toxicity and fish products. Adv. Food Nutr. Res., vol. 24, Academic Press, Inc., San Diego, CA, pp. 113-154.

Baranowski, J.D.; Frank, H.A.; Brust, P.A.; Chongsiriwantana, M. and Premaratne, R.J. 1990. Decomposition and histamine content in mahi-mahi (*Coryphaena hippurus*). *J. Food Protec.* 53: 217-222.

Bomio, M. 1998. Neutral networks and the future of sensory evaluation. *Food Technol.* 52: 62-63.

Borjesson, T.; Eklov, T.; Jonsson, A.; Sundgren, H. and Schnurer, J. 1996. Electronic nose for odor classification of grains. *Cereal Chem.* 73: 457-461.

Clifford, M.N.; Walker, R. and Wright, J. 1989. Studies with volunteers on the role of histamine in suspected scrombrotoxicosis. *J. Sci. Food Agric.* 47: 365-375.

Dainty, R.H.; Saw, B.G. and Roberts, T.A. 1983. Microbial and chemical changes in chill-stored red meats. In *Food Microbiology-Advances and Prsopects*. T.A. Roberts and F.A. Skinner (eds.), Academic Press, London, UK, 151 p.

Federal Drug Administration. 1995. Procedures for the safe and sanitary processing and importing of fish and fishery products, final rule. *Federal Register* 60: 65095-65202.

Finne, G. 1992. Non-protein nitrogen compounds in fish and shell fish. In *Advances in Seafood Biochemistry*. G. Flick and R.E. Martin (eds.), Technomic Publishing Co. Inc., Lancaster, PA, pp. 393-401.

Fletcher, G.C.; Summers, G.; Winchester, R.V. and Wong, R.J. 1995. Histamine and histidine in New Zealand marine fish and shellfish species, particularly kahawai (*Arripis trutta*). *J. Aquatic Food Prod. Technol.* 4: 53-74.

Gellert, G.A.; Ralls, J.; Brown, C.; Huston, J. and Merryman, R. 1992. Scrombroid fish poisoning underreporting and prevention among noncommercial recreational fishers. *West. J. Med.* 157: 645-647.

Hollingworth, Jr., T.A. and Throm, H.R. 1982. Correlation of ethanol concentration with sensory classification of decomposition in canned salmon. *J. Food Sci.* 47: 1315-1317.

Huss, H.H. 1994. Assurance of seafood quality. FAO Fisheries Technical Paper 334, pp. 32-35.

Ingram, M. and Dainty, R.H. 1971. Changes caused by microbes in spoilage of meats. *J. Appl. Bacteriol.* 34: 21-39.

Jacober, L.F. and Rand, Jr. A.G. 1982. Biochemical evaluation of seafood. In *Chemistry and Biochemistry of Marine Food Products*. R. E. Martin, G. J. Flick and D. R. Ward (eds.), AVI Publishing Co., Westport, CT, pp. 347-365.

Jay, J.M. 1992. *Modern Food Microbiology, 4th* (ed.), Chapman and Hall, New York, NY, pp. 221-233, 413-416.

Kramer, A. 1966. Parameters of quality. *Food Technol.* 20: 53-58.

Learson, R.J. and Ronsivalli, L.J. 1969. A new approach for evaluating the quality of fishery products. Bureau of Commercial Fisheries Technological Laboratory, Gloucester, MA, pp. 249-259.

Lerke, P.A.; Werner, B.; Taylor, S.L. and Guthertz, L.S. 1978. Scombroid poisoning – report of an outbreak. *West. J. Med.* 129: 381-386.

Liston, J. 1973. Microbial spoilage of fish and seafoods. Presented at the IVth International Conference on Global Impacts of Applied Microbiology, San Paulo, Brazil, July 23.

Luten, J.B.; Bouquet, W.; Seuren, L.A.J.; Berggraf, M.M.; Riekwel-Booy, G.; Durand, P.; Etienne, M.; Gouyou, J.P.; Landrein, A.; Ritchie, A.; Leclerq, M. and Guinet, R. 1992. Biogenic amines. In *Fishery Products: Standardization Method Within EC. Dev. Food Sci.* 30: 427-439.

Lyons, D.E.; Beery, J.T.; Lyons, S.A. and Taylor, S.L. 1983. Cadaverine and aminoguanidine potentiate the uptake of histimine *in vitro* in perfused intestinal segments of rats. *Toxicol. Appl. Pharmacol.* 70: 445-458.

Morrow, J.D.; Margolies, G.R.; Rowland, J. and Roberts, L.J. 1991. Evidence that histamine is the causative toxin of scrombroid-fish poisoning. *New Engl. J. Med.* 324: 716-720.

National Marine Fisheries Service (NMFS). Fishery Products Inspection Manual (Section 1, Chapter 18, Part II, August 25, 1975). In *Regualtions Governing*

Processed Fishery Products and U.S. Standards for Grades of Fishery Products (Chapter 2, Part 260). National Seafood Inspection Laboratory, Pascagoula, MS.

Reddy, N.R.; Schreiber, C.L.; Buzard, K.S.; Skinner, G.E. and Armstrong, D.J. 1994. Shelf life of fresh tilapia fillets packaged in high barrier film with modified atmospheres. *J. Food Sci.* 59: 260-264.

Rogers, P.R. and Staruszkiewicz, W.F. 1993. Modification of the GLC procedure for the determination of cadaverine and putrescine. FDA Laboratory Information Bulletin No. 3794, Issue 7.

Rogers, P.L. and Staruszkiewicz, W.F. 1997. Gas chromatographic method for putrescine and cadaverine in canned tuna and mahi-mahi and fluorometric method for histamine (minor modification of AOAC Official Methods 977.13): collaboratiove study. *J. Assoc. Off. Anal. Chem.* 803: 591-602.

Russel, P. 1995. Sensory analysis. *Milk Ind. Int.* 97: 11-12.

Schmitt, H. and Siebert, G. 1961. Abstract of optischer zur bestimmung der peptidase aktivat. *Biochemische Zeitschrift* 334: 96-1078.

Sikorski, Z.E.; Kolakowska, A. and Burt, J.R. 1990. Postharvest biochemical and microbial changes. In *Seafood: Resources, Nutritional Composition, and Preservation.* Z. E. Sikorski (ed.), CRC Press Inc., Boca Raton, FL, pp. 55-75.

Smart, D.R. 1992. Scombroid poisoning – a report of seven cases involving the Western Australian salmon. *Med. J. Australia* 157: 748-751.

Taylor, S.T.; Hui, J.Y. and Lyons, D.E. 1984. Toxicology of scrombroid poisoning. In *Seafood Toxins.* E.P. Ragelis (ed.), American Chemical Society, Washington, DC, pp. 417-430.

Taylor, S.L. and Sumner, S. 1987. Determination of histamine, putrescine, and cadaverine. In *Seafood Quality Determination: Procedings of an International Symposium.* D.E. Kramer, and J. Liston (eds.), Elsevier Science Publishers, Amsterdam, The Netherlands, pp. 235-245.

CHAPTER 13

STUDY OF COLD-SOLUBLE METHYLATED AGARS

Abdelwahab Riad[1], Andre Begin[2,] Marie-Rose Van Calsteren[2], and Rachid Lebar[1]

[1]Setexam, B.P. 210, Kenitra, Morocco
[2]Agricultrue et Agroalimentaire Canada, Centre de Recherché et de Developpement sur les Aliments, 3600 Boul. Casavant Ouest, Saint-Hyacinthe Quebec, J2S 8E3, Canada

Agar is widely used in microbiology, in cosmetics and in foods to control viscosity and to form gels. The solubilization temperature of the powder (Ts), the gelling temperature of the solution (Tg) and the melting temperature of the gels (Tm) determine the formulation and technological possibilities of agar. In this study, agar was partly methylated with dimethyl sulfate. Methylation was followed by hydrolysis, reduction, acetylation, and separation of the partially methylated alditol acetates by gas chromatography. The solubilization, gelling and melting temperatures were measured by rheology. Methylation occurs rapidly on C2 of the galactose residues. Solubilization, gel formation, and melting were 80, 35.6 and 85 C, respectively. Methylation with 20 g L^{-1} of DMS sharply decreased Ts, Tg, and Tm to 55, 25 and 59 C. After methylation with a 40 g L^{-1} of DMS, the Ts, Tg and Tm decreased to 48, 19.3 and 53 C, respectively.

INTRODUCTION

Agar and agarose are extracted from red algae (Rhodophyta) and are widely used in microbiology as culture media, in biochemistry as electrophoretic support, and in cosmetics and in foods as thickener and thermoreversible gels. Agar and its fractions (agarose and agaropectin) are based on linear chains of β-D-galactopyranose residues linked glycosidically through positions 1 and 3 (A units) and α-galactopyranose residues linked through position 1 and 4 (B units) (Figure 1). The galactopyranose or 3,6-anhydro-α-L-galactose.

© 2004 ScienceTech Publishing

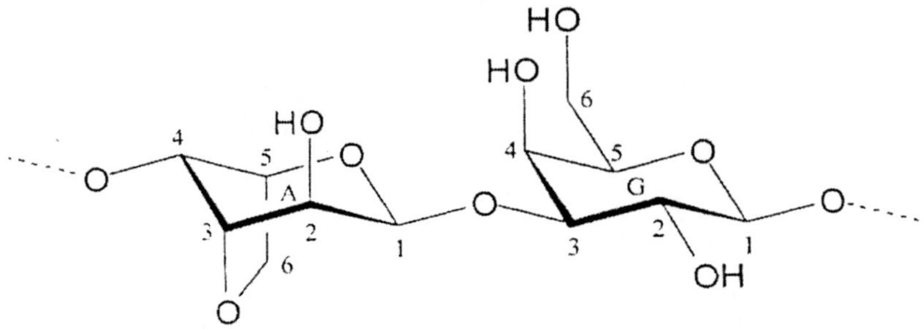

4)-3,6-anhydro-α-L-galactopyranosyl-(1   3)-β-D-galactopyranosyl-(1

Figure 1. Molecular structure of agarobiose.

For dissolving agar in water, the temperature must be maintained to at least 80 C for a long time or at water boiling temperature for 30 minutes. The temperature of gelation of agar solutions varies between 28 and 38°C, and depending on the agar variety, melting points vary between 80 and 98°C.

One serious drawback in using agars is their fixed temperature of powder solubilization (Ts), gel formation (Tg) and gel melting (Tm). For example, on food formulations agar must first be mixed with water and them boiled. In biotechnology, psychrophilic bacteria immobilization in agar is impossible because at the temperature agar gels, the bacteria are killed. To change the thermal determine the influence of methylation on agar Ts, Tg and Tm and to understand the mechanisms by which agar properties are changed.

MATERIALS AND METHODS

Agar (25 g), partially methylated on A2 in the native form, was boiled in water (850 ml) and the solution cooled at 80 C. For end-group reduction, sodium borohydride (0.15 g of NaBH$_4$ and 0.2 NaOH (16 g L^{-1}) and 20, 30 or 40 g L^{-1} of dimethylsulfate (DMS) were added. The samples were air dried at 50°C.

Hydrolysis, reduction, acetylation and sugars analysis were performed according to Stevenson and Furneaux (1991). Retention times were obtained from methylated ι-, κ-, and λ-carrageenan (Stevenson and Furneaux, 1991), FID response factor for anhydeogalactoses from Stevenson and Furneaux (1991) and the other response factors were calculated according to Sweet *et al.* (1975).

Determination of Ts was done in a M5-Osc Haake viscometer (Paramus, NJ) by following the rise in viscosity of a suspension (0.22 g of powder in 15 ml of water) in a NV system held at solution of 1.5% agar by 3.3 °C min^{-1} from 55 to 5 °C (Figure 2B). Tg determination was performed with crushed gel particles introduced between the cup and the rotor. The temperature was gradually increased from 30 to 95°C at a rate of 0.57 C/min (Figure 2C).

RESULT AND DISCUSSION

Errors related to hydrolysis and degradation products

The influence of preparation steps (hydrolysis and extraction) on the estimated degrees of methylation was estimated by calculating the G/A ratio of non-methylated agars rapidly decreased to 1.2 after 1 hour and remained almost constant (Figure 3), suggesting losses of anhydrogalactose during preparation of the samples for GC. However, the mean G/A ratio for the methylated agars was 0.993. Errors related to superposition of degradation products were considered as negligible since degradation peals were of the same order of magnitude as the G246 peak (Figure 4).

The total degree of methylation increased progressively with DMS concentration from 0.19 to 0.40 (Figure 5). At 20 g L^{-1} of DMS, 12.4% of G2 is methylated (Figure 5), representing 77% of the carbons methylated by DMS. At this concentration the reactivity of the methylation sites was G2>G4 G6 A2. At 30 and 40 g L^{-1} of DMS, the percentage of methylated G2 remained almost constant while the percentage of G6 and A2 increased. At these DMS concentrations, the order of reactivity was G6 A2>G2 G4.

The loss of reactivity of the G2 hydroxyl group with increasing concentration of DMS is probably related to the polysaccharide conformation in solution. Hayashi et al. (1978), cooled agarose solution from 85°C to 30°C, and showed (by DSC) that a transition from coil to helix occurs between 50 to 55°C. X-ray diffraction studies of agarose helix by Arnott et al. (1974) have shown that all G2 hydroxyl group are directed inside the helix. Since our products were methylated at 50°C, it suggests that parts of the molecules were rapidly methylated. The G2 hydroxyl groups in random coil portions of the molecules were rapidly methylated. The G2 hydroxyl groups inside the helix were not methylated because DMS was not able to penetrate inside.

At 20 g L^{-1} of DMS, the temperatures of solubilization, gelling and melting rapidly decreased from 80, 35.6 and 85 to 55.1, 25.6 and 59°C, respectively. At 40 g L^{-1} of DMS, Ts, Tg and Tm decreased only to 48, 19.3 and 53°C

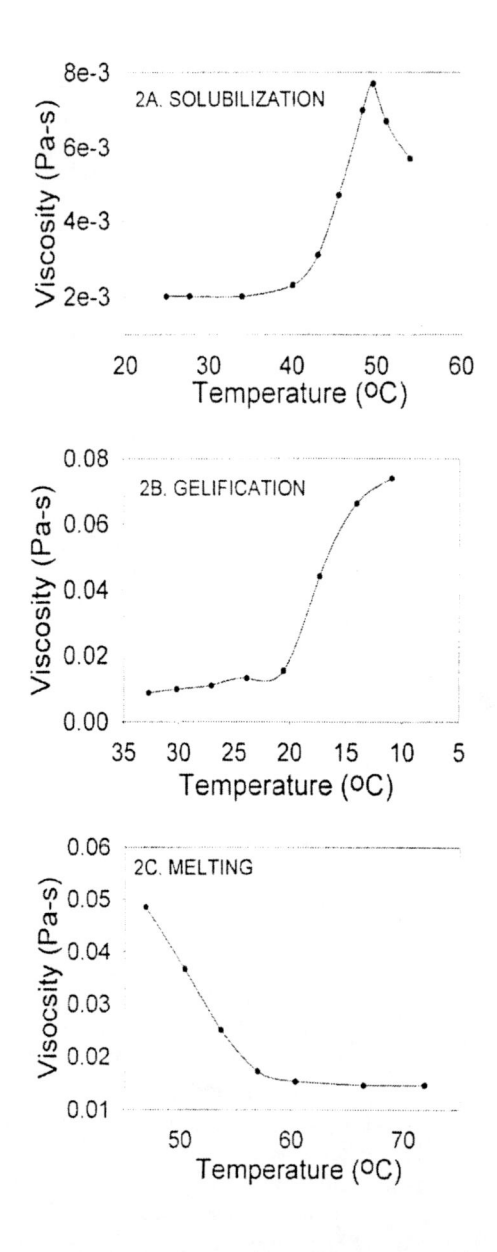

Figure 2. Determination of Ts (A), Tg (B), and Tm (C).

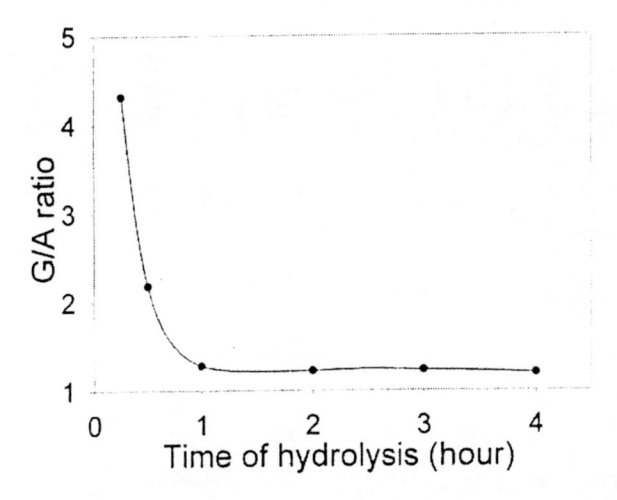

Figure 3. Influence of hydrolysis time on the recuperation of sugars

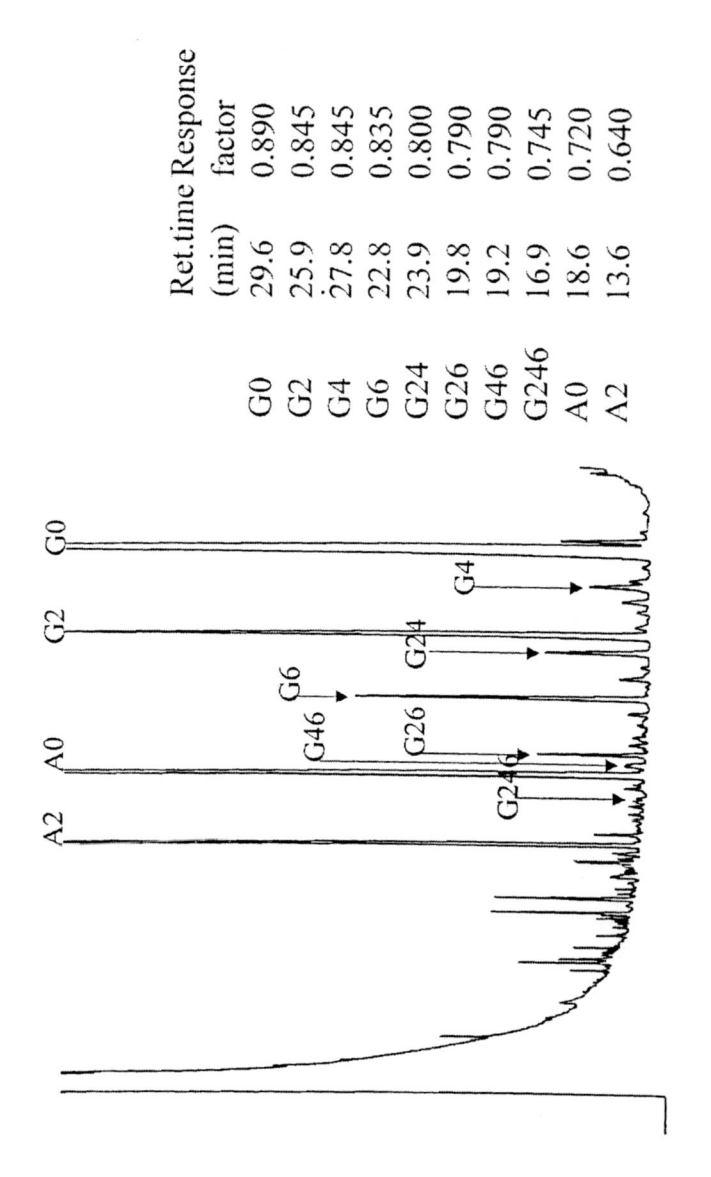

	Ret.time (min)	Response factor
G0	29.6	0.890
G2	25.9	0.845
G4	27.8	0.845
G6	22.8	0.835
G24	23.9	0.800
G26	19.8	0.790
G46	19.2	0.790
G246	16.9	0.745
A0	18.6	0.720
A2	13.6	0.640

Figure 4. Chromatogram of agar (40 g L^{-1} of DMS)

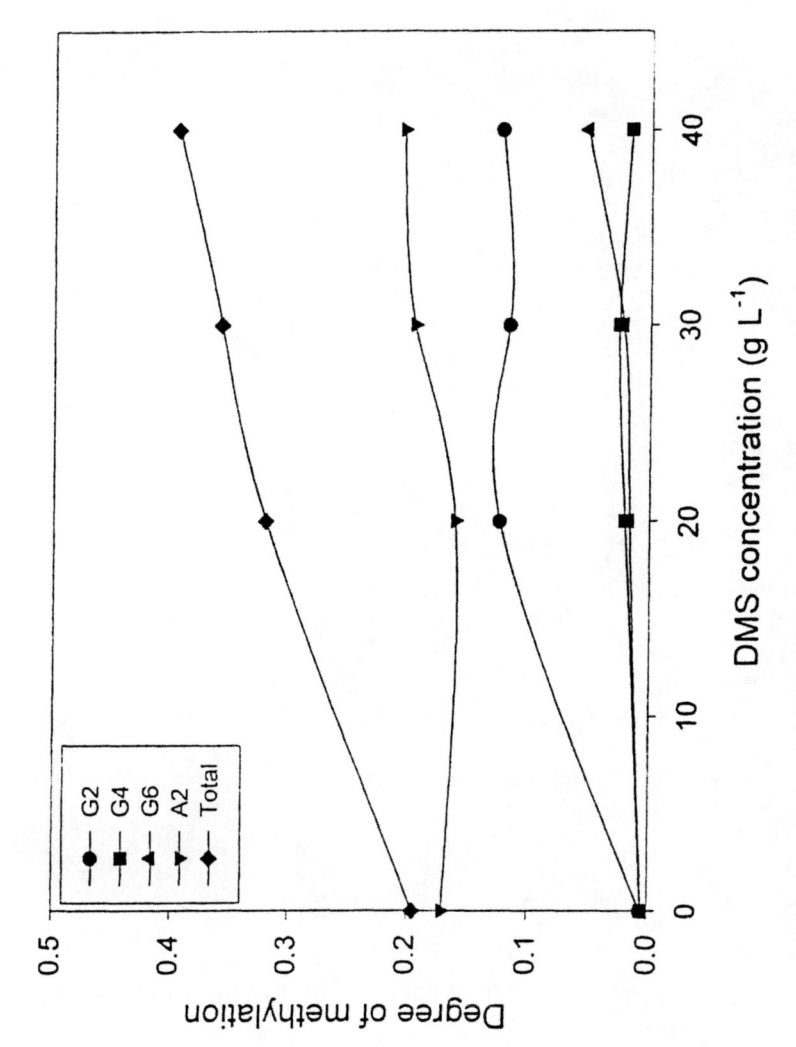

Figure 5. Effect of the DMS concentration on total methylation and methylation of the specific hydroxyl groups.

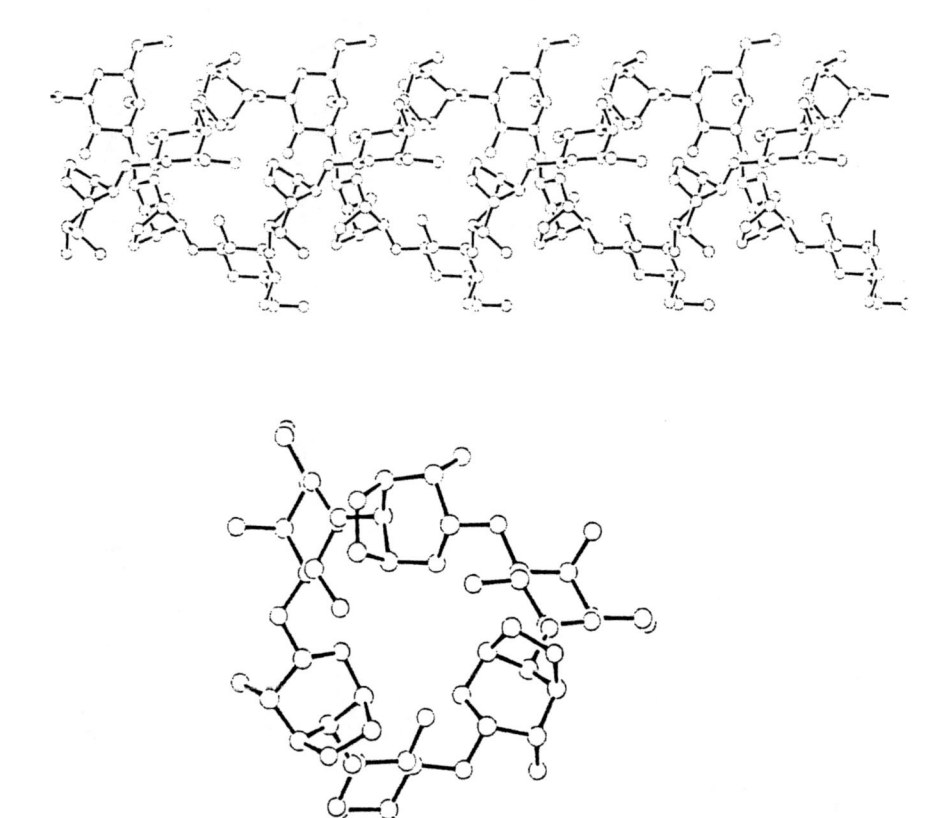

Figure 6. Mutually perpendicular projections of the agarose double helix structure. Left: perpendicular to the helix axis. Right: down the helix axis.

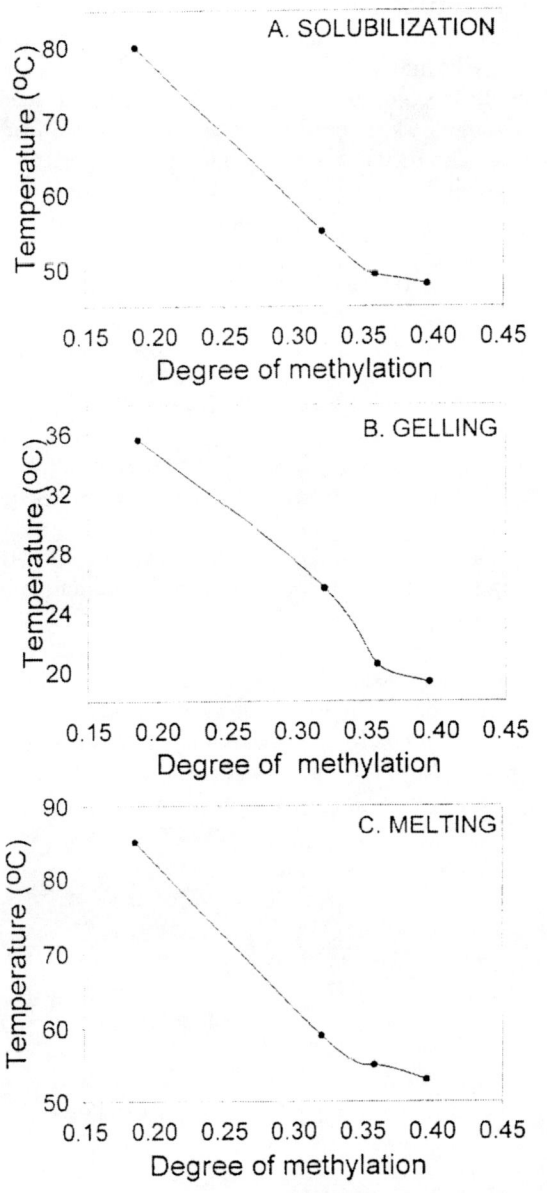

Figure 7. Influence of the degree of methylation on the temperatures of solubilization (A), of gelling (B), and of melting (C).

respectively. It suggests that thermal characteristics of agar are highly related to the methylation of G2 hydroxyl groups.

Several molecular models have been proposed to explain agar gel formation. Essentially, the gel formation process is viewed as a coil-single helix transition or a coil-double helix transition followed by crystalline aggregation of helices side by side. The methoxyl group on G2 should sterically interfere with the formation of helices. The decrease of crystallite links results in lower Ts, Tg and Tm.

ACKNOWLEDGMENT

We would like to thank Daniel Ouellet for the chromatographic analyses.

REFERENCES

Arnott, S.; Fulmer, A.; Scott, W.E.; Dea, I.C.M.; Moorhouse, R. and Rees, D.A. 1974. The agarose double helix and its function in agarose gel structure. *J. Mol. Biol.* 90: 269-284.

Hayashi, A.; Kinoshita, K.; Kuwano, M. and Nose, A. 1978. Studies of the agarose gelling system by the fluorescence polarization method. II. *Polym. J.* 10: 485-494.

Sweet, D.P.; Shapiro, R.H. and Albersheim, P. 1975. Quantitative analysis by various g.l.c. response-factor theories for partially methylated and partially ethylated alditol acetates. *Carbohydr. Res.* 40: 217-225.

Stevenson, T.T. and Furneaux, R.H. 1991. Chemical methods for the analysis of sulphated galactans from red algae. *Carbohydr. Res.* 210: 227-298.

CHAPTER 14

USE OF CHITOSAN FOR STORAGE OF MANGOES

André Bégin, Isabelle Dupuis, Marc Dufaux and Gonzague Leroux

Food Research and Development Centre, 3600 Casavant Blvd. West, St-Hyacinthe, Quebec, J2S 8E3, Canada

The ability of chitosan to control growth of virulent strains of Colletotrichum gloeosporioides, known as the main factor for commercial loss of mango fruits was studied. Initially, naturally infested fruits were coated with chitosan. All fruits coated with chitosan had a higher resistance to C. gloeosporioides. However, some fruits were partially fermented. This indicated that chitosan coatings modified the internal atmosphere of the fruits, suggesting that the antimicrobial activity of chitosan came from the change in maturity of fruits. To compare the direct antimicrobial activity of chitosan, we isolated C. gloeosporioides. On solid agar media, chitosan, Benomyl® and benzoate completely inhibited growth during 15 days. On fruits sterilized with alcohol and re-infected with C. gloeosporioides, no fungicide completely inhibited growth. However, Benomyl® and chitosan reduced mycelium growth by 46 and 74%, respectively. The results indicate that chitosan can advantageously replace chemical fungicides for storage of mango fruits.

INTRODUCTION

Fungicides are not well accepted by the consumer. Their utilization depends on the particular national legislation on pesticide residues in foods (EC, 1991) which are constantly re-evaluated (OECD, 1996; FAO, 1997; EC 1991) for all their safety aspects (consumer protection, residues in food, worker protection, fate and behavior, ecotoxicology). For an exporter, the ban of a pesticide can mean the loss of a market. In post-harvest treatments, fungicides are dangerous to workers, and residual concentrations on the product can not be checked easily. The FAO/WHO food standards (FAO, 1997) tolerate residues of benomyl, macozeb, prochloraz, propiconazole, triadimefon and triadimenol on mango fruits with specified maximum residue levels. In the USA, chitosan is accepted as a biopesticide in the production of any raw agricultural commodity and has been exempted of tolerance (USOFR, 2001).

Eighty-five percent of the commercial losses of mango fruits are related to anthracnose infection caused by *Colletotrichum gloeosporioides* (Noon, 1984). El Gaouth *et al.* (1992) have demonstrated the action of chitosan against *C. gloeosporioides* and used it as coating for the preservation of strawberries, tomatoes, bell peppers and cucumbers. Elson *et al.* (1985) used N,O-carboxymethylchitosan as coating material to generate modified internal atmospheres in several fruits.

In this study, the potential of chitosan to control anthracnose on mango fruits was evaluated by: 1) studying the relations between the host, the fungus, the environment and the fungicide, and 2) comparing the effectiveness of chitosan to that of other fungicides.

MATERIALS AND METHODS

Three coating formulations (1% chitosan, 1.5% chitosan and 1% chitosan:1% Tween®) were tested. The fruits were stored at 20 °C and their respiration rates, their respiratory quotient, the development of maturity and of anthracnose were followed. The antifungal activity of chitosan was compared to that of other commercial fungicides (oleic acid, benzoic acid, and benomyl®) on Potato Dextrose Agar plates containing different fungicides and on mango fruits infected by cutting 1 cm of the peel and replacing it by an agar plug containing growing *C. gloeosporioides*.

RESULTS AND DISCUSSION

Storage tests

In storage, coatings containing chitosan delayed ripening of fruits by 9 days (Figure 1A) and were effective to reduce anthracnose development (Figure 1B). Except at day 3, coatings reduced respiration rates (RR_{O2}) by 30 to 50% (Figure 1C). The respiratory quotients (RR_{CO2}/RR_{O2}) of the fruits coated with 1.5% chitosan and chitosan:Tween® were above 1.4 (Figure 1D), indicating internal fermentation of the fruits. The 1% chitosan coating reduced the RR_{O2} of the fruits without creating anaerobiosis for the Tommy Atkins (Figure 1C). As for modified atmosphere packaging, fermentation of coated fruits depends on gas transport

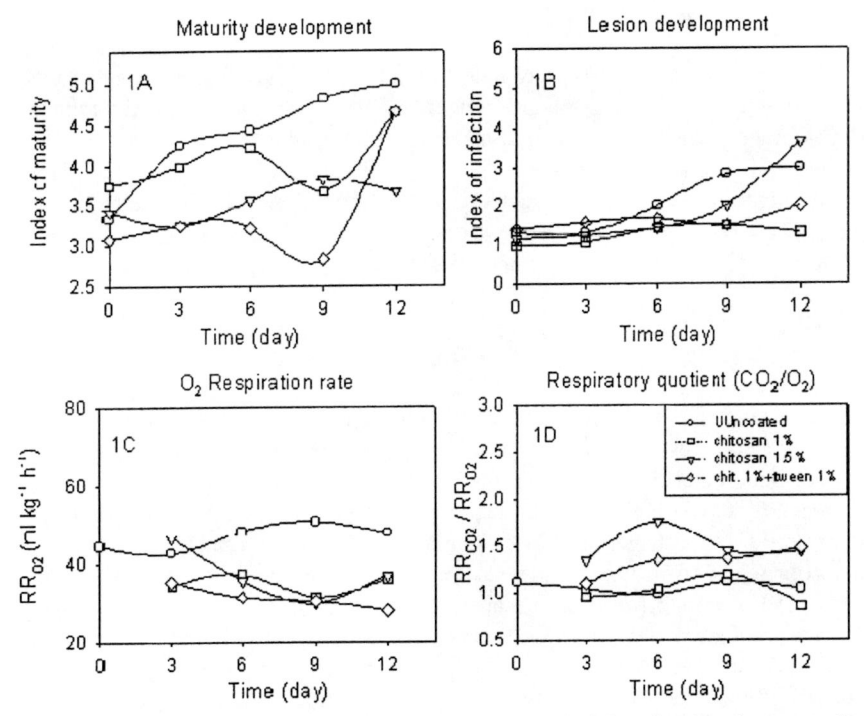

Figure 1. Tommy Atkins mangoes during storage. A. Maturity development. B. Lesion development. C. Respiration rate (O_2). D. Respiratory quotient (CO_2/O_2). ○ Uncoated, □ chitosan 1%, ◊ chitosan 1.5%, ▽ chitosan 1% and Tween 1%.

properties of membranes and tissues and on respiration rates of the tissues (Burton, 1982). To reduce the permeability of the chitosan membrane, low concentrations of chitosan or plastisizers should be used (Bégin and Van Calsteren, 1999).

Growth on agar plates

After 15 days at 12 and 20 °C, Benomyl® and benzoate at concentrations above 500 mg L^{-1} and 1%, respectively, and chitosan at 1.5% totally inhibited mycelial development on agar (Table 1). Oleic acid, benomyl at 250 mg L^{-1}, and benzoate at 0.5% were ineffective. At 12 °C, 1% chitosan totally inhibited mycelial growth, while at 20 °C the fungicidal activity was totally maintained for 6 days only, followed by growth, at 0.17 cm d^{-1}, 1.7% of the growth rate on untreated fruits. These results suggest that, for optimal antifungal activity, the chitosan concentration should be kept as high as possible to avoid anaerobiosis.

Growth on mangoes with cut peel

None of the fungicides were able to totally inhibit growth on mango fruits, indicating the importance of the cuticule of the fruits to control anthracnose development. Benomyl® and chitosan strongly reduced mycelial growth rates by, respectively, 46 and 74% in air, respectively (Figure 2A) and by 62 and 77% in modified atmosphere (5% O_2 : 5% CO_2 : 85% R.H.) (Figure 2B). By itself modified atmosphere reduced the growth rate of anthracnose by only 14% whereas oleic acid and benzoate did not show any important effect.

CONCLUSIONS

The results indicate the potential of chitosan to successfully replace chemical fungicides for post-harvest storage of mangoes. Chitosan can be considered as having fungicidal activity better than benomyl, benzoate and MA to control anthracnose development. However, as for other fungicides are concerned, chitosan does not replace refrigeration. Use of chitosan as a semi-permeable membrane to control ripening of fruits should be avoided because of the difficulty to adjust the permeability on the fruits and because of the changing RR_{O2} of the fruit during storage associated to ripening and temperature. Formulation of the coatings should be carefully adapted to each variety of fruits to avoid anaerobiosis and the concentration optimized to obtain the maximal fungicidal effect.

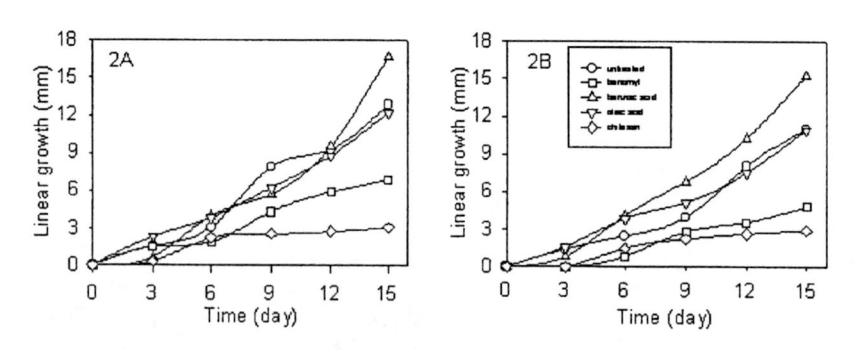

Figure 2. Development of anthracnose on mango fruits with cut peel. A. In air at 12°C. B. In modified atmosphere at 12°C.

Table 1. Effect of the different compounds tested for inhibition. of growth (time of inhibition and growth rate after the inhition period).

Fungicide	Air - 12°C Time (day)	Growth rate*	M.A.- 12°C Time (day)	Growth rate*	Air - 20°C Time (day)	Growth rate*	M.A.- 20°C Time (day)	Growth rate*
Untreated	0	2.8	0	2.63	0	10.01	0	8.95
benomyl .025%	0	2.84	0	0.4	0	0.69	0	0.56
benomyl 0.05%	15	0	15	0	15	0	15	0
benzoate 0.3%	3	0.58	6	0.59	3	0.54	6	0.4
benzoate 0.5%	12	0.05	15	0	12	0.55	12	0.41
benzoate 1.0%	15	0	15	0	15	0	15	0
oleate 0.3%	0	2.97	0	2.83	0	10.16	0	8.23
oleate 0.5%	0	2.85	0	2.89	0	9.28	0	8.51
oleate 1.0%	0	2.88	0	2.76	0	10.98	0	8.78
chitosan 1%	15	0	15	0	6	0.17	6	0.13
chitosan 1.5%	15	0	15	0	15	0	15	0

* (cm/day)

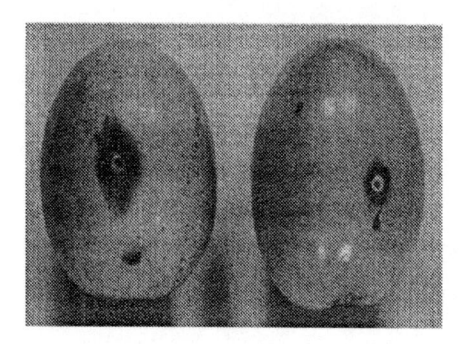

Figure 3. In vitro experiments: fruits infected by cuting the peel. Left. Control. Right. Fruit coated with chitosan.

ACKNOWLEDGEMENT

We wish to thank Claude Danis for his help with the chemical analyses and Dr. Richard Couture from A. Lassonde Inc. for his participation to the discussion which initiated this work.

REFERENCES

Bégine, A. and Van Calsteren, M.-R. 1999. Antimicrobial films produced from chitosan. *Int. J. Biol. Macromol.* 26: 63-67.

Burton, W.G. 1982. Post-harvest physiology of food crops. Longman Group Limited, New York, NY.

EC (European Commission), 1991, Directive 91/414/EEC concerning the marketing and use of plant protection products. *Off. J. Eur. Commun.* L230: 1.

El Gaouth, A.; Arul, J.; Asselin, A. and Benhamou, N. 1992. *Mycol. Res.* 96: 969-979.

Elson, C.W.; Hayes, E.R. and Lidster, P.D. 1985. In *Proc. of the 4th National Controlled Atmosphere Research Conference*. S.M. Blakenship (ed.), Deptartment of Horticultural Science North Carolina State Univ., Raleigh N.C., Hort. Report, 126, 248 p.

FAO (Food and Agriculture Organisation), 1997, FAO Manual on the Submission and Evaluation of Pesticide Residues Data for the Estimation of Maximum Residue Levels in Foods and Feed, Rome, Italy.

Noon, R.A. 1984. In *Plant diseases: infection, damage and loss*, R.K.S. Wood and G.J. Jellis (eds.), Blackwell, Oxford, UK, pp. 299-310.

OECD (Organisation for Economic Co-operation and Development), 1996, Activities to Reduce Pesticides Risks in OECD and Selected FAO Countries, Paris, Publication OCDE/GD (96)122.

USOFR (United States, Office of the Federal Register), 2001, Code of Federal Regulations, 1072: Poly-D-glucosamine (chitosan); exemption from the requirement of a tolerance, Title 40-Part 180: 531-532.

CHAPTER 15

SHRIMP ENHANCEMENT POTENTIAL

Zine-Al-Abidine Gadhi[1], Lucien Adambounou[2] and
Alain Guillou[3]

[1]Université du Québec à Rimouski, Département d'économie et gestion,
[2]Université du Québec à Rimouski, Département de biologie, chimie et sciences de la santé,
[3]Université du Québec à Rimouski, Rimouski, Quebec, Canada

Of the 2,500 species of shrimp known to exist worldwide, about thirty inhabit the Gulf of St. Lawrence. Of these, only the northern shrimp, *Pandalus borealis*, is commercially exploited in Quebec. The commercial success of this species has progressed considerably over the past ten years and is now ranked fourth among molluscs and crustaceans in total production volume, behind lobster, scallop and crab. Technological developments have added new products to existing lines of conventional primary and secondary processing products. Biotechnology has had an appreciable impact on process development, contributing to both environmental protection and to substantial gains in industry profits.

INTRODUCTION

The Canadian fisheries industry is divided into three major geographical sectors: east coast (Atlantic), west coast (Pacific) and inland (freshwater) fisheries. The Atlantic fishery (Atlantic Ocean and Gulf of St. Lawrence) processes and markets primarily three categories of catch: bottom fish, pelagic fish, and molluscs and crustaceans (Ministère de l'agriculture, 2000).

The northern shrimp (*Pandalus borealis*) is one of the main crustaceans harvested in the east coast area. Although there are some 2,500 species of shrimp known around the world and about thirty of these are found in the Gulf of St. Lawrence, *P. borealis* is the only one commercially exploited in this zone due to its extreme abundance. The shrimp industry has made great strides over the past ten years and is now ranked fourth in total production volume among and crustaceans, after lobster, scallop and crab. Its growth has been fueled by technological and technical progress, new fishing grounds and increased demand. Once fished, shrimp is processed before marketing and most of its by-products may be enhanced.

This contribution reviews shrimp processing and enhancement practices from the perspective of their adaptation to new technologies, national quality standards (quality control program or QCP) and international regulations (hazard analysis critical control point or HACCP) as well as environmental legislation. We thus present an overview of the current shrimp processing techniques and outline shrimp processing by-product enhancement processes.

SHRIMP PROCESSING

A marine products processing business is defined as "a processing or canning enterprise, for wholesale of, by the entrepreneur or by a person soliciting services in exchange for income, marine products intended for human consumption. This activity requires a processing or canning permit corresponding to the product type" (Ministère de l'agriculture, 1996). In the case of shrimp processing systems, the major finished or semi-finished product categories are described below.

Peeled and cooked shrimp, chilled or frozen (in blocks or individually) or canned

The major shrimp products forms include the following (Fisheries and Oceans Canada, 1996).
a) Fresh or frozen as whole in the shell, headless, peeled with tail, peeled without tail, peeled and deveined, as well as wrinkled or broken meats; and b) Canned as peeled, peeled and deveined, cocktail, as well as salad or broken meats.

The various shrimp products are a) In season as raw, whole and chilled shrimp or cooked, whole and chilled shrimp; and b) Off season as cooked, peeled and frozen shrimp or canned shrimp.

Freeze-dried shrimp

Freeze-drying is a preservation method defined as dehydration of materials at low temperature (below the freezing point) and vacuum sublimation (Fisheries and Oceans Canada, 1993). Applied to northern shrimp, this method provides: a) Preservation of product organoleptic (sensory) and nutritional properties; b)

Preservation of the product in its packaging during 9 months at ambient temperature.

Breaded shrimp

The production of this type of product is due to shrimp flesh restructuring process (Purdel Inc, 1991). The starting material for this is small shrimp and processing residues such as broken meats, thin fresh slices, debris and so on. The finished product is sold frozen in the form of nuggets.

Shrimp cooked meats

Fish and seafood cooked meats are defined as products manufactured from the shredded or crushed flesh of these animals, mixed with various savoury, nutritional and functional ingredients, and generally cooked (Boyer et al., 1995). They include a wide variety of products:
a) Products using bowels such as sausage;
b) Products to be spreaded such as butter, cream and foam; and
c) Products to be sliced such as pâté and terrine.

SHRIMP ENHANCEMENT

By definition, fishery industry by-products are processing residues and non-marketed accidental catches. In the case of shrimp, by-products refer primarily to peeling residues (shells, internal organs, legs), which are generally estimated to make up 48-68% of the fresh biomass (Bélanger et al., 1990).

The principal chemical components of shrimp shells (on a dry-mass basis) are as follows:

i) Chitin, 15-25%; ii) Protein, 25-40%; iii) Minerals, 40-55%; and iv) Lipids and pigments, < 1 %.

In Eastern Quebec, annual shrimp residues production is nearly 9 to 10 thousand tons. In the past, all of these residues were either composted, simply buried with domestic wastes or dumped in the ocean, since the industry was unable to make profitable use of it. The shrimp processing industry was thus confronted with a two-pronged difficulty. These were i) Environmental management pressures, which translated to processing wastes pollution problem; and ii) An economic burden, as the fixed cost of waste treatment and disposal decreased profit margins and competitiveness.

Today, the profitability of this industry has been improved as a result of developing alternative uses for these interesting raw materials. Shrimp by-products uses may be directed towards three product categories: foods for human consumption, animal feeds and various non-food uses.

Shrimp by-products contain numerous valuable components as described below.

Proteins

Examples for the use of proteins recovered are given below.
a) Aromatic raw material for food preparation such as soups, bisques or broths.
b) Raw material for high-proteined foods (i.e. sporty, effort and elderly supplements, reconstituting foods during convalescence, infant formulas, etc).
c) Condiments as in Japan to enhance food flavor or used in the surimi.
d) Raw material for appetite stimulators (e.g. in European markets) as ingredients in sauces or pastas.

Carotenoid pigments

Examples for the use of carotenoproteins are both in feed and the food industries.
a) Natural pigments to replace synthetic forms used in salmonidae feeds to emphasize the flesh coloring of farmed fish.
b) Pigments added to shrimp mousses, heat-and-serve foods or dry products, such as in coatings (chips, crackers, etc.) to impart pleasing colour.

Chitin and its derivative chitosan

Potential uses of chitin/chitosan include the following.
a) Nematocidal agent;
b) Thickener and stabilizer in foods and pharmaceutical products;
c) Analgesic bandage and wound healing agent (e.g. for burns);
d) Glucosamine salts precursor.

Potential uses of chitosan may be extended from those from chitin/chitosan.
a) Flocculating agent for wastewater;
b) Coagulant in the purification of drinking water;
c) Edible protective film for fruits and vegetables;
d) Microfilter membranes for sterilizing air in aseptic (e.g. surgical) environments.

ECONOMIC ASPECTS

The budget cuts due to moratorium crisis have forced fisheries operators to adopt a new outlook and to search for alternatives to ensure their economic, ecological and social viability. The following aspects have been emphasized. a) Diversification of harvested species; b) Diversification of processed products; and c) Global market context and exportability.

Several facts are in favor of this orientation, as given below.

The significance of the fisheries sector

In Quebec, harvesters collectively represent about 1,200 commercial fishing companies and nearly 4,000 full-time jobs as fishermen and assistants (Ministère de l'agriculture, 2000). It should be noted that these fishermen reach only the zones of the Gulf of St. Lawrence. They operate on 1,508 vessels, including 51 shrimp boats.

Analysis of the evolution of Quebec fisheries landings (fish, crustaceans, molluscs) from 1985 through 2000 indicates that:
a) Volume has dropped sharply by 35 %, from 90,189 tons in 1985 to 58,209 tons in 2000;
b) Value has increased by 135.5 % from $67.6 million in 1985 to $158.8 million in 2000.

The year 2000 was the second most profitable one for commercial fisheries during the past two decades, regarding the value of landings (all species combined). The record was established in 1995 at $177 million. The three pillars of the excellent performance in 2000 were crab, lobster and shrimp.

The significance of shrimp landings volume

In the year 2000, shrimp represented about 30 % of the total catch by Quebec fishermen (Ministry of Agriculture, 2000). This level has been stable over recent years. The quantity of shrimp landed (17,089 tons) is much larger than the 7,431 tons of bottom fish as well as the 9,135 tons of pelagic fish and approaching half (41.07 %) of the molluscs and crustaceans catches (41,607 tons).

In 1999, the value of shrimp landings in the Gulf of St. Lawrence reached $35.5 million and the average price paid to the fishermen was $1.52 per kilogram.

Shrimp processing industry development

A hundred fishery product processing establishments operate in Quebec, about half of them in maritime areas (Bélanger *et al.*, 1990; Ministry of Agriculture, 2000). These plants employ 4,133 workers (estimated for the year 2000 in monthly maximum employment).

Shrimp processing is a well-developed biofood sector, thanks largely to technological progress, which has added new processed products to existing lines of conventional primary and secondary processing products. The processes currently in use in this industry are of two types:

a) Those based on conventional methods (freezing, canning);
b) Those based on newer methods (freeze-drying, flesh restructuring, manufacture of cooked meats).

Appreciable input of biotechnology to shrimp by-product enhancement

Industrial processing of crustaceans (lobster, crab, and shrimp) generates considerable quantities of residues, making proper biomass disposal relatively expensive ($10 to $40 per ton in 1994, depending on establishment location) (Bélanger *et al.*, 1990; Gadhi, 2001; Martin, 1994).

The most widely practiced method for these residues remains to be in-land dumping and burying. This represents a significant cost for processors; up to $50,000 per year for a typical plant. Today however, the biotechnology sector can be proud of the high-value-added products to which shrimp by-products can be enhanced.

These processes, which represent significant biotechnological innovations, may be divided into two types.
a) Low skills processes, such as composting, chemical silage and hydrolysis;
b) Technologically sophisticated processes such as extraction and purification of chitin and its derivative chitosan and of carotenoid pigments.

The significance of these developments is primarily from an environmental perspective and clearly demonstrates the know-how of Quebec businesses specialized in marine biotechnologies. The products obtained represent great economic potential due to their numerous applications (biomedical, pharmaceutical, food-processing, etc.), making possible to protect the environment while realizing substantial economic gains. These include:

Flavors

In 1991, the world market for flavors was estimated at about US$3 billion with an annual growth of 30%. Market research carried out in 1993 estimated the gross value of concentrated flavor extract at the factory to be in the $14.00 to $17.70 per kg range.

Carotenoid pigments

The yield of pure astaxanthin, a carotenoid pigment, from crustacean shells is about 1 kg per 33 tons, its price being evaluated in 1991 at $0.33 per gram. In that year, the world demand for pure pigments was 50 tons, which wasn't met.

Chitin

According to a study carried out in 1993 by a US west coast industrial group involved in crab processing, it is estimated that the world demand for chitin is about 900 tons, of which at least two thirds (600-700 tons) is consumed in Japan. The Japanese are already very advanced in the marketing of chitin-based products.

An evaluation done in 1987 showed that the potential American market for chitin and its derivatives reached $335 million US per year, including $190 million in the health field alone.

CONCLUSIONS

Quebec businesses that have carried out biotechnological research and development projects focused on shrimp residues have been successful in finding high-value-added products. There is a consensus that marine biotechnology will be a very fertile area of scientific and technological endeavor over the coming decades. We believe that businesses supporting this emerging science start collecting benefits.

Among the dedicated marine biotechnology companies that have, after many years of research supported by the Quebec Ministry of Agriculture, Fisheries and Food (MAPAQ), reached the product marketing stage and may be thus regarded as successful businesses:
a) *ABK-Gaspésie*, located in Matane as the subsidiary *Aqua-Biokem BSL Inc.*
b) *Marinard Biotech*, located in Rivière-au-Renard as the subsidiary *Pêcheries Marinard Inc.*

REFERENCES

Bélanger, C.; Chevrier, R.J. 1990. Transformation des produits marins. Québec, Canada: La Revue Maritime L'Escale, 278 p.

Boyer, J.; Frentz, J-C.; Michaud, H.; Aubert, G. 1995. La charcuterie de poisson et fruits de mer. Québec, Canada: Erti Éditeur, 328 p.

Gadhi, Z. 2001. Potentialités de valorisation de la crevette. Mémoire de recherche en gestion des ressources maritimes. Rimouski, Canada: Université du Québec à Rimouski, 86 p.

Martin, A.M. 1994. Fisheries processing: Biotechnological applications. Londres, Grande-Bretagne: Chapman & Hall, New York, NY, 494 p.

Ministry of Agriculture, Fisheries and Food, Quebec. 1996. Politique ministérielle de délivrance et de renouvellement des permis d'exploitation, d'établissement, de préparation et de conserverie de produits marins. Québec, Canada: MAPAQ, 44 p.

Ministry of Agriculture, Fisheries and Food, Quebec. 2000. Pêches et aquaculture commerciales au Québec en un coup d'œil - Portrait statistique: Édition 2000. Québec, Canada: MAPAQ, 20 p.

Fisheries and Oceans Canada. 1996. Normes et méthodes des produits du poisson - Crevettes en conserve. Ottawa, Canada: MPO. Chapitre 2, norme 3, 8 p.

Fisheries and Oceans Canada. 1996. Normes et méthodes des produits du poisson - Crevettes fraîches et congelées. Ottawa, Canada: MPO. Chapitre 3, norme 2, 8 p.

Fisheries and Oceans Canada. 1993. Lyophilisation de crevettes. Québec, Canada: MPO. Fiche 27 (nouvelles technologies, No. de cat. Fs 41-39/88-1993.

Purdel inc. et Université Laval. 1991. Étude sur la restructuration de la chair de la crevette nordique. Phase III: expérimentation à l'échelle semi-industrielle. Québec, Canada: MPO. Programme d'essai et d'expérimentation halieutiques et aquicoles, 82 p.

CHAPTER 16

HEAVY METAL CONTAMINATED LEACHATE REGENERATION WITH CHITOSAN FLAKES

Benedictus Hope Soga[1], Shiv O. Prasher[2] and
Benjamin Kofi Simpson[1]

[1]Department of Food Science and Agricultural Chemistry, McGill University (Macdonald Campus), 21,111 Lakeshore Road, Ste Anne-de-Bellevue, Quebec, H9X 3V9, Canada
[2]Department of Agricultural and Biosystems Engineering, McGill University (Macdonald Campus), 21,111 Lakeshore Road, Ste Anne-de-Bellevue, Quebec, H9X 3V9, Canada

Temperature, pH, and chitosan amounts were evaluated for optimum chelation of heavy metal with chitosan flakes from citrate solution, using response surface methodology (RSM). Chitosan was used to polish zerovalent metal treated soil leachates at optimum conditions. pH mostly influenced metal chelation. Extraction pH was optimal at 6.1 for Cu, Pb - 6.6, Zn - 6.69 and Cd – 6.71. Average chitosan quantity needed was 1.321% (w/v). Chitosan polished zero-valent metal regenerated leachate, pH 6.62, effectively reduced Pb and Zn by 99% with Mg treatment, while removing 64% of Cu and 44% of Zn and Pb, respectively, from Fe treated leachate with chitosan alone.

INTRODUCTION

Significant amounts of waste containing both inorganic and organic contaminants such as pesticides, dyes, suspended solids, proteins and heavy metals are discharged from various industries. Wastewaters containing a variety of heavy metals originate from many sources such as mining operations, metal-plating facilities, electroplating facilities or electronic device manufacturing operations and are often toxic at low concentrations, besides not being biodegradable. These contaminated wastewaters may pose serious environmental and health problems due to their potential transformation, activity and toxicity. The contaminants must thus be removed to meet the increasing stringent environmental quality standards (Peniston and Johnson, 1978; Rorrer et al., 1993). The nontoxic and biodegradable biopolymer, chitosan, has been recognised for its potential use to regenerate these wastewaters.

Chitosan is a partially acetylated polymer of glucosamine, produced by deacetylation of chitin. Chitin, the parent material for chitosan is the major component of crustacean shells such as crab, shrimp, lobster, and krill. The shells comprise more than 70% of the waste material from the crustacean processing industry. The waste material itself poses an environmental concern, since most of these industries were established without consideration of reclaiming by-product from the waste (Hansen and Illanes, 1994). However, the recovery of chitin, may help in the regeneration of other industrial waste and the recycling of some wastes such as protein recovered from diary, poultry and fish processing with chitosan. Chitin is an easily obtainable and renewable natural polymer, and forms 14-35% of shells on a dry weight basis.

Most commercial grade chitosans contain 75-95% glucosamine and 5-25% N-acetylglucosamine monomers (Domszy and Roberts, 1985). Chitosan has a high molecular weight, and when subjected to dilute acids, the primary amine groups of the chitosan molecule are protonated, thus acquiring a positive charge. The solvated molecule is polycationic at pH below 6.0 with a high charge density (NH_3^+), increasing the degree of polymer to polymer repulsion and porosity of the polymer. Chitin and chitosan have ion exchange capabilities as well as chelating properties. They form complexes with many transition metals as well as metals from Groups 3-7 of the periodic table (Muzzarelli, 1973). The metal complexation involves the donation of non-bonding pair of electrons from the nitrogen and /or the oxygen of the hydroxyl groups to the heavy metal ion. The rate of formation and the stability of these complexes are heavily dependent on counterions, competing heavy metal ions, temperature and pH of solution, as well as particle size, crystallinity, hydrophilicity and the degree of acetylation of the chitin and chitosan (Winterford and Sandford, 1995). Temperature plays an important role in the sequestration of metals from solution, however there is no general rule for predicting the influence of temperature on the sequestration process. Hence, one should expect variations in both directions, but a better result is observed at 4 °C than at room temperature (Muzzarelli, 1977).

The current major application of chitosan is in wastewater treatment, where it acts as coagulants or flocculants, though it has received increased attention for commercial applications in the biomedical, food and chemical industry in recent years (No and Meyers, 2000). Chitosan has been applied as chelating polymer chelatin of toxic metals such as copper, lead, mercury and uranium, and its effectiveness in the removal of transition metals from waste waters has been reported by many workers (Muzzarelli, 1977; Bassi et al. 1999). The polymer's selectivity towards binding virtually all Group III transition metal ions at low concentrations, but not to Groups I and II alkali and alkali earth metal ions makes it very attractive for the separation of heavy metals from wastewater.

The precipitation of metal ions with Fe oxides can occur in soil solution and wastewaters to play a significant role in limiting heavy metal solubility, toxicity and bioavailability. Redox processes in the soil and water environment lead to the dissolution and reformation of the hydrous ferric or aluminum oxides and in the process coprecipitate, adsorb or absorb other metal ions such as Pb, Zn, Cu, and Cd along with it in noncrystalline forms. The immobilizing process may be influenced by the type and concentration of organic acids present and is affected by pH, metal ion load, and contact time (aging).

Surfaces of finely divided solids such as iron have high sorption capabilities for solutes due to excess surface energy. The hydrated surfaces may lose or gain H^+ that can sorb metal ions, Mt^{z+}, by complexation to the surface,

$M\text{-}OH + Mt^{z+} \rightleftharpoons M\text{-}OMt^{z-1} + H^+$

A metal complexed with a ligand, L, may likewise be sorbed by loss of either H^+ or OH^-:

$M\text{-}OH + MtL^{z+} \rightleftharpoons M\text{-}OMtL^{(z-1)} + H^+$

$M\text{-}OH + MtL^{z+} \rightleftharpoons M\text{-}(MtL)^{(z+1)} + OH^-$

The reduction process is normally accomplished by aggregation of particles. Another process, cementation, involving electroactive metals such as Cu and Cd sedimenting in the elemental state on iron particle surfaces, thus oxidizing the zero-valent iron (Fe^0). The process involves the electrochemical reduction of the dissolved metal ions to the zero-valent state, subsequently getting sorbed to the iron surface.

$$Fe^0 + M^{2+} \rightarrow Fe^{2+} + M^0$$

Reports of this process being used to remove heavy metals from acid rock drainage suspensions and uranium from ground water to nondetectable levels have been published (Morrison et al., 2001; Xie, 2000). Martinez and Mcbride (2001) coprecipitated heavy metals from a solution with Fe by raising the pH to 6 and 7. They reported the formation of Fe hydroxy polymers [$Fe_x(OH)_y$] with increasing pH, a condition which favored nucleation and growth of floc aggregates in an amorphous precipitate containing high levels of heavy metals.

Xie (2000), used Mg^0 metal with a higher redox potential (>2 V) than Fe^0 (~0.44V) to strip heavy metals from their bis (2-hydroxyethyl)dithiocarbamate (HEDC) complexes to trace level of Fe, and Mn ions to nondetectable levels. Thus, polishing the wastewater or reagent with chitosan may remove the residual amount of heavy metals to trace levels. Because of the expected high cost of chitosan sequestration of heavy metal laden soil leachate, a zero-valent metal (M^0) pretreatment to reduce the level of contaminants in the leachate may be implemented.

The objectives of this work were to optimize sequestration of Pb, Zn, Cu, and Cd from an aqueous citrate solution under variable conditions of temperature, pH and amount of chitosan required. Evaluation of the efficiency of zero-valent Fe^0 and Mg^0 in regenerating citrate leachate from heavy metal contaminated soil and post treatment polishing of the zero-valent metal regenerated leachate with chitosan flakes was also intended.

MATERIALS AND METHODS

Crab chitosan flakes, purchased from Sigma Chemical Co. (St. Louis, MO) was sieved through a 2mm mesh sieve. Citrate solutions were prepared by spiking buffer with standard solutions of Pb, Zn, Cu, and Cd to concentration of 10.0, 3.0, 1.0 and 0.5 µg/g, respectively. Standard metal solutions, citric acid and tri-sodium citrate used for buffer preparation were obtained from Fisher Scientific (Nepea, ON) and all solutions were prepared with distilled deionized water.

The central composite rotatable design (CCRD) was used as the experimental design to optimize the sequestration at three levels of operating variable factors as summarized in the Table 1. This consisted of 19 combinations in five buffer pH buffers at five temperatures and five quantities (g) of chitosan. The amount of each metals sequestrated from the citrate buffer was evaluated using a second degree polynomial (see below) including quadratic terms to afford better precision. All calculations and plots were carried out with Microsoft Excel.

$$Y = \beta_0 + \beta_1 X_1 + \beta_2 X_2 + \beta_3 X_3 + \beta_{12} X_1 X_2 + \beta_{13} X_1 X_3 + \beta_{23} X_2 X_3 + \beta_{11} X_1^2 + \beta_{22} X_2^2 + \beta_{33} X_3^2 + \varepsilon$$

Where, Y = yield of metal ion per gram of chitosan and β_0 β_1 β_2 β_3 = coefficients determined from the results of the experiment variable after regression analysis of the data.

Various quantities of chitosan flakes were incubated in 25 ml of buffer at the various temperatures for 24 and 36 h to allow for equilibration. The process was carried out without agitation as this criterion was previously determined not to be significance on the sorption of heavy metals by chitosan. Supernatants were analyzed using a GBC 903 single beam flame atomic absorption

Table 1. Coded and uncoded factors and their levels for the CCRD experimental design

RUN No.	Coded (X_1)	Actual(X_1)	Coded (X_2)	Actual(X_2)	Coded (X_3)	Actual(X_3)
1	-1	5.0	-1	20	-1	0.125
2	1	7.0	-1	20	-1	0.125
3	-1	5.0	1	40	-1	0.125
4	1	7.0	1	40	-1	0.125
5	-1	5.0	-1	20	1	0.375
6	1	7.0	-1	20	1	0.375
7	-1	5.0	1	40	1	0.375
8	1	7.0	1	40	1	0.375
9	-1.682	4.32	0	30	0	0.250
10	1.682	7.68	0	30	0	0.250
11	0	6.0	-1.682	13.2	0	0.250
12	0	6.0	1.682	7.68	0	0.250
13	0	6.0	0	30	-1.682	0.04
14	0	6.0	0	30	1.682	0.46
15	0	6.0	0	30	0	0.250
16	0	6.0	0	30	0	0.250
17	0	6.0	0	30	0	0.250
18	0	6.0	0	30	0	0.250
19	0	6.0	0	30	0	0.250

Where,
X_1 = pH
X_2 = Temperature
X_3 = quantity of chitosan (g)

spectrometer (FAAS), GBC Scientific Equipment PTY LTD, Dandenong, Victoria Australia, for residual Pb, Zn, Cu, and Cd.

Leachate obtained from flushing a column of heavy metal contaminated soil with 0.5M sodium citrate buffer pH 6.0 was used as a source of heavy metal contaminated wastewater/solution. Zero-valent Fe^0 (40 mesh) and Mg^0 (-12+50 mesh) metal granules were used for pretreatment of leachate to reduce the total metal burden before chitosan flake polishing. The target metal burden and pH of the leachate are indicated in Table 2 (untreated leachate).

Table 2. Residual metal levels of soil leachate before, and zero-valent metal and chitosan treatments.

Sorbent	pH	Pb	Zn	Cu	Cd
Untreated	8.45 ± 0.02	136.06 ± 0.96	4.13 ± 0.07	0.17 ± 0.008	0.09 ± 1.3E-09
Fe	9.15 ± 0.02	129.39 ± 0.96	4.05 ± 0.05	0.50 ± 0.030	0.09 ± 1.3E-09
Mg	12.0 ± 0.06	13.83 ± 2.80	0.11 ± 0.008	0.10 ± 0.010	0.07 ± 0.003
CF	6.63 ± 0.01	15.28 ± 1.94	1.95 ± 0.001	0.18 ± 0.007	0.08 ± 0.002
CM	6.62 ± 0.02	1.97 ± 0.27	0.04 ± 0.27	0.11 ± 0.008	0.06 ± 0.00

Heavy Metal (µg/ml)

Fe- Fe granule treated leachte, Mg- Mg granule treatment, CF- chitosan treated Fe regenerate, CM- chitosan treated Mg regenerate.

Zero-valent metal regeneration

Soil leachate (100 mL) was added to 0.5g of either Fe^0 or Mg^0 in acid washed 400 ml beakers in triplicates. The mixture was agitated by shaking on Innova 2100 model platform shaker, from New Brunswick Scientific Co. (Edison, NJ) for 6 h at 300 rpm and allowed to settle for another 18 h at room temperature. The supernatant was centrifuged (MSE Mistral 2000 John's Scientific, City, UK) at 4350g for 20 min in 50ml polyethylene centrifuge tubes. The supernatant was analyzed for its content of Pb, Zn, Cu and Cd on GBC model 903 FAAS.

Polishing with Chitosan flakes

Twenty milliliters of the zero-valent regenerated leachates were treated with 1g of chitosan flakes in Mg^0 treated leachate and 3g per tube of Fe^0 treated leachate. The amount of chitosan flakes used was dependent on the residual amount of heavy metal in the zero-valent treated leachate. The leachate was left in contact with the chitosan flakes for 24 h at 24±2 °C after which the supernatant was decanted and centrifuged for 20 min at 3000g at ambient temperature. The residual amounts of Pb, Zn, Cu and Cd were then measured using FAAS.

RESULTS AND DISCUSSION

The heavy metal uptake per gram of sample was calculated as:
$q = (C_e/C_i)*V$

Where: q = metal uptake per gram (μg/g), C_e = equilibrium concentration of metal, C_i = initial concentration of metal, and V = volume of metal solution.

Regression analyses of the data to evaluate their linear, quadratic and interactive influence on metal sorption shows the following results and interpretations:

Cadmium sequestration with chitosan flakes was significant influenced by pH and its corresponding doubling effect ($P < 0.05$ and 0.1, respectively). Copper sequestration from solution was significantly affected by chitosan amounts used ($P < 0.05$), but its doubling effect was not found to be significant. The effect of pH on Cu uptake was slightly significant at the quadratic level ($P < 0.10$). Zinc sequestration was significantly influenced by pH ($P < 0.05$).

Lead (Pb) sequestration was significantly influenced by pH at both linear and quadratic levels, and also by the amount of chitosan (CA), pH and temperature, as well as pH and the amount of chitosan which showed interactive influences ($P < 0.05$). However, the interactive influence of temperature and chitosan amounts was significant at ($P < 0.10$) while all other factors and their interactive levels were not significant.

Increasing the incubation time to 36 h, the regression statistics of the experimental data showed no significant influence of temperature and chitosan amounts ($P < 0.05$) on Pb, Zn, Cu and Cd sorption from the citrate solution. The influence of pH on Pb sorption was significant at both linear and quadratic levels ($P < 0.05$), Zn sorption was slightly significant under the quadratic influence of pH ($P < 0.10$). Copper, however, was not influenced by pH, whereas Cd was significantly influenced at the linear level ($P < 0.05$). Bassi et al. (1999) indicated that contact time between the metal solution and chitosan flakes had little influence on the sorption capacity of chitosan flakes, however, they did not evaluate the influence of contact time on other variables.

The influence of paired factors on chitosan sorption of heavy metals from citrate buffer

1. The influence of temperature and pH

The response surface curves predicted that temperature had no significant influence on the sorption of heavy metals from citrate buffer. Contrary to this finding, MaKay et al. (1986, 1989) measured the sorption isotherms of Cu, Zn, Ni, and Hg ions on chitosan and concluded that sorption capacity decreased with increasing temperature. However, the sorption efficiency tends to increase with pH above 6.3. Chitosan becomes predominantly positively charged below pH 6.0 (Roller and Covill, 1999) due to the presence of unshared electron pair of the amine nitrogen that makes the functional group nucleophilic, hence metal ions will compete with hydronium ion for the amine reactive site. Thus, the mechanism of metal uptake can be described to be complexing or by nodule formation, ion sorption and ion absorption at the surface of the flakes (No and Meyers, 2000) above pH 6 for Zn, and ion exchange in character for Cu below pH 6.0. The influence of pH on metal sequestration capacity of chitosan has been reported (Rorrer et al. 1993; Udaybhaskar et al. 1990)

Lead (Pb) had optimal sorption above pH 6.3, ($R^2 = 0.7839$) and minimum sorption pH range of 5-6.3, sorption being minimal at the critical pH of 5.65 (Figure 1), based on the response surface curve constructed from equation 2.

$$Y = 170.864 + 36.130 X_1 + 40.326 X_1^2, \tag{Eq.2}$$

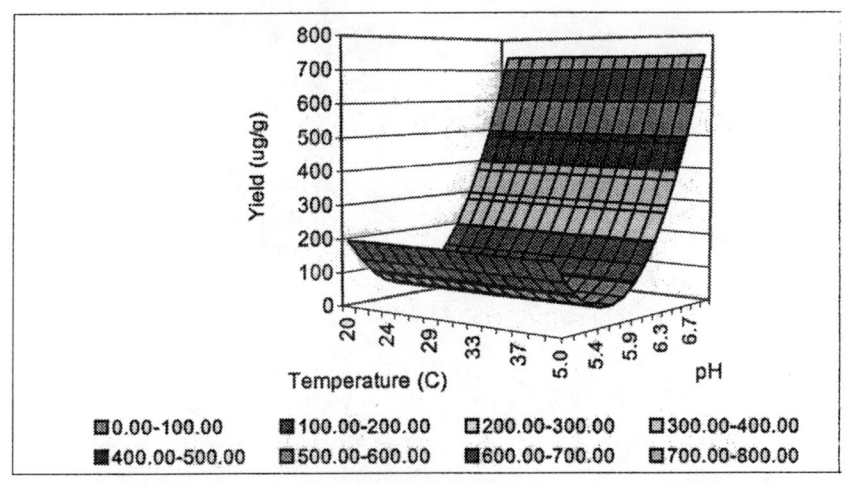

Figure 1. The pH and temperature effect on chitosan sorption of Pb.

Copper sorption was optimal at pH 6.1, ($R^2 = 0.6571$) and minimal in the pH range of 5.2-6.14. Above pH 6.14 and below pH 5.2 sorption increased, suggesting both ion exchange and chelation/complexation mechanisms (Figure 2), as derived from equation 3.

$$Y = 8.726 + 5.668 X_1 + 5.902 X_1^2, \qquad (Eq.3)$$

Zinc on the other hand was optimal at pH 6.8 ($R^2 = 0.8969$) and exhibited minimum sorption in the pH range of 5.2-6.26, above which sorption tended to increase with increasing pH (Figure 3), as derived from equation 4.

$$Y = -36.741 - 20.879 X_1 + 29.155 X_1^2, \qquad (Eq.4)$$

The sorption of Pb and Zn was minimal at pH 5.6 after 36 h incubation period and began to increase with decreasing pH, a condition that favors the exchange of metal ions and hydronium ions at the amino sites of the polymer. Above this critical value, Pb sorption tended to increase with increasing pH. The sorption capacity of both metal ions was more pronounced above pH of 6.0 when the amine nitrogen was not protonated and chelation/complex or nodule formation

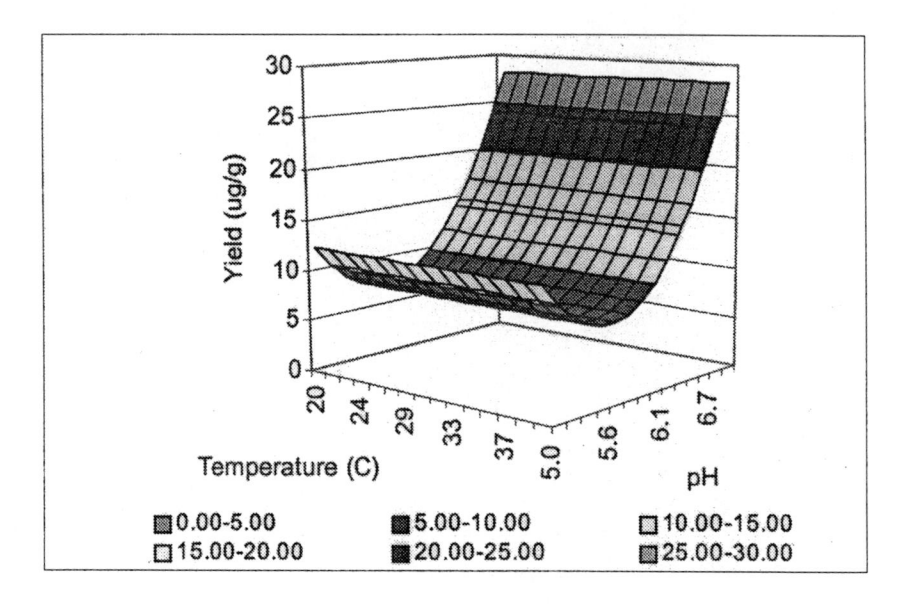

Figure 2. The pH and temperature effect on chitosan sorption of Cu.

was the prominent mechanism for metal sequestration. Thus, the chelation/complexing mechanism was more effective for Pb sorption than ion exchange and was optimum in the pH range of 6.6 - 7.0 (700-800ug/g). Zinc sorption was optimum in the pH range of 6.69-7.0 (300-350µg/g).

Cadmium sorption increased with increasing pH and was estimated to be optimal at pH > 6.71, (R^2 = 0.9265). Sorption was minimal in the pH range of 5.0-6.3, above which sorption tended to increase with increasing pH, (Figure 4) as derived from equation 5. This confirms the report of Maruca *et al.* (1982), who found that Cd sorption decreased with decreasing pH, thus suggesting that the exchange of hydronium ions at the available amino sites of the chitosan flakes at lower pH was not efficient due to competition. Thus, metal uptake was mainly due to complex or chelate formation.

$Y = 7.355 + 4.170 X_1 + 3.556 X_1^2$ (Eq.5)

Copper and Cd sequestration trends showed that minimal sorption occurred between pH 5.0 and 5.45 for Cu and pH 5.0 and 5.28 for Cd. Metal sorption

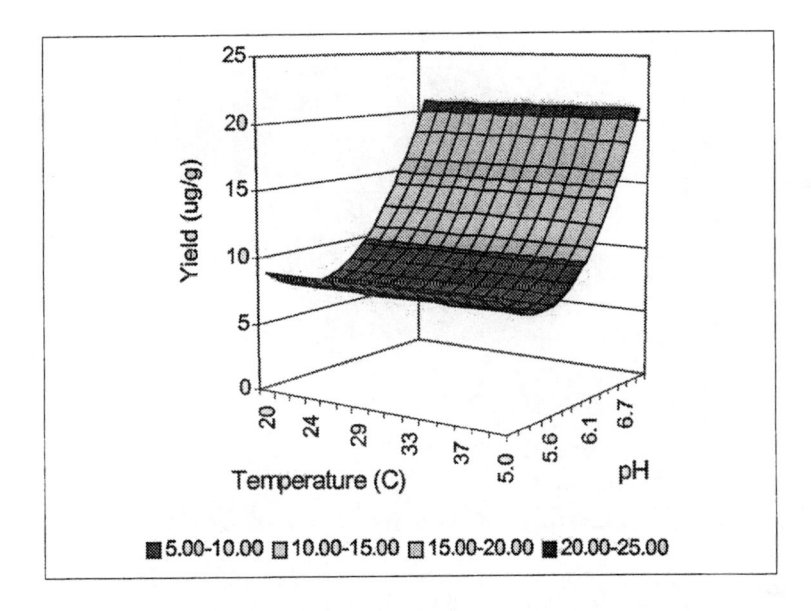

Figure 3. The pH and temperature effect on chitosan sorption of Zn.

increased with increasing pH and was optimal in the pH range of 6.95-7.0 for Cu (60-80 ug/g) and pH of 6.55-7.0 for Cd (80-100 ug/g). Muzzarelli *et al.* (1980) while studying chitosan complexed with Cu (II) concluded that an increase in pH during complex formation increased the number of amino groups involved in union from 1 at pH<5 to 3 at pH 6-8. Thus it may be assumed that a similar complexing mechanism was operative for Pb, Zn and Cd in the present study.

The predominant mechanism for Cu and Cd sorption was complexation/chelation where the polymer was not in the polycationic state above pH 6.0. Temperature had no significant influence on metal sorption in all cases.

2. The effect of pH and chitosan amounts

Zinc ion removal from solution was significantly influenced by both the temperature and the amount of chitosan used at both linear and quadratic levels, but not by their interactive effects ($P < 0.05$). Surface response curve was with coefficients of significant factor/variables as shown by equation 6.
$Y = -51.1345 + 4.587735 - 22.1757X_1 - 198718X_3 + 33.67278X_1^2 + 22.97411X_3^2$
(Eq.6)

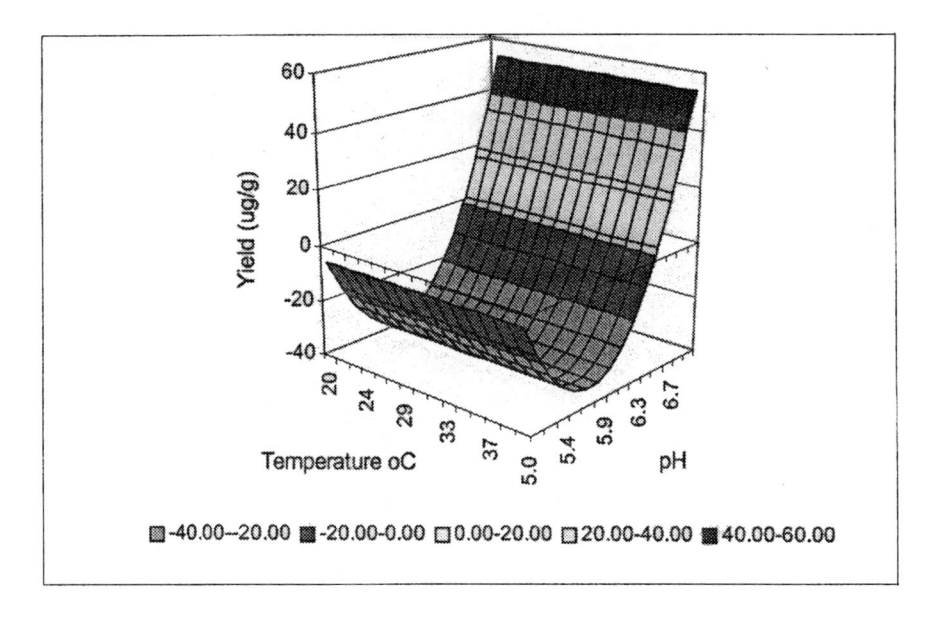

Figure 4. The pH and temperature effect on chitosan sorption of Cd.

Minimum sorption was observed in the pH range of 5.55-6.00 with 0.18-0.25g chitosan. Sorption tended to increase in all directions at pH range between 5.0 and 5.1 on the lower end and 6.4-7.0 on the upper end. Chitosan quantities were in the range of 0.13-0.17g and 0.25-0.33g per 25 ml solution on the lower and upper limits, respectively (Figure 5). It was assumed that two sorption mechanisms were involved, with ion exchange occurring at lower pH values (around 5.0) where the polymer was predominantly positively charged and chelation or complexation above pH 6.0 where the polymer in the neutral state. The complexation mechanism was however more efficient than the ion exchange. Zinc tended to have minimal sorption at lower pH values, and increased with increasing pH. It is suggested that chelation and complexation are more important over ion exchange as a mode of sorption.

Lead sequestration was influenced by both factors, however, unlike Zn was not influence by the quadratic effect of chitosan quantities, but was influenced by the interactive effects of both variables.

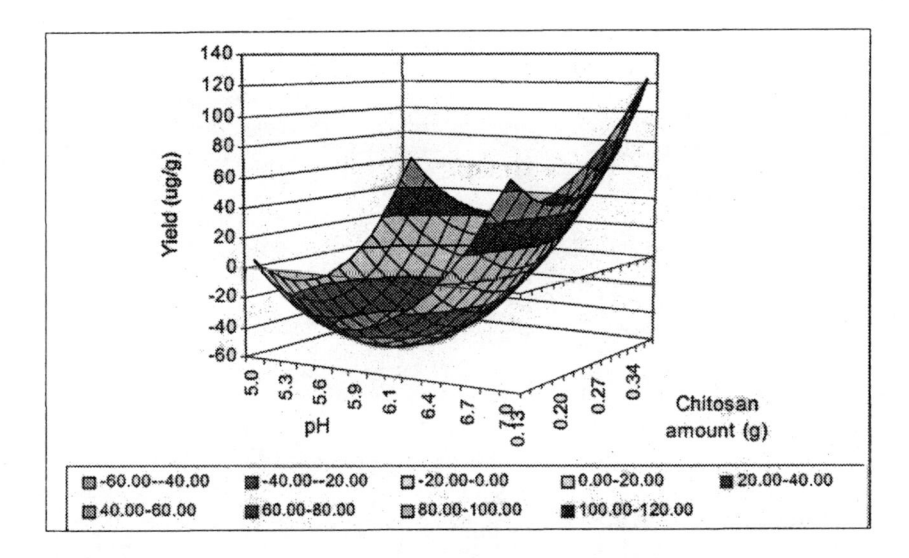

Figure 5. Effect of pH and chitosan amount used on Zn sequestration from solution.

The surface response curve (Figure 6) for Pb sorption was based on equation 7.

$Y = 232.5529 - 3.6244 + 9.10405X_1 - 46.8805X_3 + 18.8522X_1^2 + 42.7350X_1.X_3$ (Eq.7).

Lead sorption was high at pH above 6.0, optimal at pH 6.78 and minimal in the pH range of 5.0-5.5. Sorption was lowest when chitosan quantities in the range of 0.18-0.27g at pH above 6.0, but generally were lowest in the pH range of 5.0-6.0, irrespective of the quantity of chitosan used. Optimal sorption was estimated to occur with a minimum of 0.33g per 25 ml of spiked solution. Lead sorption was essentially favored by complexation/chelation. Lead sequestration was minimal in the pH range of 5.54-6.5 after 36 h incubation. Outside this pH range either a decrease or increase in pH resulted in uniform sorption capacity of the chitosan flakes. Chitosan quantities had no significant influence on Pb sorption. A different sorption mechanism might have been responsible for this observation besides chelation or ion exchange when longer contact time made the influence of chitosan quantities insignificant. Due to the hydrophilicity of chitosan, a longer contact time may lead to swelling that reduces the crystallinity of the polymer. A slow diffusion of hydrated ion into the pores of the sorbing occurs during the equilibration period and it binds to the inner surface. The

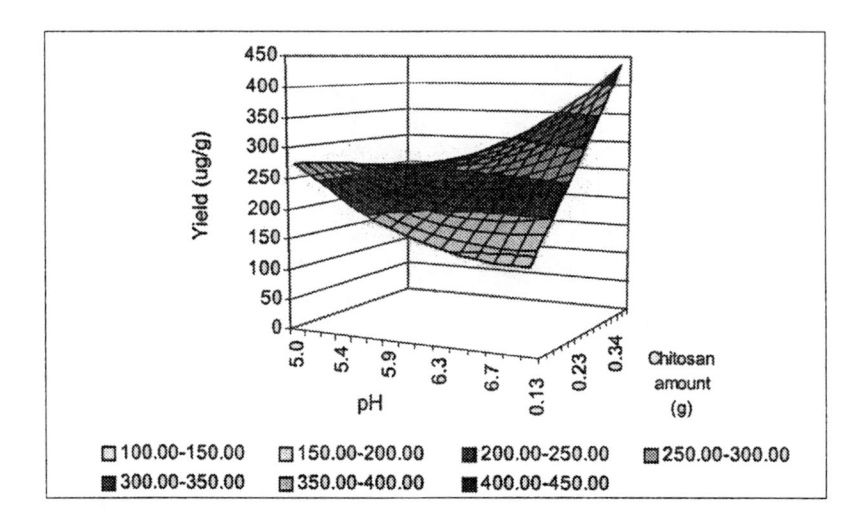

Figure 6. Effect of pH and chitosan amount used on Pb sequestration from solution.

diffusion rate and binding may also depend on hydrated size of the ion, thus influencing the sorption capacity of the ionic species. The nodular formation mechanism observed by Sudar *et al.* (1983) might have influenced sorption over such extended equilibration period. Nucleation on the polymer surface is rapid initially, increasing its surface energy which, can only be reduced by forming larger nodules as a result of metal complexes being sorbed on to the nuclei points. The intra-particle diffusion is the rate determining step as observed by Yang *et al.* (1984) and Piniche-Covas *et al.* (1992). However, the difference in the concentration of metal ions in the multi-element system could have influenced the sorption of other metal ions.

Copper sequestration was under the influence of chitosan quantities used at both linear and quadratic levels, but was not affected by either pH or the interactive effect of both variables. The surface response curve was generated from equation 8.

$$Y = -3.12694 + 2.6795 - 14.6063X_3 + 13.23207X_3^2 \quad \text{(Eq.8)}$$

Chitosan sorption capacity was minimal in the range of 0.155-0.245g in 25 ml spiked buffer. Optimum sorption occurred with chitosan amounts of 0.325g

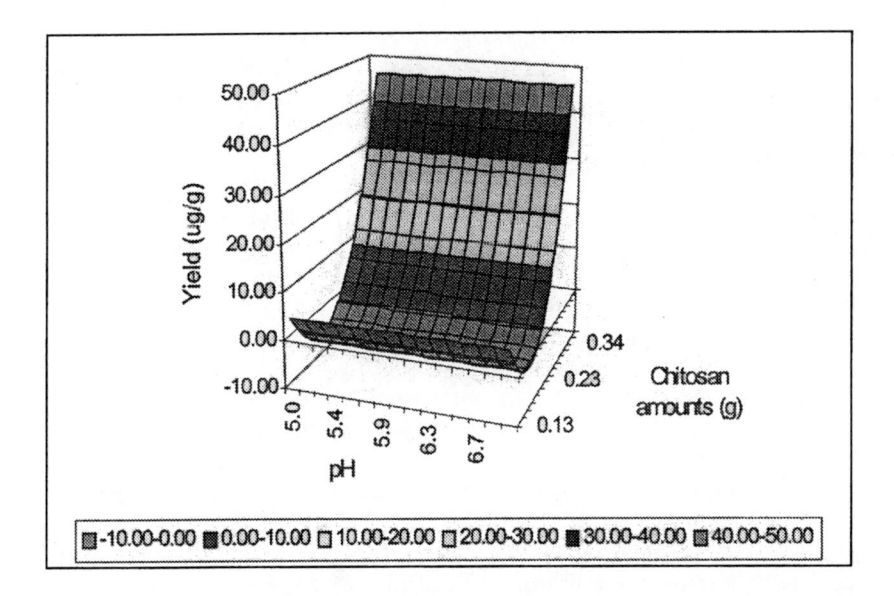

Figure 7. Effect of pH and chitosan amount used on Cu sequestration from solution

or greater, irrespective of pH (Figure 7), confirming the lack of pH significance on Cu uptake by the flakes as observed by Park and Park (1984). The predicted yields of metal ion sorption per gram chitosan was 400, 120 and 40 µg/g Pb, Zn, and Cu, respectively.

Cadmium sequestration was influenced only at the linear level of pH, whereas chitosan quantities and the interactive effects of both variables had no influence on Cd sorption. The influence of incubation or contact time can not be ignored if maximum sorption is to be achieved.

Zero-valent metal treatment of soil leachate

Treatment of the soil leachate with zero-valent metal resulted in an increase in pH of the solution. The pH of the soil leachate regenerated with iron $-Fe^0$ granules increased slightly by 0.7±0.02 compared to Mg^0 regenerated leachate that increased to pH 12±0.06 as indicated in Table 2. Metal precipitation or removal capacity of the Fe^0 granules was not efficient (5% for Pb, Zn and Cd). Copper was preferably complex into solution than to be deposited. This might be due to the higher stability of Cu-citrate than the ability of Fe to strip metals from the organic complex. Little precipitation was observed in the solution, which suggests cementation of the metals as zero-valent elements on the granule surface with a resultant oxidation of Fe^0 to Fe (II or III) state, releasing limited amounts of Fe ions into solution. On the other hand, Mg^0 granules effectively reduced the level of Zn and Pb by 97 and 90%, respectively, and Cu and Cd to a lesser extent of 40 and 24%, respectively, probably due to their initial low concentration in solution. The high pH of the regenerated solution coupled with large deposit of white precipitate indicated that Mg^0 was hydrolyzed into solution and the oxyhydroxide group acted as inorganic ligands in sorbing the various metal ions or their complexes out of solution. If left for a longer contact time of 24 h, magnetite (Fe_3O_4) formation and corrosion could lead to improved reduction in the level of contaminants in the use of both zero-valent metals. Thus, with time, the hydrated Fe will form ferric oxyhydroxides. The reactive surface of the floc tends to adsorb free metal ions and complexes from solution. As the size of the aggregate increases some metals get coprecipitated, occluded, and absorbed to aid further reduction of metal content of the leachate. Excessive accumulation of the oxides and hydroxides at the surface of the Fe^0 surface might increase the distance between the metal ions in solution and the Fe^0 to limit contact between them with a resultant slowing of sorption rate.

The inefficient removal of Cd could have been influenced by the presence of other metal and inorganic ligands in solution, which interfere with its reaction with the zero-valent metal. The assertion of Xie (2000), that Mg would be a better sorbent to use was confirmed in this study, since Mg performed more effectively than Fe in reducing the leachate contaminant levels. Chitosan flakes were able to reduce the level of residual Pb by 80%, but could not remove any of Cu from Mg^0 regenerated solution after adjusting the pH to 6.62 (Figure 8). The ionic strength of the leachate and the presence of other organic components from the soil environment could have compromised the selective sorption of the polymer flakes.

Chitosan treatment of Fe^0 regenerated leachates on the other hand removed 44% of residual Zn and Pb as well as 3% of Cd and 64% of Cu. These results reiterate the dependency of chitosan sequestration on the initial concentration of the metal ion species. Metals of higher concentration compete more favorably for the reactive sorption sites than those of lower concentration. There is the possibility of chitosan exhaustion due to sorption of other metal ions not evaluated in the soil leachate by competing with target metals for the chitosan sorption sites, leading to poor sorption efficiencies of some target metals.

From Figure 9 it can be deduced that the combined effects of zero-valent metal and chitosan flake treatments was efficient in removing Zn and Pb from

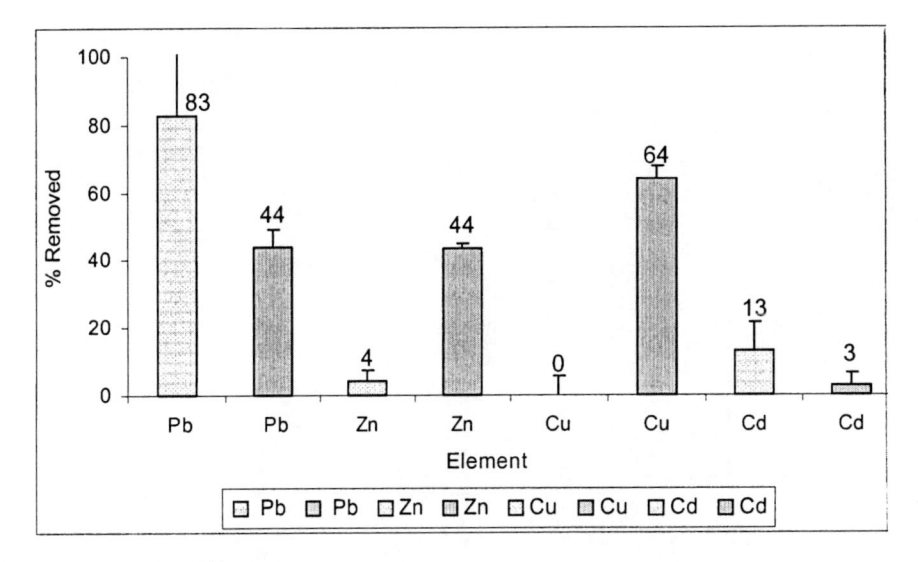

Figure 8. Chitosan treatment of post zero-valent metal treated leachate.

the soil leachate, whereas Cd removal from solution was not very favorable when using both Mg and Fe granules. Copper sequestration was efficient when Fe^0 was used. If adequate quantities of chitosan were utilized, Cu could be effectively removed from the solution.

CONCLUSIONS

The stability of the chitosan-metal complexes was higher than the citrate-metal buffers. The uptake capability of the metal ions depended on the type of element and its initial concentration in the buffer. The model shows that all the metals in this study tended to favor the complexation or chelation mechanism, averaging in the range of pH 6.45-6.73 for optimal sorption from citrate solution, rather than ion exchange. However, ion exchange was favored at lower pH values, the efficiency compared to chelation and related mechanisms above pH 6.0 was lower. The use of sequential pH adjustments followed by chitosan treatments might be more useful for efficient element sequestration with chitosan flakes.

The contact time between the chitosan flakes and the metal solution affected metal sorption capabilities of chitosan. Contact time did not influence pH effect on Pb, Cu and Cd sorption. Temperature did not significantly influence the sorption of metals studied from citrate buffer irrespective of the contact time.

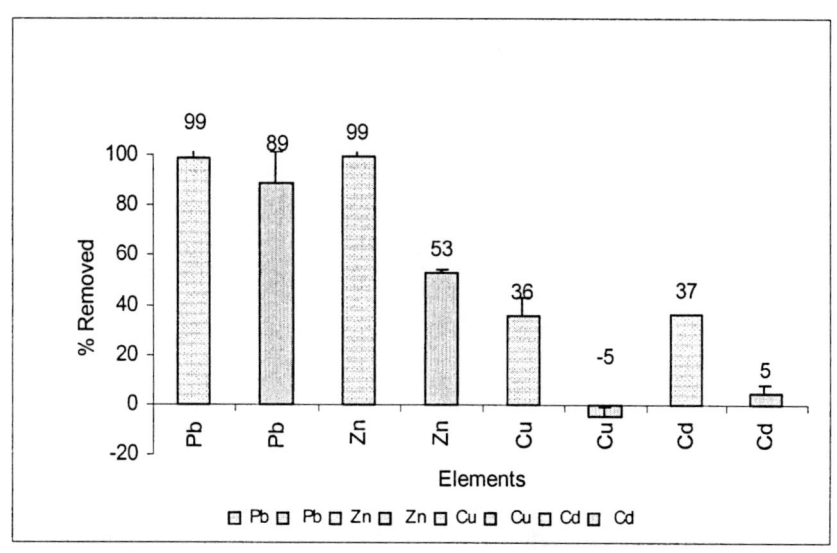

Figure 9. Heavy metal removed after leachate treatment with zero-valent metal and chitosan.

The pretreatment of soil leachate with Mg^0 granules effectively reduced Zn and Pb from leachate through both processes of cementation and oxyhydroxide induced precipitation of the metal. Fe^0 was not effective in the stripping of metals from the leachate within the experimental time frame. Mg granules thus demonstrated their ability to perform better than the Fe^0 in stripping the metals (Zn, Pb and part Cu and Cd) from the complex. This is due to its relatively higher redox potential. However, it is more conceivable that Cu was preferentially complexed by citrate rather than being reduced to the elemental solid state.

The combined effect of both Mg and chitosan treatment virtually reduced Zn and Pb levels to 1%, but Cu and Cd levels were not reduced sufficiently due to the final pH of 6.63 employed in the chitosan treatment. A stepwise pH adjustment of leachate to the optimal range of individual metal ions followed by chitosan treatment may offer an appropriate treatment approach. Slightly larger amounts of chitosan would have to be used to cater for other metals not evaluated in the leachate.

ACKNOWLEDGEMENT

Financial support from the National Science and Engineering Research Council of Canada (NSERC) is gratefully acknowledge. We also acknowledge the Dr. W.D. Marshall of the Food Science and Agricultural Chemistry Department for making available his laboratory and equipment for the metal analysis.

REFERENCE

Bassi, R.; Prasher, S.O. and Simpson, B.K. 1999. Remediation of contaminated leachate using chitosan flakes. *Environ. Technol.* 20: 1172-1182.

Domszy, J.G. and Roberts, G.A.F. 1985. Evaluation of spectroscopic techniques for the analyzing chitosan, *Makromol. Chem.* 186: 1671.

Hansen, M.E. and Illanes, A. 1994. Applications of crustacean waste in biotechnology. In *Fishery Processing Biotechnological Applications*. A.M. Martin (ed.) Chapman and Hall, London, UK, pp. 174-205

Martiner, C.E. and Mcbride, M.B. 2001. Cd, Cu, Pb and Zn coprecipitation in Fe oxide formed at different pH: Aging effect on the metal solubility and extractability by citrate. *Environ. Toxicol. Chem.* 20: 122-126.

Mckay, G.; Blair, H.S. and Findon, A. 1986. Kinetics of copper uptake on chitosan. In *Proceedings of the Third International Conference on Chitin and Chitosan*. R.A.A. Muzzarelli, C. Jeuniaux and G.W. Gooday (eds.) Senigallia, Italy, pp. 559-565.

Mckay, G.; Blair, H.S. and Findon, A. 1986. Equilibrium studies for the sorption of metal ions onto chitosan. *Indian J. Chem.* 28(A): 356-360.

Morrison, S.J.; Metzler, D.R. and Carpenter, C.E. 2001. Uranium precipitation in a permeable reactive barrier by progressive dissolution of zerovalent iron. *Environ. Sci. Technol.* 35: 385-390.

Muzzarelli, R.A.A. 1973. *Natural Chelating Polymers*. Pergamon Press Ltd, Oxford, UK.

Muzzarelli, R.A.A. 1977. *Chitin*, Pergamon Press, Oxford, UK.

Muzzarelli, R.A.A.; Tanfani, F.; Emanuelli, M. and Gentile, S. 1980. The chelation of capric ions by chitosan membranes. *J. Appl. Biochem.* 2: 380-389.

No, H.K. and Meyers, S.P. 2000. Application of chitosan for treatment of wastewaters. *Rev. Environ. Contamin. Toxicol.* 163: 1-28

Park, J.W. and Park, M.O. 1984. Mechanism of metal ion binding to chitosan in solution. Cooperative inter and intramolecular chelations. *Bull. Korean Chem. Soc.* 5(3): 108-112.

Peniston, Q.P. and Johnson, E.L. 1978. Process for determination of chitin from crustacean shells. US Patent 4006 735.

Piniche-Covas, C.; Alvarez, L.W. and Arguelles-Monal, W. 1992. The adsorption of mercuric ions by chitosan. *J. Appl. Polym. Sci.* 46: 1147-1150.

Roller, S. and Covill, N. 1999. The antifungal properties of chitosan in laboratory media and apple juice. *Int. J. Food Microbiol.* 47: 67-77.

Rorrer, G.L.; Hsien, T.Y and Way, J.D. 1993. Synthesis of porous-magnetic chitosan beads for removal of cadmium ions from wastewater. *Ind. Eng. Chem. Res.* 32: 2170-2178.

Suder, B.J. and Wightman, J.P. 1983. Interaction of heavy metals with chitin and chitosan. II. Cadmium and Zinc. In *Adsorption from solution*. R.H. Ottewill, C.H. Rochester and A.L. Smith (eds.) Academic press, London, UK, pp. 235-244.

Udaybhaskar, P.; Iyenger, L.; Prabhakara, A. and Rao, A.V.S. 1990. Hexavalent chromium interaction with chitosan. *J. Appl. Polym. Sci.* 39: 739-747.

Winterford, J.G. and Sandford, P.A. 1995. Chitin and Chitosan. In *Food polysaccharides and their applications*. A.M. Stephens (ed.) Marcel Dekker, New York, NY, pp. 441-462.

Xie, T. 2000. Heavy metal removal from soil by complexing reagents and recycling of chelating reagents. MSc Thesis, Department of Food Science and Agricultural Chemistry, McGill University, Montreal, Canada, pp. 74-80.

Yang, T.C. and Zall, R.R. 1984. Absorption of metals by natural polymers generated from seafood processing waste. *Ind. Eng. Chem. Prod. Res. Dev.* 23:168-172.

CHAPTER 17

CHITOSAN FILM IN SEAFOOD QUALITY PRESERVATION

Fereidoon Shahidi

Department of Biochemistry, Memorial University of Newfoundland,
St. John's, NL, A1B 3X9, Canada

The effect of different deacetylation times for production of chitosan was reflected in the viscosity of the products; the longer periods gave lower viscosity chitosans. The chitosans so produced were employed as a protective edible film on raw cod and herring. The quality of products, as reflected on their oxidation and microbial load, was enhanced in the presence of chitosan coating.

INTRODUCTION

Processing of crustaceans, including crabs, leads to the production of a large volume of by-products. These by-products are composed mainly of chitin, minerals and proteins. The name "chitin" is derived from the Greek word "chiton" meaning a coat of nail and was first isolated from mushrooms and then from cuticles of insects. Chitin is the second most abundant natural polymer on earth after cellulose and it is structurally similar to cellulose in that it is a (1 4) linked polymer of N-acetyl-D-glycosamine with a lesser proportion of D-glucosamine. When the degree of deacetylation is 70%, the product is known as chitosan. Chitosan was first isolated by Rouget in 1859 and is the simplest product that could be obtained from chitin. Presence of positively charged amino groups regularly located along the chitosan chain allows it to bind to negatively charged surfaces via ionic or hydrogen bonding.

Interest in chitin and chitosan began in the 1930's and early 1940's when over 50 patents were issued. However, industrial exploitation of chitin and chitosan began to expand since the 1970's. The advancement in research and small-scale production of chitin and chitosans has resulted in an expansion in the number and variety of potential applications. The unusual multifunctional properties, including high tensile strength, bioactivity and biodegradability are among other characteristics that make both chitin and chitosan attractive specialty products (Ikejma and Inove, 2000). Extensive toxicity tests have shown lack of toxicity of chitosan (Rao and Sharma, 1997). Chitosan is an approved feed additive by the US Food and Drug Administration and its use for potable water purification has also been approved by the US Environmental Protection Agency, up to 10 mg/L. According to Japan's Health Department, chitin and its derivatives are listed as functional food ingredients.

In this contribution, the production of chitosan under different deacetylation conditions is reported. Application of chitosan for quality preservation of cod and herring was also attempted. Finally, use of chitosans for purification of water was tested. Results indicate that chitosan could serve as a novel ingredient for quality enhancement of food and water.

EXPERIMENTAL

Materials

Fresh samples of crab processing by-product as well as cod and herring were procured from local sources in Newfoundland. Cod and herring were immediately cleaned, gutted, filleted and deskinned. Processed fillets were vacuum packed in Whirl pack plastic bags and frozen at -60°C until used. All chemicals and reagents were purchased from Sigma-Aldrich Canada (Oakville, ON). Plate count agar and peptone were obtained from Difco Laboratories (Detroit, MI).

Methods

Chitin/chitosan. Chitin was recovered from crab processing by-products by deproteinization, demineralization and subsequent washing and air-drying, as explained elsewhere (Shahidi and Synowiecki, 1991). Alkali treatment of chitin, using 10 volumes of 50% (w/v) sodium hydroxide in distilled water at 100 C, was carried out for 4, 10 and 20 h under a nitrogen atmosphere. The resultant product was subsequently filtered and washed with hot-deionized water to a neutral pH. The chitosan so obtained was then lyophillized for 72 h at -49 C at 62×10^{-3} mbar (Freezone 6, Model 77530, Labconco Co., Kansas City, MO).

Chitosans were prepared at a 1% level in 1% solution of acetic acid containing glycerol as a plasticizer. Each fillet (5 x 15cm) was immersed in the chitosan solution (5°C) for 30s and after a 2-min period immersed again. Coated samples were allowed to dry at 40°C for 2 hours and then stored at 4±1°C.

Lipids were extracted into a mixture of chloroform and methanol as described by Bligh and Dyer (1959). The peroxide value of the extracted lipids was determined as described in the AOCS (1990) methods. The 2-thiobarbituric acid reactive substances in the samples were determined according to the method of Siu and Draper (1978) as explained by Shahidi and Hong (1991).

Total volatile basic-nitrogen (TVB-N) was determined according to the method described by Cobb III *et al.* (1973) using a micro-Kjeldahl distillation procedure. This method determines the total amount of trimethylamine, dimethylamine, ammonia and other volatile basic nitrogen compounds generally associated with seafood spoilage.

For assessment of microbial spoilage, 11 g fish were put into sterilized plastic bag (Seward Medical Stomach "400" bags) along with 99 mL of sterilized peptone water (10 g peptone and 5 g NaCl in 1L of distilled water, pH 7.2-7.3). The mixture was then homogenized in a laboratory stomacher (Type BA 7021, Seward Medical, London, UK) for 30s. From this mixture 3, 4, and 5 dilutions were obtained by mixing with peptone water. A 1 mL of each diluent was applied on a sterilized standard plate count agar (23.5 g agar in 1 L distilled water). The plates were triplicated and incubated at 20°C for 72 h. Afterwards, the colonies grown on the plates were counted using a colony counter and total aerobic psychrotropic plate count values were indicated as colony forming units (cfu) per gram of fish.

RESULTS AND DISCUSSION

Production of chitosans

The processing by-products of crab contained 50.48% moisture. The content of solids in the by-products, on a dry weight basis, was 0.33% lipid, 32.49% chitin, 40.62% ash and 20.08% protein. The chitin isolated, following washing, deproteinization, demineralization and decolorization (Shahidi and Synowiecki, 1991) was subjected to 50% caustic solution for different time periods at 100 C. The chitosans produced showed variations in their viscosity, 360, 57 and 14 cP, respectively, which appears to be related to the deacetylation period of 4, 10 and 20 h (Table 1). Thus, nitrogen contents of chitosans were 7.55, 7.63 and 7.70%, respectively, for samples deacetylated over 4, 10 and 20h at 100 C. These correspond to a degree of deacetylation of 86.4, 89.3 and 91.3%, respectively, and for sake of simplicity the corresponding chitosan samples are referred to as chitosan I, II and III.

Table 1. Characteristics of chitosans prepared from crab shell chitin and as affected by deacetylation time.

Characteristics	Chitosan I	Chitosan II	Chitosan III
Deacetylation time (h)	4	10	20
Moisture (%)	4.50	3.95	3.75
Nitrogen (%)	7.55	7.63	7.70
Ash (%)	0.30	0.25	0.30
Apparent viscosity (cP)	360	57	14
Deacetylation (%)	86.4	89.3	91.3
Mv (Dalton)	1.8×10^6	9.6×10^5	6.6×10^5

Effect of chitosan coatings on fish quality

The composition and characteristics of fillets of cod and herring are summarized in Table 2. The relative moisture loss of chitosan-coated cod and herring fillets stored at 4 1 C is summarized in Table 3. The pattern of relative moisture loss in cod was different from that of herring under the same storage conditions; the values being lower for herring samples compared to cod fillets regardless of the type of chitosan used. Perhaps the higher fat content in herring is responsible for a lower moisture loss in the fillets examined. The coating with chitosans I and II was more effective than chitosan III in terms of control of moisture loss for both types of fillets.

Table 2. Compositional characteristics of cod and herring fillets

Component	Cod	Herring
Moisture	80.92 ± 0.93	75.53 ± 0.38
Protein	15.44 ± 0.07	13.21 ± 0.02
Lipid	1.25 ± 0.04	12.43 ± 0.13
Ash	0.52 ± 0.01	0.35 ± 0.01

Table 3. Effect of chitosan coating on moisture loss of cod and herring stored at 4±1°C.

Chitosan	Storage Period (days)		
	2	8	12
Cod			
Uncoated	6.82	14.60	9.81
360 cP (I)	6.70	10.50	7.00
57 cP (II)	6.75	11.43	8.10
14 cP (I)	6.80	12.15	10.21
Herring			
Uncoated	4.86	7.10	8.60
360 cP (I)	4.02	4.93	6.98
57 cP (II)	4.10	5.50	6.85
14 cP (I)	4.43	5.89	7.94

The effect of chitosan coating on changes in peroxide values (PV) and 2-thiobarbituric acid reactive substances (TBARS) of cod and herring samples (Tables 4 and 5, respectively) was examined after extraction of lipids from fillet samples. As reflected in the data, PV and TBARS values of the uncoated samples increased progressively during the entire storage period. The PV of chitosan coated samples was 48-63% lower than that of the control. Generally, chitosans I and II were more effective in inhibiting the formation of hydroperoxides as compared to chitosan III. However, the existing differences were not always significant ($P>0.5$).

Chitosan films show extremely good barrier effect against oxygen permeation, thus chitosan coating of fish fillets retards diffusion of oxygen to fish meat surface. Presence of a large number of ionic functional groups creates strong polymer interactions which restrict chain motion in high viscosity-chitosans which results in a good oxygen barrier properties. Chen and Hwa (1996) observed that the tensile strength, tensile elongation and enthalpy of the membrane prepared from high molecular weight chitosans were higher than those of low molecular weight chitosans. They also demonstrated that permeability of high molecular weight chitosans (high viscosity) was lower than those of their low molecular weight counterparts (low viscosity). The results of the present study indicate that chitosan coating is effective in retarding oxidation of fish fillets stored at 4 ± 1°C. These results are in agreement with those of Stuchell and Krochta (1995) who reported the effectiveness of why protein isolates as edible coating for reduction of oxidation of salmon.

The post-mortem metabolism of nitrogenous compounds in fish flesh is mainly responsible for its gradual loss of fresh quality (Sikorski et al., 1990).

Thus, the effect of chitosan coating on total volatile basic nitrogen (TVB-N) content of cod and herring was monitored upon storage at 4 ± 1°C. The TVB-N of treated cod samples (Table 6) increased by about 3-4 fold as compared to 6-fold increase for the control sample at the end of a 12-day storage period. Reduction of TVB-N contents of herring samples treated with chitosans was in the range of 26-51% after the same 12 days of storage. Treatment of both cod and herring samples with chitosan I and II reduced TVB-N formation throughout the entire storage period. After day 6, the TVB-N levels in uncoated cod sample exceeded the acceptable level of 30 mg N/100 g of flesh suggested for fish and shellfish (Cobb III et al., 1973; Connell, 1990). For herring, this level was reached after 8 days for uncoated samples. Meanwhile, 14 cP chitosan-treated samples (chitosan III) exceeded this acceptable level after 8 and 10 days for cod and herring, respectively.

The metabolic processes of the microflora contribute, in part to the gradual loss of taste substances in iced fish and ultimately lead to spoilage due to partial proteolysis and accumulation of unpleasant metabolites (Barile et al., 1985; Sikorski et al., 1990). The surface bacteria of cold-water fish are mainly psychotropic (Huss, 1995).

The total aerobic psychrotropic bacterial count in both cod and herring coated with chitosans was compared with those treated with a 1% acetic acid solution which was also used for dissolution of chitosan samples, and their uncoated counterparts. Fresh cod and herring fillets had initial total aerobic psychrotropic count of 3-3.7 $\log_{10}$ cfu/g (Table 7). Samples of cod coated with chitosan contained less than 10^6 cfu/g fish during the entire 12 days of storage period. Meanwhile, uncoated and 1% acetic acid-treated cod samples exceeded this level after 6 and 10 days, respectively. The acceptability limit of 10^6 cfu/g has been proposed for fresh fish (ICMSF, 1986). Stenstrom (1985) observed that cod fillets stored at 2°C exceeded the maximum acceptable level of total plate count value of 10^6 cfu/g after 6 days of storage. For herring fillets, however, this acceptable limit was reached much faster and exceeded it after 4 and 6 days of storage for uncoated and 1% acetic acid-treated samples, respectively. While chitosan I and II were nearly equally effective, chitosan III was somewhat less effective. The data obtained showed that treatment of herring and cod samples with chitosan resulted in reductions of 10^3 and 10^2 total plate counts, respectively, after 12 days of refrigerated storage.

Table 4. Inhibition of formation of peroxides (%) in chitosans in coated cod and herring fillets during storage at 4±1°C.

Chitosan	Storage period (days)			
	0	2	8	12
Cod				
360 cP (I)	1.4	7.9	46.3	45.1
57 cP (II)	5.0	2.6	46.7	33.9
14 cP (III)	3.6	1.3	22.8	13.4
Herring				
360 cP (I)	0	48.7	57.4	59.4
57 cP (II)	0	46.4	58.4	59.9
14 cP (III)	0	42.0	53.4	49.6

Chitosans were evaluated for their capacity to remove metal ion contaminants from waste water samples obtained from a zinc mining site in Newfoundland. Table 8 shows the elements present and their concentration in the waste water tested.

Table 5. Inhibition (%) of TBARS (2-thiobarbituric acid reactive substances) formation of chitosan-coated cod and herring during storage at 4±1°C.

Chitosans	Storage Period (days)			
	0	2	8	12
Cod				
360 cP (I)	-	28.7	70.1	84.2
57 cP (II)	-	27.0	63.8	83.8
14 cP (III)	-	8.2	58.7	52.5
Herring				
360 cP (I)	4.8	31.5	73.2	54.8
57 cP (II)	0.5	21.9	70.0	55.3
14 cP (III)	8.3	6.3	47.2	48.0

Table 6. Content of total volatile basic nitrogen (mg-N/100g fish) of chitosan-coated samples stored at 4 ± 1°C.

Chitosan	Storage Period (days)			
	0	2	8	12
Cod				
Uncoated	9.83	11.98	37.10	53.39
360 cP (I)	8.65	9.80	20.23	21.94
57 cP (II)	8.13	11.55	20.88	25.33
14 cP (III)	10.05	15.88	31.93	38.10
Herring				
Uncoated	8.65	10.70	27.50	48.91
360 cP (I)	7.50	10.63	17.53	24.33
57 cP (II)	9.60	9.83	20.75	24.19
14 cP (III)	8.09	14.85	23.18	36.35

Table 7. Effect of chitosan coating on microbial load ($\log_{10}$ cfu/g fish) of cod and herring fillets stored at 4±1°C.

Chitosan	Storage period (days)			
	0	2	8	12
Cod				
Uncoated	3.30	4.17	6.70	6.80
1% acetic acid	3.20	4.22	5.10	6.08
360 cP (I)	3.05	3.60	4.80	4.95
57 cP (II)	3.30	4.00	4.65	4.75
14 cP (III)	3.10	4.03	5.03	5.06
Herring				
Uncoated	3.08	4.90	7.15	8.03
1% acetic acid	3.10	3.95	6.15	7.22
360 cP (I)	3.20	4.00	5.10	5.10
57 cP (II)	3.40	4.40	5.10	5.10
14 cP (III)	3.0	3.80	5.20	5.22

Table 8. Major metal ions in the waste water samples from a zinc mining site in Newfoundland.

Element	Concentration (ppb)
Ca	14748
Zn	1330
Mg	1191
Fe	162
Mm	159
Pb	33.0
Cu	15.4
Cd	5.0
Hg	0.6
Co	0.5

The waste water sample contained excessive amounts of Zn, Mn, Pb, Cu, Cd, Hg and Co, based on the Canadian Water Quality Guidelines (1999) for protection of fresh water. Removal of these metal ions was best achieved with chitosan 360 cP (chitosan I) at pH 7.0. Concentration of Zn was reduced up to 50% and similarly other metal ions were removed to a large extent. However, multiple stage chitosan treatment was necessary to bring the level of metal ions present below the recommended limits.

REFERENCES

AOCS. 1990. Official methods at recommended practices of the American Oil Chemists' Society. 4th ed. American Oil Chemists' Society. Champaign, IL.

Barile, L.E.; Milla, A.D.; Reilley, A. and Villadsen, A. 1985. A spoilage pattern of mackerel *Rastrelliger faughni Matsui*. 2. Mesophilic and psychrophilic spoilage. In: *Spoilage of Tropical Fish and Product Development*. FAO Fish. Rep. No. 317 (Supp.). Food and Agricultural Organization, Rome, Italy, pp. 146-154.

Bligh, E.G. and Dyer, W.J. 1959. A rapid method of total lipid extraction and purification. *Can. J. Biochem. Physiol.* 32: 911-917.

Canadian Water Quality Guideline. 1999. Canadian Council of Ministers of Environment. Winnipeg, Canada, pp. 1-8.

Chen, R.H. and Hwa, H.D. 1996. Effect of molecular weight of chitosan with the same degree of deacetylation on the thermal, microbial and permeability properties of the prepared membrane. *Carbohydr. Polym.* 29: 353-358.

Cobb III, B.F.; Alaniz, I. And Thompson, C.P. 1973. Biochemical and microbial studies of shrimp: volatile nitrogen and amino nitrogen analysis. *J. Food Sci.* 38: 431-436.

Connell, J.J. 1990. Control of Fish Quality. 3rd ed. Fishing News Books, Oxford, UK.

Huss, H.H. 1995. Quality and Quality Changes in Fresh Fish. FAO Fisheries Technical Paper 348. Food and Agricultural Organization of the United Nations. Rome, Italy.

ICMSF. 1986. Microorganisms in Foods. 2: Sampling for Microbiological Analysis: Principles and Specific Applications. International Commission on Microbiological Specifications for foods of the International Union of Microbiological Societies. University of Toronto Press, Toronto, Canada.

Rao, S.B. and Sharma, C.P. 1997. Use of chitosan as a biomaterial: studies on its safety and hemostatic potential. *J. Biomed. Nat. Res.* 35: 21-28.

Shahidi, F. and Hong, C. 1991. Evaluation of malonaldehyde as a marker of oxidative rancidity in meat products. *J. Food Biochem.* 15: 97-105.

Shahidi, F. and Synowiecki, J. 1991. Insolation and characterization of nutrients and value-added products from snow crab (*Chinoecetes opilio*) and shrimp (*Pandalus borealis*) processing discards. *J. Agric. Food Chem.* 39: 1527-1532.

Sikorski, Z.E.; Kolakowska, A. and Burt, J.R. 1990. Post-harvest biochemical and microbial changes. In: *Seafood: Resources, Nutritional Composition, and Preservation.* Ed. Z.E. Sikorski, CRC Press, Boca Raton, FL, pp. 55-75.

Siu, G.M. and Draper, H.H. 1987. A survey of the malonaldehyde content of retail meat and fish. *J. Food Sci.* 43: 1147-1149.

Stuchell, Y.M. and Krochta, J.M. 1995. Edible coatings on frozen king salmon: effect of whey protein isolate and acetylated monoglycerides on moisture loss and lipid oxidation. *J. Food Sci.* 80: 28-31.

CHAPTER 18

EMULSION STABILIZING PROPERTIES OF CHITOSAN

Serge Laplante, Sylvie L. Turgeon and Paul Paquin

Dairy Reseach Center (Centre STELA), Faculté des Sciences de l'Agriculture et de l'Alimentation, Université Laval, Québec, G1K 7P4, Canada

The effect of various factors, namely pH, ionic strength (μ), chitosan (CN) content and whey protein isolate (WPI) concentration on the stabilizing properties of CN with various characteristics in a model emulsion containing 10% canola oil and whey protein isolate (WPI) as emulsifier was examined. Results from phase separation measurements revealed that pH and ionic strength were most important factors affecting emulsion stability. No significant difference between CN preparations. At pH 4.0-5.0, a rapid wheying off (WO) with flocculation between droplets was observed, but small gradient creaming phenomenon with fine droplet dispersions at pH 5.5-6.0 occurred. With μ=0.3 M or low CN:WPI ratio ($\leq$1:10), some creaming conditions at pH=5.5 destabilized towards WO. WO conditions at pH<pI of proteins ($\cong$5.1) may be explained by the incompatibility between CN and adsorbed proteins, favoring depletion flocculation and wheying off. The more stable gradient creaming conditions at pH 5.5-6.0 are due to CN-protein compatibility, thus permitting stabilizing effect of CN by coadsorption with protein.

INTRODUCTION

Proteins and polysaccharides play important roles in various food systems. For instance, under suitable conditions and processes, protein-polysaccharide (P-PS) mixtures may control and improve the texture and stability of lipid dispersions in various food emulsions (homogenized milk, ice cream, whipping cream, dressing, mayonnaise, etc.) (Cao et al., 1990). In such oil/water emulsions, the major functional role of proteins (eg.: casein, whey proteins) is to emulsify lipid phase, by forming interfacial protein film with elastic and steric-repulsive properties, which prevents flocculation and coalescence mechanisms leading to creaming (Leman and Kinsella, 1989) and wheying off (Prajapati et al., 1990). In the case of PS, their main stabilizing role is by gelation or viscosification of the continuous phase, which prevents destabilizing mechanisms leading to creaming (Dickinson, 1988a,b). However, in most food systems, particularly at low PS concentrations (≤0.1% (w/w)), the overall stability is less dependent on the specific role of each biopolymer, but becomes rather more dependent on interactions among themselves, thus affecting emulsion stability (Dickinson, 1995, 1996, 1998; Dickinson and Euston, 1991; Syrbe et al., 1998).

According to the structure of biopolymers and environmental conditions in a particular food emulsion (pH, µ, PS/P ratio, concentrations. of PS and P), P-PS interactions depend on many attractive and repulsive forces, mainly electrostatic. When attractive interactions predominate over P-P or PS-PS, both biopolymers are compatible and can coadsorb at oil/water interface, thus the resulting emulsion stability could be increased by steric and charge repulsivity between droplets. On the contrary, incompatibility and destabilization towards phase separation occurs.

PS studied to date as stabilizers in low viscosity model emulsions containing protein emulsifiers have either been neutral (hydroxyethylcellulose, dextran) or anionic (xanthan, carrageenan, guar gum, alginate, carboxymethylcellulose) (Cao et al., 1990, 1991; Dickinson, 1996; Dickinson and Galazka, 1991; Samant et al., 1993; Ward-Smith, et al., 1994). However, a distinctive cationic PS, namely chitosan (linear co-polymer of D-glucosamine and N-acetyl-D-glucosamine units connected through ß-(1-4) linkages), has never been studied for its stabilizing role in model emulsions containing protein emulsifier, in spite of its numerous patented food applications (Winterowd and Sanford, 1995). Food applications of chitosan are of interest because of the availability (from crustaceous shells) and hypocholesterolemic/ hypolipidaemic properties of chitosan (Winterowd and Sandford, 1995; Muzzarelli, 1996, 1997; Nauss and Nagyvary, 1983). Moreover, emulsifying properties of chitosan (Schulz et al., 1998; Del Blanco et al., 1999), as well as their coadsorption properties on lipid droplets covered with anionic lipid emulsifiers (Nauss and Nagyvary, 1983; Fäldt et al., 1993; Magdassi et al., 1997; Alamelu and Panduranga Rao, 1990) have been reported. Other studies have shown that emulsifying properties would be influenced by characteristics such as ash content (Cho et al., 1998) and the degree of deacetylation of chitosan (Del Blanco et al., 1999). By comparing the effect of various environmental factors (pH, µ, %WPI and %CN, characteristics of CN) on emulsion stability with a model emulsion containing whey protein isolate (WPI), known for its good emulsifying properties in many food systems (Leman and Kinsella, 1989), it would be possible to show that cationic chitosans, as opposed to anionic PS, would favor P-PS interactions and stability by interfacial coadsorption at pH > protein pI. This may allow for a better evaluation of potential applications of chitosan in food emulsions containing protein emulsifiers, and knowledge about P-PS emulsions using cationic PS.

MATERIAL AND METHODS

Material

Chitosans, produced from shrimp chitin (*Pandalus borealis*), were kindly donated by Les Pêcheries Marinard Ltée (Rivière-au-Renard, QC, Canada). The characteristics of CN are presented in Table 1. Whey protein isolate (WPI) was supplied by Davisco International (Le Sueur, MN). The sample contained 94.85% proteins (3.2% bovine serum albumin, 13.0% α-lactalbumine, 66.1% ß-lactoglobuline and 17.7% other proteins), 1.95% ash, and 6.15% moisture. Canola oil was purchased from a local supermarket. A lipophilic dye (Red Oil O, Sigma Co., St. Louis, MO) was added (0.06%, w/v) to the oil. Reagents were purchased from Fisher (Pittsburgh, PA) and Sigma. Distilled water was used in all preparations.

Experimental design and statistical analysis

For each CN, the selected experimental design was a central composite rotatable design made up of 3 factors x 5 levels: (%(w/w) WPI= 0-0.25-0.5-0.75-1.0; %(w/w) chitosan= 0-0.075-0.15-0.225-0.3; pH= 4.0-4.5-5.0-5.5-6.0), making a total of 19 experiments (14 treatments and 5 repeats of the central point treatment). While the ionic strength (μ) was not adjusted in the first experiment, in the second experiment, μ was adjusted to a constant value, equivalent to 0.3 M NaCl=([Na$^+$]+[Cl$^-$])/2 by the addition of sodium chloride. Statistical analysis was performed using SAS software.

Emulsion preparation

Each emulsion was made using 180 ml of solution containing the required concentrations of CN and WPI in 0.2 M acetic acid and 0.02% (w/v) sodium azide long with 20 ml of canola oil, in order to obtain a complete emulsion containing 10% (v/v) oil. The pH adjustments were made with sodium hydroxide. Once the oil was added, a pre-mixing was done with Ultra-Turrax (Janke-Kunkel, GmbH) over a 30s period with an electrical input of 30 Volts.

The homogenization was carried out using an Emulsiflex-C5 homogenizer (Avestin Co., Ottawa, ON) in 2 passes (6000 and 3000 psi).

Particle size determination

Immediately after emulsification, the volume-weighted average diameter of droplets in emulsions was determined by photon correlation spectroscopy (PCS), using a Nicomp 370 submicron particle sizer (Pacific Scientific, Hiac-Royce Instruments Division, Silver Spring, FL).

Stability of emulsions during storage

Immediately after emulsification, two glass tubes were filled with 15 ml of emulsion, capped, and kept at room temperature (23°C). The evolution of stability for each treatment was done over a period of 20 days by daily measurements of thickness (centimeter units) of serum layer (lower distinct clear or semi-transparent phase) or creamed phase (red layer gradient at the upper phase). In the latter case, the presence of a red lipophilic dye in the oil enabled us to distinguish and follow thickness evolution of creamed phase. For each experimental design, statistical analysis for comparison of stability profiles between treatments, where serum layer appeared (wheying off treatments), was established from their maximal speed of phase separation (Smax), evaluated by the second derivative of non-linear fitting curves for each evolution profile of thickness of phase separation versus time. Treatments performing gradient creaming were compared by LSD multiple comparison test between their average thickness of phase separation over a period of 20 days.

Protein load in lipid phase

A known amount of each emulsion was centrifuged at 40000g for 1 h (23°C). Protein assay (BCA assay method, Pierce Co., Rockford, IL) was carried out using serum, by taking acetic acid buffer with the same concentration of chitosan as in the formulation for blank. By subtracting the amount of protein in serum from the total protein amount in the formulation, protein concentration in the lipid phase (in g protein/100 g lipid phase) was calculated.

Description of methods:

1) %DD: Acido-basic titration (Tan *et al.*, 1998)
2) Brookfield viscosity: Brookfield model LV-DVII, spindle#4, 60 rpm, 25°C.

Table 1. Characteristics of chitosan preparations

Name of chitosan	CNI[1]	CNHI[2]	CNHK[3]
Degree of Deacetylation (%)	78.1	78.5	67.7
Brookfield Viscosity (1% CN/1% AcOH) (Cps)	4413	490	1030
Viscosity average molecular weight (KDa)	1494	694	749
Weight Average molecular weight (KDa)	1299	918	1284
Polydispersity index (Mw/Mn)	2.39	4.91	6.00
% Ashes*	0.244	0.403	0.142
% Protein*	0.254	0.099	0.225
% Humidity	9.010	8.624	11.152

(*) Calculated on a dry weight basis.
[1] CNI preparation provided by Les Pêcheries Marinard Ltée.
[2] CNHI preparation was produced from CNI by acidolysis
[3] CNHK was obtained by acidolysis of a chitosan of high molecular weight produced in our laboratory.

3) Viscosity average Mw: Capillary viscosimetry (Terbojevich and Cosani, 1997)
4) Weight average Mw: Size exclusion chromatography, with TSK-4000 and 5000 GWXL as column system, pullulan as standard, mobile phase= 0.2M CH_3COONH_3 + 0.1M HCl (pH 4.5, 35°C), flow rate=0.3ml/min. (Denuzière et al., 1995; Ottoy et al., 1996).
5) Protein content: Total nitrogen (%) – nitrogen corresponding to %DD.

RESULTS AND DISCUSSION

Observations of phase separation phenomenon in tubes

Figure 1a shows the assembly of tubes corresponding to each condition of the central composite design for CNI, after 40 days at room temperature (others CN giving similar results). Mainly two types of creaming phenomena occurred: "wheying off" as shown by a lipid-rich upper phase and a clear lower serum

238

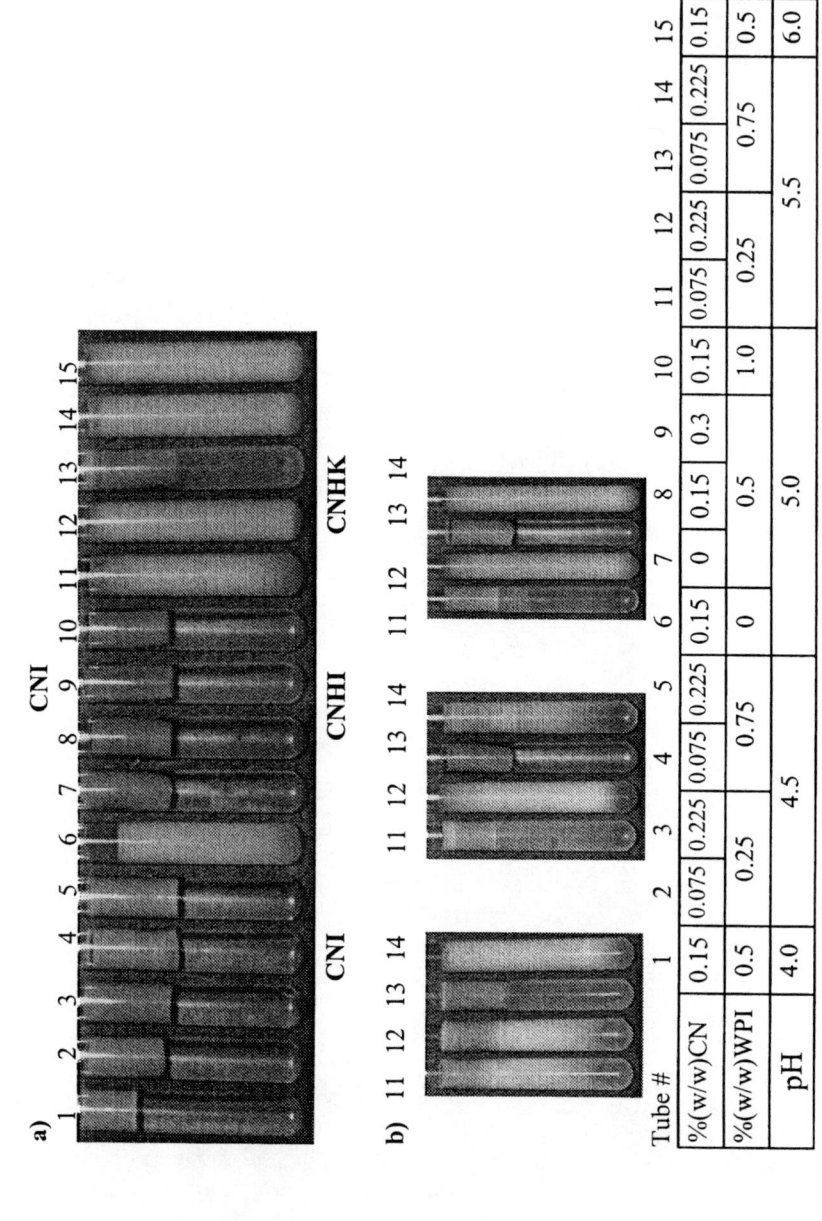

Tube #	1	2	3	4	5	6	7	8	9	10	11	12	13	14	15
%(w/w)CN	0.15	0.075	0.225	0.075	0.225	0.15	0	0.15	0.3	0.15	0.075	0.225	0.075	0.225	0.15
%(w/w)WPI	0.5	0.25		0.75		0		0.5		1.0		0.25		0.75	0.5
pH	4.0	4.5					5.0					5.5			6.0

Figure 1. Stability tube test with a) CNI after 40 days at room temperature; and b) all 3 chitosans at pH 5.5 with $\mu=0.3M$, b; The chart below describes corresponding treatment for each tube number.

phase, and "gradient creaming", as shown by a gradient on top of the emulsion due to higher concentrations in oil phase marked with a red lipophilic dye.

Figure 1a reveals that most conditions at pH≥5.5 give the most homogenous and stable emulsions, even if little gradient creamed phase occurred on top. The only condition at pH 5.5 where wheying off occurred (tube#13), corresponded to a minimum ratio of CNI/WPI=1:10. At pH≤5.0, all treatments containing CNI-WPI mixtures exhibited wheying off.

At pH 5.0, however, gradient creaming occurred for, the control without WPI (tube#6), on the top layer and a light lipid droplet suspension remained in the lower phase, without flocculation. Microscopic observations of each phenomenon (not shown) confirmed that wheying off corresponds to a flocculated droplet network expulsing the serum phase, whereas gradient creaming corresponds to a dispersion of individual droplets with heterogeneity in size.

These results indicate that of all factors examined and levels selected, pH was the most important factor affecting the stability in this emulsion system. According to the literature (Dickinson, 1995; 1998; Syrbe *et al.*, 1998), this could be explained by the predominance of electro-attractive forces (compatibility) occurring between adsorbed WPI proteins (mostly β-lactoglobulin, mainly negatively charged) and chitosan (mainly positively charged) at pH values above the isoelectric point of WPI (pI≅5.1). This condition may lead to interfacial coadsorption of both kinds of biopolymers, thus preventing destabilization (i.e., through flocculation, coalescence or creaming) by creating steric and electrostatic repulsion between lipid droplets. However, at pH 5.5, when CNI:WPI ratio was too low (1:10), wheying off with flocculation occurred. This type of flocculation associated with compatibility at low polysaccharide concentrations is known as bridging flocculation (Dickinson, 1995; 1998; Syrbe *et al.*, 1998).

At pH<5.5, due to a net electro-repulsive force between adsorbed WPI emulsifying proteins (mainly positively charged) and chitosan in the aqueous phase, the incompatibility between both types of biopolymers would favor adsorbed protein-protein association among droplets, resulting in lipid flocculation and wheying off. As reported in the literature (Dickinson, 1995; 1998; Syrbe *et al.* 1998), this type of flocculation associated with incompatibility is known as depletion flocculation. Protein-protein associations between droplets would also be favored at pH 5.0 without CNI, due to the combined effect of neighborhood of WPI pI in the presence of the AcOH buffer (at μ=0.136 M), which indicates instability in the dispersing effect of WPI proteins alone. However, in spite of a low stability against gradient creaming with the chitosan control devoid of protein, residual stable lipid dispersion are observed, thus confirming, the reported emulsifying property of chitosan (Schulz et al., 1998; Del Blanco et al., 1999).

Figure 1b shows differences in the stability of chitosans preparations at pH5.5 and μ=0.3M, where the increase in μ reduces the stability under most conditions with CNI and CNHI as shown by transition from previous gradient creaming phenomenon to flocculation with wheying off (appearance and progression of clearer lower phases). Under all other pH condition, (not shown), and in comparison with Fig. 1a, no visual change was observed. However, with CNHK, the stability of the emulsion appeared unaffected at 0.225% CN (tubes #12 and 14, Fig. 1b), suggesting a stabilizing effect of a lower deacetylation degree of chitosan against wheying off when μ was increased. According to the literature (Dickinson, 1995; 1998; Syrbe et al. 1998), the destabilizing effect upon increase in μ may be due to the loss of electro-attractive interactions and compatibility between adsorbed protein and polyelectrolyte due to charge screening effect, which enhances adsorbed protein-protein attractions between droplets, thus resulting in the formation of a flocculated upper lipid phase rich in adsorbed proteins. In the case of chitosan as polyelectrolyte, a lower deacetylation degree would be less affected by the charge screening effect (Anthonsen et al., 1993). Consequently, the structural rigidity and extension that favors viscosity-stabilizing role would be less affected, and favorable hydrophobic interactions of chitosan with oil phase would permit better adsorption at the interface even if compatibility with proteins is decreased. However, more repeats and experimental conditions for comparison of chitosans at μ=0.3M would be necessary in order to validate a significant stabilizing effect of the lower deacetylation degree of CNHK. Therefore, μ is another important factor affecting stability, in addition to pH, in the model system employed in this work.

Comparison of the effect of treatments producing wheying off phenomenon by response-surface analysis

The final product of destabilization after 40 days was observed so far, and the predominating effect of pH on stability was demonstrated. However, a more accurate comparison of stability between treatments must be based on relative evolution of destabilizing phenomena, in order to better discriminate the effect of factors other than pH and μ, if possible. However, because of the subjectivity in the graphical comparison of evolution profiles (not shown), possible relationships between factor effects was validated by experimenting with a quantitative approach for easier validation of our conclusions, through response-surface analysis between wheying off profiles. In order to do so, a non-linear regression analysis for each wheying off evolution profiles was performed by fitting the following mathematical model (Seber and Wild, 1989) relating thickness in centimeters of serum phase (cms) vs. days of phase separation (t):

$$Cms = \alpha(1 - e^{(-k \cdot t)})^\beta$$

Where α, β, and k represent fitting constants at p=0.05 for each curve fitting.

From each fitted wheying off profile from the experimental design, a new parameter of comparison of instability, Smax (maximum Speed of serum phase in crease, in cm/day units) was calculated by considering the second derivative as $\delta^2 cms/(\delta t)^2 = 0$. The response-surface analysis was achieved by combining Smax as dependent variable, and setting Smax=0 for treatments performing gradient creaming because of their independence from wheying off.

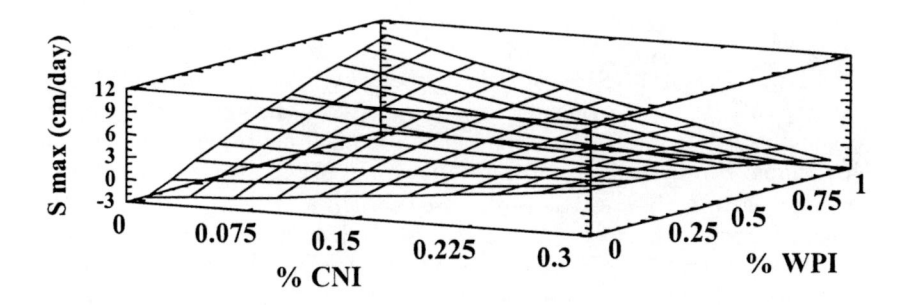

Figure 2. Response-surface curve for interaction effect between %CNI-%WPI on Smax.

As expected, results of wheying off response-surface analysis revealed that quadratic effect of pH on Smax was the most significant factor for all CN preparations without adjustment of µ (results not shown), where the maximal instability (Smax) was reached at pH 5.0, confirming our previous observations and discussions. However, another significant effect was the %CNI-%WPI interaction (Figure 2), where a general tendency towards maximal instability was seen when CN/WPI ratio was minimized. The same tendancy was observed with others CN, but it wasn't significant in the response-surface regression model at p=0.05.

As reported with other protein-polysaccharide model systems (Dickinson, 1995; 1998; Syrbe et al., 1998), the conditions of compatibility between WPI proteins and CN (pH≥5.5), with low CN/WPI ratio (low [CN]) would favor emulsion destabilization by bridging flocculation, whereas in conditions of incompatibility (pH≤5.0), it would destabilize the emulsion by the excluded volume effect of CN, with its lower ability to adsorb to neutral lipids (vegetable oil) when pH is decreased due to loss of hydrophobicity (higher charge density) (Nauss *et al.*, 1983). Since WPI alone in emulsion formulation did not resist wheying off; Fig. 2 indicates that an increase in [CN] would slow down wheying off, due to the enhanced viscosity effect, as reported for other gums and systems (Dickinson, 1995; 1996; 1998). Finally, about the experimental designs at $\mu=0.3$ M, no significant factor effect was observed, due to comparable rapid destabilization in most wheying off conditions.

Comparison of the effect of μ and type of chitosan on Smax values for each wheying off treatment

Comparison of the effect of μ and type of chitosan on Smax for each treatment (Figure 3), indicates a general increasing tendency for Smax when $\mu=0.3$ M. The charge screening effect on the protein coating around lipid droplets (Dickinson, 1995; 1996; 1998; Syrbe *et al.*, 1998) that increase Smax of wheying off by decreasing the electrorepulsivity between lipid droplets (containing WPI proteins and/or chitosan) may be responsible for this observation. However, the μ effect was lowest at pH 5.0, the lowest stable pH condition against wheying off, as discussed earlier. Moreover, at pH≥5.5, the μ destabilizing effect is most effective because, as reported in literature (Syrbe *et al.*, 1998), increases of μ could convert compatibility into incompatibility between adsorbed protein and polysaccharide, that converts a gradient creaming towards a wheying off.

About the effect of CN preparations on Smax, there is no important difference, particularly without μ adjustment (Fig. 3a). At μ=0.3M (Fig. 3b), even if some conditions at pH 5.0 seem to show lower Smax for CNHK, more repeats would be needed for each treatment in order to validate such interpretation of results, because in fact, the control without CN represents a true repetition of the same treatment and shows a significant variability.

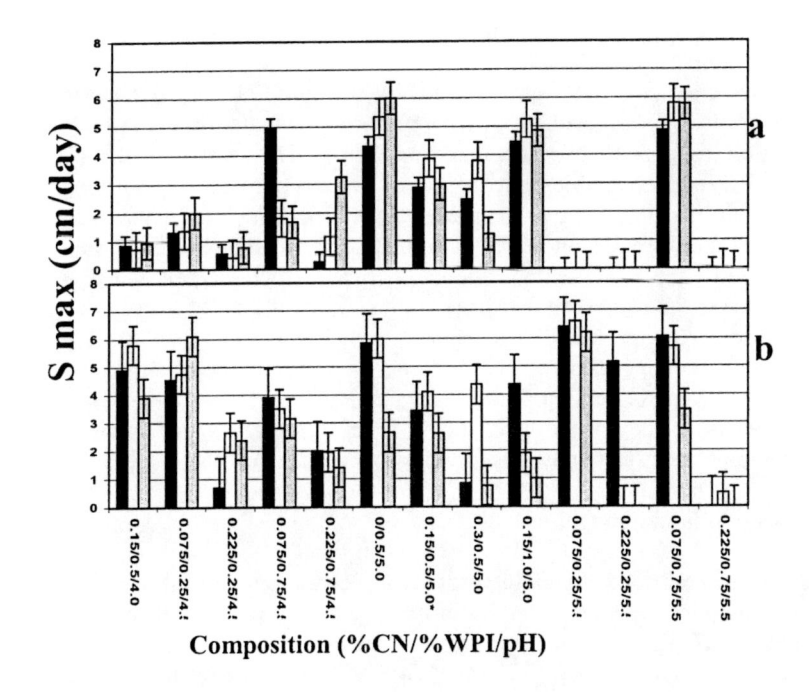

Figure 3. Effect of type of chitosan and µ on Smax for wheying off
(= CNI■ = CNHI ☐ = CNHI◯) a) without, and b) with µ= 0.3M.
Error bars represent the standard error of estimate at p=0.05

Comparison of the effect of treatments producing gradient creaming phenomenon

In order to better discriminate the generally close and slow gradient creaming profiles (not shown), multiple comparison tests (LSD test) between average creaming thickness for each treatment over a 20 day period were done. Graphical results are presented in Figure 4 with or without µ=0.3M. Comparison of the effect of chitosan characteristics and composition on the stability against gradient creaming without µ adjustment (Fig. 4a), shows mostly the same creaming thickness at pH5.5 and 6.0, with a small tendency towards lower values at higher %CN (0.225%), as a result of the combined stabilizing effects of higher CN/WPI ratios and viscosities. However, with µ=0.3M (fig. 4b), we see that some value stripes representing gradient creaming are disappeared due to the destabilization towards wheying off.

Among the remaining treatments at µ=0.3M, condition at pH 6.0 was the only one insignificantly affected by µ. Therefore, we conclude it is the most stable condition from our experimental conditions. This result could be explained by this more favorable pH condition for interaction and coadsorption of chitosan with proteins at the interface, associated with an increase of electrostatic and steric repulsion between droplets due to a more cationic and hydrophilic state of the lipid coating, but also because µ without salt added at pH 6.0 (=0,19M) is the nearest to 0.3M. About the effect of type of chitosan on gradient creaming at µ=0.3M, CNHK seems to give slightly higher stability against gradient creaming than other chitosans, because of higher structural hydrophobicity and emulsificating power, as explained in an earlier section. As a general rule, as in the case of wheying off, the stability against gradient creaming phenomenon is also mostly affected by pH and the ionic strength.

Results of protein load determination

Results of protein load for gradient creaming conditions are presented in Table 2, along with corresponding values of mean droplet diameters. If we keep our attention at pH 5.5, we mainly observe a decrease in protein load when %CN is increased from 0.075 to 0.225% (%WPI= 0.25), whereas it increases when %WPI is increased from 0.25 to 0.75% (%CN= 0.225). Also, similar protein loads between CN preparations either with or without µ =0.3M was observed. However, at µ=0.3M, there is no change in protein load when %CNHI and %CNHK was increased from 0.075 to 0.225% at constant %WPI=0.25%. A general tendency towards slightly higher protein loads at µ=0.3M was also observed when %CN was at the maximal value of 0.225%. The results with unadjusted µ were indicative of interfacial coadsorption of chitosan with protein that would be responsible for the stabilizing properties of chitosan in conditions of compatibility (pH>pI WPI). It also suggests competitive adsorption between both components for the interface. However, when µ was increased to 0.3M, interactions (mainly electrostatic) between CN and adsorbed proteins would become unfavorable (incompatibility) due to charge screening effects. Therefore, chitosan cannot coadsorb efficiently with the already adsorbed protein, explaining the absence of change in protein load when %CNHI or CNHK was increased at constant %WPI. Examining the droplet size measurements, one can see lower sizes at pH5.5 for higher concentrations of CN (0.225%) and at pH6.0, but most of treatments gave higher sizes when µ=0.3M. Those results are in good accordance with the higher stabilities against gradient creaming (Fig. 4) when droplet sizes were decreased.

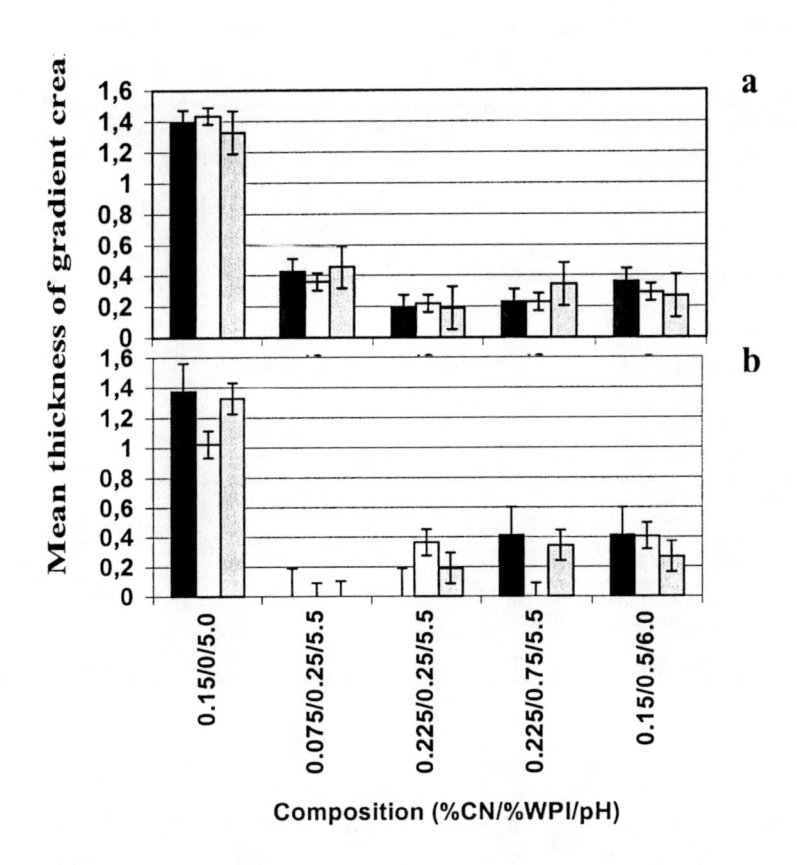

Figure 4. Comparison of treatments producing gradient creaming a) without, and b) with µ=0.3M.
(■ = CNI ; ☐ = CNHI ; ☐ = CNHK). Error bars represents the least significant difference (LSD) at p=0.05

Table 2
Results of emulsion characterization for conditions producing gradient creaming at $\mu = 0.3M$

Treatment composition (%CNI/%WPI/pH)	Protein load (g Protein/ 100g lipid) low	Protein load μ^1	Mean droplet diameter (μm) low	Mean droplet diameter μ^1
CNI				
0.15/0/5.0	0	0	1.498	5.805
0.075/0.25/5.5	1.17	1.06*	1.110	2.346*
0.225/0.25/5.5	0.73	0.84*	0.959	1.970*
0.225/0.75/5.5	1.58	1.67	0.987	1.043
0.15/0.5/6.0	1.80	1.45	0.983	0.963
CNHI				
0.15/0/5.0	0	0	2.253	2.057
0.075/0.25/5.5	1.25	1.07*	2.084	n.d.*
0.225/0.25/5.5	0.99	1.06	0.750	1.288
0.225/0.75/5.5	1.86	2.04*	0.972	1.250*
0.15/0.5/6.0	1.86	1.47	0.836	0.869
CNHK				
0.15/0/5.0	0	0	2.114	2.853
0.075/0.25/5.5	1.16	0.98	1.285	1.213
0.225/0.25/5.5	0.88	0.98*	0.892	0.965*
0.225/0.75/5.5	1.44	1.48	0.773	0.931
0.15/0.5/6.0	1.70	1.57	0.809	0.903

(* indicates an experimental condition where wheying off was observed during the first 20 days of tube observation).
(n.d.= not detectable because of rapid droplet flocculation during measurements)
[1] unadjusted ionic strength (0.088M at pH 5.5; 0.095M at pH 6.0)

In fact, results show that droplet sizes are less affected by μ effect at pH6.0 condition, confirming the highest stability at this pH condition, as previously observed (Fig.4).

CONCLUSIONS

In a model system containing WPI as emulsifier and low concentrations of chitosan as stabiliser, two kinds of phase separation phenomena were observed (wheying off, mainly at pH≤5.0 and gradient creaming at pH≥5.5). Among the various factors studied, pH and µ affected the stability most, demonstrating the predominance of electrostatic interactions between WPI proteins and chitosan. In fact, due to its cationic charge, chitosan is distinctive among other anionic gum stabilizers already used in industrial food emulsions by the way that it could perform interfacial stabilisation by coadsorption with WPI proteins at pH>pI, provided a CN:WPI ratio>1:10 and µ<0.3 M. These results suggest suitable applications of chitosan in low acidity milk products (whipping and ice creams, milk chocolates, etc.)

ACKNOWLEDGEMENTS

The financial support from Conseil de Recherche en Pêche et Agro-alimentaire du Québec (CORPAQ) and Les Pêcheries Marinard Ltée. is gratefully acknowledged.

REFERENCES

Alamelu, S. and Panduranga Rao, K. 1990. Effect of surfactants on the stability of modified egg-yolk phosphatidylcholine liposomes. *J. Microencapsulation* 7: 541-551.

Anthonsen, M.W.; Varum, K.M. and Smidsrod, O. 1993. Solution properties of chitosans: conformation and chain stiffness of chitosan with different degrees of N-acetylation. *Carbohydr. Polym.* 22: 193-201.

Cao, Y.; Dickinson, E. and Wedlock, D.J. 1990. Creaming and flocculation in emulsions containing polysaccharides *Food Hydrocolloids* 4: 185-195.

Cao, Y.; Dickinson, E. and Wedlock, D.J. 1991. Influence of polysaccharides on the creaming of casein-stabilized emulsions. *Food Hydrocolloids* 5: 443-454.

Cho, Y.I.; No, H.K. and Meyers, S.P. 1998. Physicochemical characteristics and functional properties of various commercial chitin and chitosan products. *J. Agric. Food Chem.* 46: 3839-3843.

Del Blanco, L.F.; Rodriguez, M.S.; Schulz, P.C. and Agullo, E. 1999. Influence of the deacetylation degree on chitosan emulsification properties. *Colloid Polym. Sci.* 277: 1087-1092.

Denuziere, A.; Yagoubi, N.; Baillet, A. and Ferrier, D. 1995. An improved statistical parameter allowing elaborate comparison of polymer molecular weight distribution by gel filtration chromatography: application to chitosan. *S.T.P. Pharma Sci.* 5: 481-485.

Dickinson, E. 1988a. The structure and stability of emulsions. In *Food Structure-Its Creation and Evaluation*. J.M.V. Blanshard and J.R. Mitchell (eds.) Butterworths Publ., London, UK, pp. 41-57.

Dickinson, E. 1988b. The role of hydrocolloids in stabilizing particulate dispersions and emulsions. In *Gums and Stabilisers for the Food Industry*, 4th Edn. G.O. Phillips, D.J. Wedlock and P.A. Williams (eds.) IRL Press, Oxford, UK, pp. 249-263.

Dickinson, E. 1995. Emulsion stabilisation by polysaccharide and protein-polysaccharides complexes. In *Food Polysaccharides and their Applications*. A.M. Stephen (ed.) Marcel Dekker, New York NY, pp. 501-515.

Dickinson, E. 1996. Biopolymer interactions in emulsion system: influences on creaming, flocculation and rheology. In *Macromolecular Interactions in Food Technology*, ACS series (No. 650). N. Parris (ed.) American Chemical Society, Washington D.C., pp. 197-207.

Dickinson, E. 1998. Stability and rheological implications of electrostatic milk protein-polysaccharide interactions. *Trends Food Sci. Technol.* 9: 347-354.

Dickinson, E. and Euston, S.R. 1991. Stability of food emulsions containing both protein and polysaccharide. In *Food Polymers, Gels and Colloids*. E. Dickinson (ed.) Royal Society of Chemistry, Cambridge, UK, pp. 132-146.

Dickinson, E. and Galazka, V.B. 1991.Emulsion stabilization by ionic and covalent complexes of beta-lactoglobulin with polysaccharides. *Food Hydrocolloids* 5: 281-296.

Faldt, P.; Bergenstahl, B. and Claesson, P.M. 1993. Stabilization by chitosan of soybean oil emulsions coated with phospholipid and glycoholic acid. *Colloids and Surfaces A: Physicochem Eng. Aspects* 71: 187-195.

Leman, J. and Kinsella, J.E. 1989. Surface activity, film formation, and emulsifying properties of milk proteins. *Crit. Rev. Food Sci. Nutr.* 28: 115-138.

Magdassi, S.; Bach, U. and Mumcuoglu, K.Y. 1997. Formation of positively charged microcapsules based on chitosan-lecithin interactions. *J. Microencapsulation* 14: 189-195.

Muzzarelli, R.A.A. 1996. Chitosan-based dietary foods. *Carbohydr. Polym.* 29: 309-316.

Muzzarelli, R.A.A. 1997. Human enzymatic activities related to the therapeutic administration of chitin derivatives. *CLMS Cell. Mol. Life Sci.* 53: 131-140.

Nauss, J.L. and Nagyvary, J. 1983. The binding of micellar lipids to chitosan. *Lipids* 18: 714 -719.

Ottoy, M.H.; Varum, K.M.; Christensen, B.E.; Anthonsen, M.W. and Smidsrod, O. 1996. Preparative and analytical size-exclusion chromatography of chitosans. *Carbohydr. Polym.* 31: 253-261.

Prajapati, P.S.; Gupta, S.K.; Patel, A.A. and Patil, G.R. 1990. In *Brief Communications of the XXIII International Dairy Congress.* Montreal, October 8-12, vol. II, International Dairy Federation, Brussels, Belgium, pp. 531.

Samant, S.K.; Singhal, R.S.; Kulkarni, P.R. and Rege, D.V. 1993. Protein-polysaccharides interactions: a new approach in food formulations. *Int. J. Food Sci. Technol,* 28: 547-562.

Seber, G.A.F. and Wild, C.I. 1989. *Nonlinear Regression,* John Wiley & Sons, New York, NY, pp. 327.

Schulz, P.C.; Rodriguez, M.S.; Del Blanco, L.F.; Pistonesi, M. and Agullo, E. 1998. Emulsification properties of chitosan. *Colloid Polym. Sci.* 276: 1159-1165.

Syrbe, A.; Bauer, W.J. and Klostermeyer, H. 1998. Polymer science concepts in dairy systems. *Int. Dairy J.* 8: 179-193.

Tan, S.C.; Khor, E.; Tan, T.K. and Wong, S.M. 1998. The degree of deacetylation of chitosan: advocating the first derivative UV-spectrophotometry method of determination. *Talanta* 45: 713-719.

Terbojevich, M. and Cosani, A. 1997. In *Chitin Handbook.* R.A.A. Muzzarelli and M.G. Peters (eds.) European Chitin Society, pp. 87-101.

Ward-Smith, R.S.; Hey, M.J. and Mitchell, J.R. 1994. Protein-polysaccharide interactions at the oil-water interface. *Food Hydrocolloids,* 8: 309-315.

Winterowd, J.G. and Sandford, P.A. 1995. Chitin and chitosan. In *Food Polysaccharides and their Applications.* A.M. Stephen (ed.) Marcel Dekker Inc., New York, NY, pp. 441-462.

CHAPTER 19

CHITOSAN MODIFICATIONS FOR PHARMACEUTICAL APPLICATIONS

Le Tien Canh[1,2], Monique Lacroix[2], Pompilia Ispas Szabo[1] and Mircea-Alexandru Mateescu[1]

[1]Department of Chemistry and Biochemistry, Université du Québec à Montréal, Montréal, Québec, H3C 3P8, Canada
[2]Research Centre in Microbiology and Biotechnology, Canadian Irradiation Centre, INRS-Institut Armand-Frappier, Laval, Québec, H7V 1B7, Canada

Chitosan is a natural polymer derived from chitin. Its biocompatibility, biodegradability and antimicrobial action make it suitable for pharmaceutical and medical use. The new technology described here consists of i) acylation with palmitoyl chloride, or ii) acylation following a previous cross-linking with glutaraldehyde. These derivatives were used for drug controlled release. Monolithic forms (tablets) containing 20% acetaminophen as tracer were obtained by direct compression of dry powders. The kinetics of drug release showed an excellent control up to 180 h and good mechanical behavior of tablets during swelling and drug liberation. No significant differences in drug release time ($t_{90\%}$) were found between the two types of derivatives, but X-rays diffraction patterns of the corresponding films were different. FT-IR band at 1700 cm^{-1} increased with increasing amounts of palmitoyl chloride used for derivatization.

INTRODUCTION

Chitosan is a natural non-toxic polymer (Sashiwa et al., 1990) derived from partial deacetylation of chitin that is widely present in shells of crustaceans. Recently, different applications were reported for chitosan including drug delivery systems and as a hypocholesterolemic agent (Bodmeier et al., 1989; Illum et al., 1994). The current work reports the incorporation of fatty acids in

© 2004 ScienceTech Publishing

the chitosan molecule and its application as pharmaceutical excipients for drug controlled release.

METHODS

Two formulations were studied, palmitoyl chitosan (PC) and cross-linked palmitoyl chitosan (CLPC). PC was obtained according to Le Tien *et al.* (2001), by acylation with palmitoyl chloride in aqueous medium, giving amidic derivative, involving part of the amino groups of the 2-aminoglucose units. For CLPC, the same acylation was done but the chitosan was first cross-linked with glutaraldehyde. After chemical modification, the preparations were neutralized, filtered and dried with acetone solutions (at gradually increasing concentrations) to obtain the corresponding powders. Monolithic forms (tablets of 500 mg) containing modified chitosan and 20 % acetaminophen as tracer were made by direct compression of powders. The kinetics of drug release were determined in aqueous medium (pH 7.0, 37 °C, 50 rpm) with a dissolution apparatus (Distek^{TM}) using a USP XXII method. FT-IR, X-ray diffraction and electronic microscopy were used for structural analysis of the polymers so obtained.

RESULTS AND DISCUSSION

The kinetics of drug release showed an excellent control up to 160 h and good mechanical behavior of tablet during swelling and drug liberation for modified chitosan tablets, whereas non-modified chitosan tablets disintegrated in few hours (Figure 1). No significant differences in drug release time ($t_{90\%}$) were found between the two types of derivatives (Figure 1). Although release times of longer than 24h are less useful for oral administration, the result is of great interest in the use of modified chitosan for implant forms. Interest in implants is very high for human and veterinarian therapeutics, such as in the sub-cutaneous administration of antibiotics, steroids, peptide hormones, contraceptives, etc. For both compounds (PC and CLPC), the FTIR analysis showed an increase of the band at 1650 cm^{-1} (elongation of C=O groups) due to acylation of chitosan with palmitoyl chloride (Figures 2A, B). The same increase was observed at 2900 cm^{-1} that might be due to the presence of C-H groups from alkyl chains of fatty acids. Minor differences between the two compounds found at this frequency were ascribed to the contribution of glutaraldehyde C-H groups of the cross-linked polymer (CLPC). Concerning the cross-linked palmitoyl chitosan, the formation of imine bond (C=N) as a result of reaction of the amine group (from chitosan) and of carbonyl group (from glutaraldehyde) is observed in the spectral region of 1720-1650 cm^{-1} (Figure

[Graph: Mt/M inf vs Temps (h), comparing Palmitoyl chitosan and Cross-linked palmitoyl chitosan]

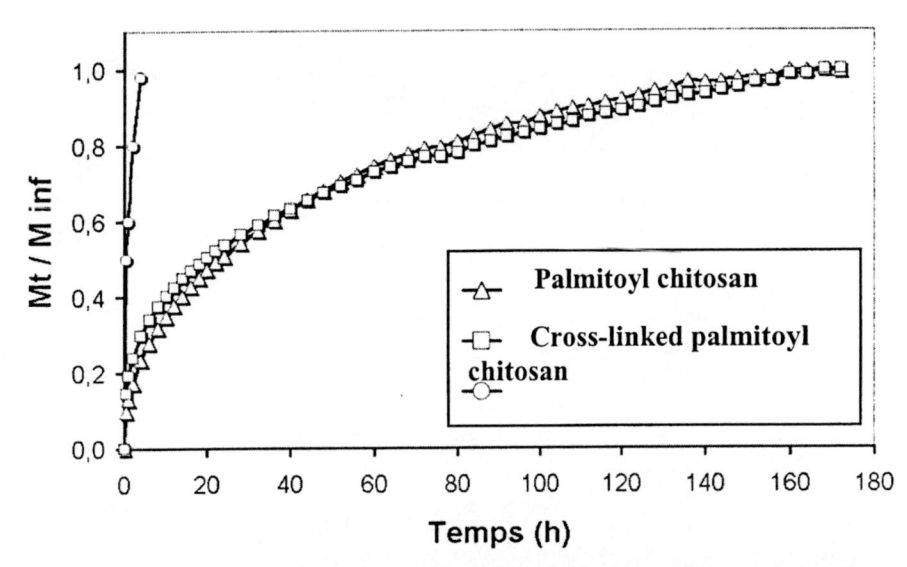

Figure 1. Release profiles of acetaminophen from tablets (500 mg) based on palmitoyl chitosan, cross-linked palmitoyl chitosan and non-modified chitosan containing 20 % drug. All the dissolution experiments were carried out in phosphate buffer (pH 7.0) at 37 °C and 50 rpm.

2B). Moreover, X-ray diffractograms showed that the maximum intensity at 4.37 Å decreased by modification of chitosan with fatty acids and a new peak appeared at 6.66 Å (Figure 3A). Electron microscopy analysis revealed that unmodified chitosan morphology is characterized by a regular and smooth surface, whereas palmitoyl chitosan derivative possesses a rough surface with a fibrous aspect (Figure 3B). The observations from X-ray diffraction and of electron microscopy indicated that palmitoyl chitosan structure was less crystalline and more disorganized. The natural chitosan is essentially characterized by hydrogen associations (Figure 4A) whereas the palmitoyl chitosan presents more hydrophobic interactions (Figure 4B). These hydrophobic contributions should be involved in the access of water and in drug release. Thus, the drug liberation can be better modulated and an extended application of these polymers could be envisaged.

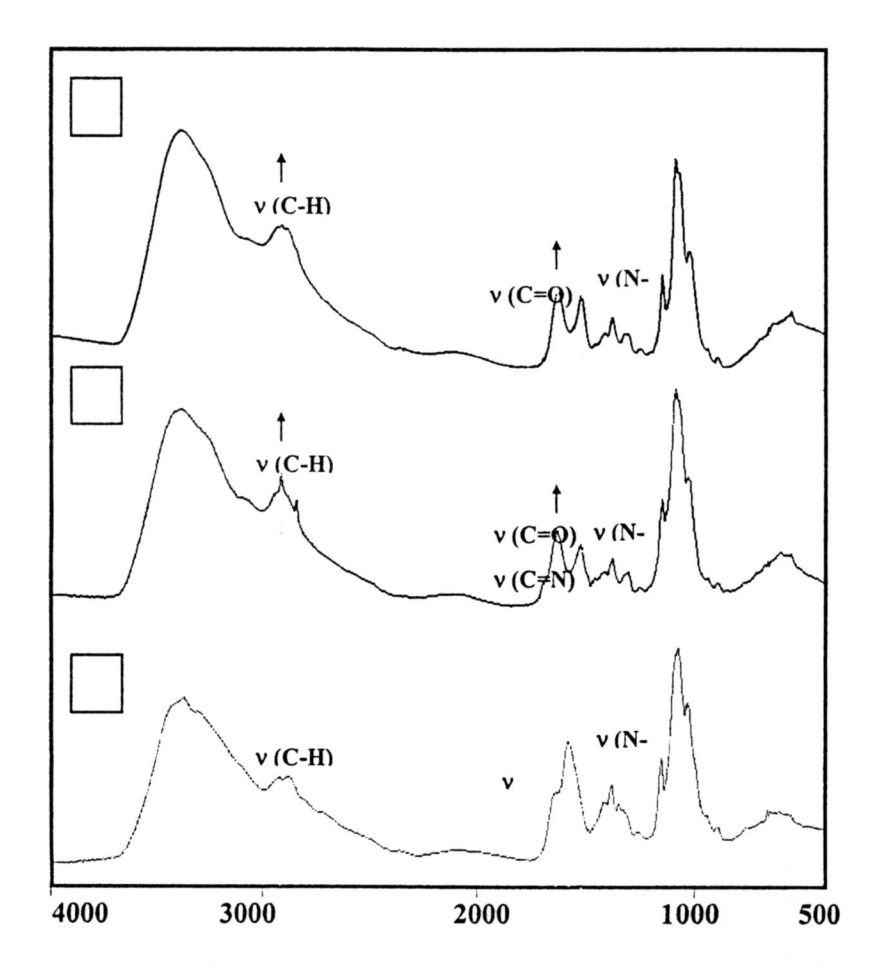

Figure 2. FT-IR spectra of A) palmitoyl chitosan, B) cross-linked palmitoyl chitosan and C) unmodified chitosan.
[1] N-H bending vibrations of non-acylated 2-aminoglucose primary amines (1560 cm^{-1})
[2] N-H bending vibrations of secondary amides (1525 cm^{-1})

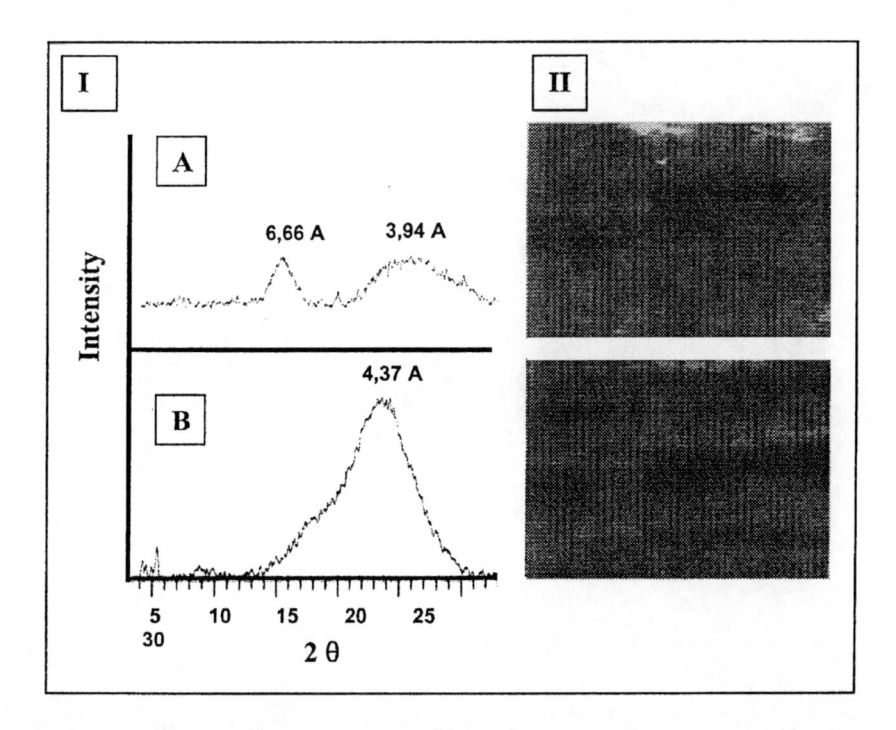

Figure 3. X-ray diffraction (I) and electronic microscopy (II) of palmitoyl chitosan (A) and non-modified chitosan (B).

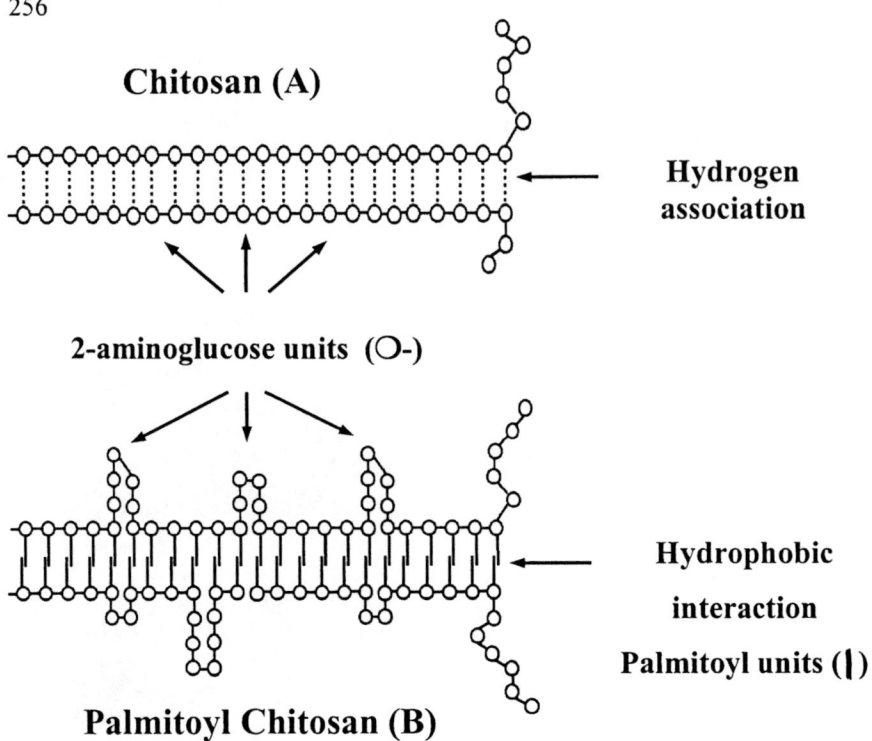

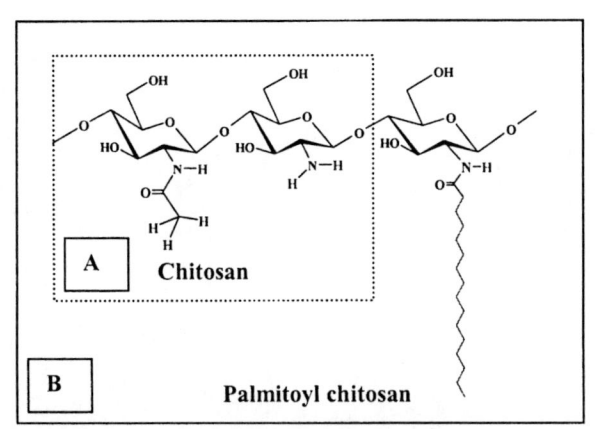

Figure 4. Schematic presentation of hypothetical structures of the native chitosan (A) and palmitoyl chitosan (B).

CONCLUSIONS

The new technology, based on modified chitosan, appears very interesting for the drug controlled formulation. The obtained polymers, which are biocompatible, may be employed under different forms (implants, films, beads and tablets) with a large potential application in pharmaceutical and in biomedical fields.

ACKNOWLEDGMENT

An FCAR University-Industry graduate studentship granted to LTC is gratefully acknowledged.

REFERENCES

Bodmeier, R.; Chen, H.G. and Paeratakul, O. 1989. A novel approach to the oral delivery of micro- or nanoparticles. *Pharm. Res.* 6: 413-418.

Illum, L.; Farraj, N.F. and Davis, S.S. 1994. Chitosan as a novel nasal delivery system for peptide drugs. *Pharm. Res.* 11: 1186-1190.

Le Tien, C.; Lacroix, M.; Ispas-Szabo, P. and Mateescu, M.A. 2004. Biocompatible compositions as carriers or excipients for pharmaceutical and nutraceutical formulations and for food protection. International Patent pending.

Sashiwa, H.; Saimoto, H.; Shigemasa, Y.; Ogawa, R. and Tokura, S. 1990. Lysozyme susceptibility of partially deacetylated chitin. *Int. J. Biol. Macromol.* 12: 295.

CHAPTER 20

ECONOMICS OF CHITIN, CHITOSAN AND CAROTENOPROTEIN PRODUCTION FROM SHRIMP WASTE

K. Gunjal[1], H. Teftal[2], B.K. Simpson[3] and P. Goldsmith[4]

[1]Department of Agricultural Economics, McGill University
[2]Suncor Energy Inc.
[3]Department of Food Science and Agricultural Chemistry, McGill University
[4]Department of Agriculture and Consumer Economics, University of Illinois

More than 12,000 metric tons of shrimp waste is produced in Quebec each year. This waste presents a source of raw material for making high-value biotechnology products such as chitin, chitosan and carotenoprotein. The simulated production costs at the industrial level are estimated at $0.65 for chitosan, $0.26 for chitin and $0.07 per gram for carotenoprotein and show gross margins over 80%. Based on the related data of cellulose derivatives (the closest substitute for chitosan), the Bass model is used to forecast the sales of high-grade chitosan in Quebec. The results show that the potential market for chitosan could be worth $37 million in Quebec and $59 million in Canada cumulative for the next 20 year period.

INTRODUCTION

The importance of seafood in the Canadian diet has been increasing over time. In 1997, global crustacean landing was 1,428,926 metric tons (MT). Canada and the U.S. produced 58% of that catch. Shrimp and prawn landing averaged 107,980 MT, while lobster and spider crab averaged 306,392 MT in the North American fishing area as given by FAO. The per capita fish consumption was 8.65 kg in 1988 and reached 9.97 kg in 1999 (Statistics Canada, 1999). Many people are choosing fish and shrimp as an alternative to red meat. The seafood industry has responded positively to meet this demand. Atlantic coast shrimp landings have tripled from a level of 43,660 MT in 1989 to 126,477 MT in 2000 (Figure 1). Quebec landings have also increased during this period as they have doubled. Consequently there has been an increase in waste material from shrimp processing industry in this region.

The waste material from the seafood processing industry represents more than 70% of the landing. Since most of these industries were established without consideration for by-product reclamation (Hansen and Illanes, 1994) this presents environmental concerns. However, these waste materials have valuable recoverable by-products such as protein, carotenoid pigments, seafood peptones and flavor compounds which may be recovered for use as feed supplement for farmed fish, as well as other biotechnology products such as chitin and chitosan.

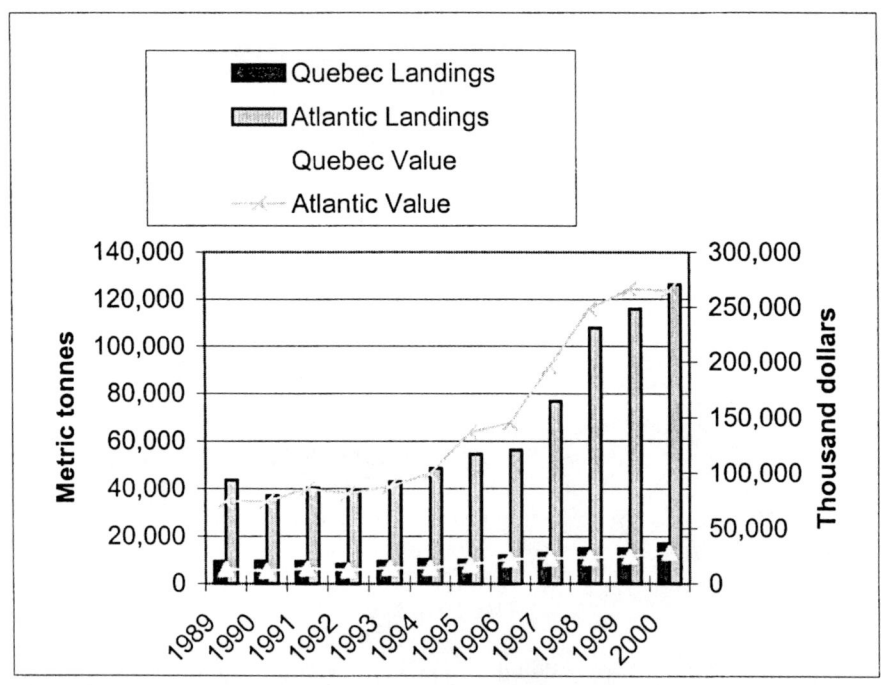

Figure 1. Commercial landings and value of shrimp for Atlantic coast and Quebec, 1989-2000.

Though considerable research has been carried out in the production of chitin, the chemical processing technology remains widely used to deproteinize and demineralize the raw material, leading to products of varied qualities and properties. In order to recover high-value biotechnology products from shrimp waste with more consistent physiochemical properties, Simpson et al. (1994) used biological means. Although the use of biological methods are well known, processing cost involved has not been fully explored.

This study, therefore, investigated the potential of chitin and chitosan industry with particular reference to the situation in Quebec. Specifically the objectives of the study were 1) to estimate the quantity of waste generated by the shrimp processors in Quebec, 2) to estimate the cost of production at the semi-pilot plant level and the scaled up processing plant level, and 3) to investigate the potential market for high grade chitosan in Quebec.

MATERIALS AND METHODS

Methods

The main steps followed in this study were:
1) Forecasting the quantity of waste generated by the shrimp processors in Quebec based on the production function of the shrimp processing industry and available data.
2) Based on the economic engineering theory, an evaluation of the production cost was performed at a semi-pilot level and then an estimate of the scaled-up process plant was determined by using the cost capacity factor.
3) Determining the potential market for high-grade chitosan in Quebec by using the Bass model to forecast the sales.

Data Sources

The data used for the study was obtained from earlier work done by Simpson *et al.* (1994). In addition, other information was obtained from government agencies, utility companies, pharmaceutical companies and from published sources as described below. Further details are provided by Teftal (1999).

Estimation of shrimp waste in Quebec

The quantity of shrimp waste is linearly dependent on the quantity of the processed shrimps. Therefore, the quantity of waste can be inferred from the quantity of the processed shrimp. Ideally, the best way to forecast this quantity is by estimating the production function of the shrimp processing industry. A production function is mathematical model relating the maximum output that can be produced from given amounts of various inputs (McGuigan and Moyer, 1975). Data on total allowable catch (TAC) and the of volume of the shrimp landings in Quebec (CATCH) were obtained from Fisheries and Oceans Canada

(FOC). The data on the quantity of shrimp processed in Quebec (PROC) was obtained from MAPAQ. The data covered the period 1985 to 1996.

Due to the scarcity of shrimp predators in the Gulf, the TACs was assumed to remain constant for the period of study (FOC, 1997). A multiple regression model was estimated to determine the statistical relationship among the variables with PROC as the dependent variable. The regression model was estimated as:

$$LnPROC = 5.20 + 0.445 LnCatch + 0.927 LnTAC$$

The details of the regression analysis used in this study are given in Teftal (1999).

Processing Costs Estimation

Production costs (operating costs) are the expenses necessary to maintain and operate a plant, processing line and equipment used in production (Zugarramurdi and Parin, 1995). A profit maximizing firm sets production at a rate where marginal revenue equals marginal cost, with the condition that total revenue covers variable costs. Alternatively, a firm focuses on a target production level and minimizes costs when it has little control over prices (Georgianna et al., 1986). Accordingly the production cost of three byproducts were analyzed to find the levels of production at which the pilot plant could operate profitably. Annual cost was used for the basis of comparison in order to even out seasonal variations and to factor for infrequently occurring large expenses. The pilot plant was assumed to be located in a rental facility at Macdonald Campus of McGill University.

The direct costs associated with the production process were the cost of raw materials, labor and utilities. For example the cost of raw material was the cost of handling and transporting the shrimp waste from the processing plant to the laboratory at the plant facility. Indirect cost were those that involved depreciation, rent, insurance and plant overhead. Estimates of all cost items were obtained from the laboratory experiments conducted by Simpson et al. (1994). Simulations based on economic engineering approach (Horton, 1994) were used to estimate the production cost of the value added products.

The annualized cost of capital model was then employed to compute the equipment cost. The machines were assumed to last 25 years with no salvage value. The annual equivalent cost of capital (AECC) was obtained by using the following formula (Lovett, 1995):

$$AECC = P \frac{i(1+i)^n}{(1+i)^n - 1}$$

Where:

P is the purchase cost of the equipment,
i is the minimum acceptable rate of return (a discount rate, i=10%)
n is the equipment's expected life (number of years)

Finally, the net realizable value approach was used to distribute joint costs among three joint products, namely chitin chitosan and pigments. The portion of the fixed cost (FCM) related to each piece of equipment was estimated as:

$$FCM = (Cost\ Portion\ Allocated * AECC)$$

Details of list of equipment and cost portion allocated to each process and joint cost distribution are available in Teftal (1999).

Scaled-up process plant

The concept of cost-capacity factor based on the economies of scale was used to estimate production cost of the value-added products at the scaled-up level. The total investment (I_1, I_2) for two plants with different capacities (Q_1, Q_2) producing the same product are linked by a certain cost-capacity factor (Williams, 1960). Thus:

$$I_2 = I_1 (Q_2 / Q_1)^x$$

Where x is the cost-capacity factor and I_i is the cost of the equipment required to produce quantity Q_i.

A cost-capacity factor of 0.7 was used and the plant was assumed to operate at it's full capacity. Pikulik and Diaz (1977) and Remer and Idrovo (1990) have used a similar value of cost-capacity factor to estimate the production cost for chemical plants. Black (1982) also used it to estimate operating cost for plants of different sizes.

Market potential

There are several methods identified in literature for estimating sales of a new product prior to its introduction. Thomas (1985) reported difficulties in modeling the growth of new product without the existence of historical data. Although market research has been used often to forecast the demand of a new product, it is preferable to use subjective managerial and/or experience with analogous products in order to estimate parameters prior to the launch of a new product (Lawrence and Lawton, 1984).

In this study, cellulose was used as an analogous product for chitosan. Time series data on value of cellulose for the entire manufacturing industry in Quebec were obtained from the "Bureau de la statistique du Quebec[1]". The data covered the period 1978 to 1993. The data for entire cellulose industry was used since data specific to the pharmaceutical industry in Quebec was not released. The current dollar values were adjusted for inflation using the industrial product price index (1986 = 100%).

The Bass model estimated by using the Nonlinear Square Estimation procedure was employed to forecast the sales of high-grade chitosan in Quebec. The assumptions of this model are similar to the theoretical concepts in literature on new product adoption and diffusion (Bass, 1969).

The Bass Model defines sales as follows:

$$S_t = pm + (q-p)Y_{t-1} - (q/m)Y_{t-1}^2$$
OR
$$S_t = a + bY_{t-1} + cY_{t-1}^2$$

Table 1 shows the definition of the different variables and their analogue definition in the original Bass model.

Bass model contains three parameters (p, q, and m), theoretically it is necessary to utilize only three data points to estimate these parameters. However, the parameter estimates and the predictions are very unreliable when only a few data points are used to calibrate the Bass model (Srinivasan and Mason, 1986). Heeler and Hustad (1980) found that their predictions were improved if the data used include the peak.

---

[1] Publication entitled, "Les produits utilizes par les manufacturiers Quebecois". Statistics Canada produces these data for the Quebec Ministry of Industry and Technology.

Table 1. Definitions of the Model's Variables and their Analogues in Bass Model

Variable	Definition in the Bass Model	Definition in the Model used for Chitosan in this study
Y_{t-1}	The cumulative number of adopters at time t-1.	The value of the cumulative sales at time t-1.
M	The total number purchasing during the period for which the density function was constructed i.e. the maximum number of potential adopters.	The maximum purchases that can be made by the potential buyers. The potential sales during the life cycle of the product
p	The coefficient of innovation	Percentage of sales that can be made by innovative firms (i.e. early buyers).
q	The rate of imitation	Percentage of sales made by the follower firms.

RESULTS AND DISCUSSION

Raw material supply

Table 2 shows a summary of the correlation matrix. LnPROC, the dependent variable, was found to be highly correlated to LnTAC and LnCATCH. No autocorrelation was found in the estimated model since the Durbin-Watson d statistic value was within the acceptable range. The results showed that the relationship was statistically significant at 5%. R-square statistic indicates that 64% of the variation in the dependent variable was explained by the variation of LnCATCH and LnTAC. Using the stable values of the two independent variables, the regression model predicted the availability of 4,208 MT of processed shrimps in Quebec in 1999 generating 12,624 MT of waste for the biotechnology industries.

Table 2. Correlation Matrix for the Logarithm of the Total Processed Shrimp (LnPROC), Total Allowable Catches (LnTACs) and of the Shrimp Landings (LnCATCH)

	LnPROC	LnTACs	LnCATCH
LnPROC	1		
LnTACs	0.745	1	
LnCATCH	0.731	0.722	1

dL=0.812 and dU=1.579

Results of production cost estimation

Table 3 shows a summary of the production cost structure and the unit gross margin of the three by-products at the pilot plant level. The lowest cost of all value-added products occurs when the pilot plant is functioning at its full capacity. The estimated per gram production costs at the pilot plant level are $0.72 for chitosan, $0.32 for chitin and $0.15 for carotenoprotein under the assumption of 100% capacity utilization. An increase in use of 20% (from 80 to 100), results in the following decreases in the production costs: 8% for pigments, 1.25% for chitin and 1.6% for chitosan. The pilot plant can be efficiently exploited at 80% capacity. Figure 2 shows that the gross margin of all products increased with an increase in capacity use. However, the marginal change of the gross margin starts slowing down after the capacity has reached 60%, and remains almost unchanged when the plant operates between 80 and 100% of its capacity.

Table 3. Production cost structure and the Unit Gross Margin of 1g of the Three By-Products at the Pilot Plant Level

Plant Capacity Use (%)	20%	40%	60%	80%	100%
Production Cost (1g)					
Chitin	$0.36	$0.33	$0.32	$0.32	$0.32
Pigments	$0.36	$0.23	$0.19	$0.17	$0.15
Chitosan	$0.87	$0.77	$0.74	$0.73	$0.72
Reference Price ($/g)					
Chitin	$6.67	$6.67	$6.67	$6.67	$6.67
Pigments	$3.58	$3.58	$3.58	$3.58	$3.58
Chitosan	$7.00	$7.00	$7.00	$7.00	$7.00
Gross Margin (%)					
Chitin	94.56%	95.05%	95.16%	95.21%	95.28%
Pigments	89.85%	93.53%	94.73%	95.33%	95.71%
Chitosan	87.54%	88.99%	89.37%	89.56%	89.73%

The gross margin of the pilot plant exceeds 80% despite using discounted selling prices (the price of chitin, chitosan and carotenoprotein). These high margins are mostly due to the price reference used to compute them. The reference prices were obtained from the price catalogue of Sigma-Aldrich Canada Ltd., Pronova and other laboratory suppliers. Chitin is one of the inputs used to make chitosan, thus explaining its higher gross margin when compared to chitosan. However, the production cost of chitosan is almost double that of chitin. The total production cost of chitin is a part of chitosan cost of production. The analysis was based on a reference price of $6.67/g for chitin and $7/g for chitosan in order to present a conservative scenario and to compensate for the risk of underestimating the production cost of these value-added products.

Figure 3 depicts the production cost curve of each product plotted against the quantity of dried shrimp waste processed. The increased volume of production activities spreads fixed costs over a larger quantity produced. Hence the production cost of each of the value added products decreases with an increase in the capacity use of the pilot plant. The cost of producing one gram of chitin, chitosan and carotenoprotein (pigments) is respectively $0.32, $0.72 and $0.15 at the pilot plant level. However an industrial scale plant processing 2,000 MT of shrimp waste a year, using the biological process to recover the value added products from the waste would result in the production of 91 MT of chitin, 55 MT of chitosan and 12 MT of carotenoprotein. The respective production costs for one gram of each product is $0.26, $0.65 and $0.07. Slight economies of scale exist up to the processing of 500MT shrimp waste. Thus, costs of producing chitosan are almost double that of chitin. Table 4 shows the

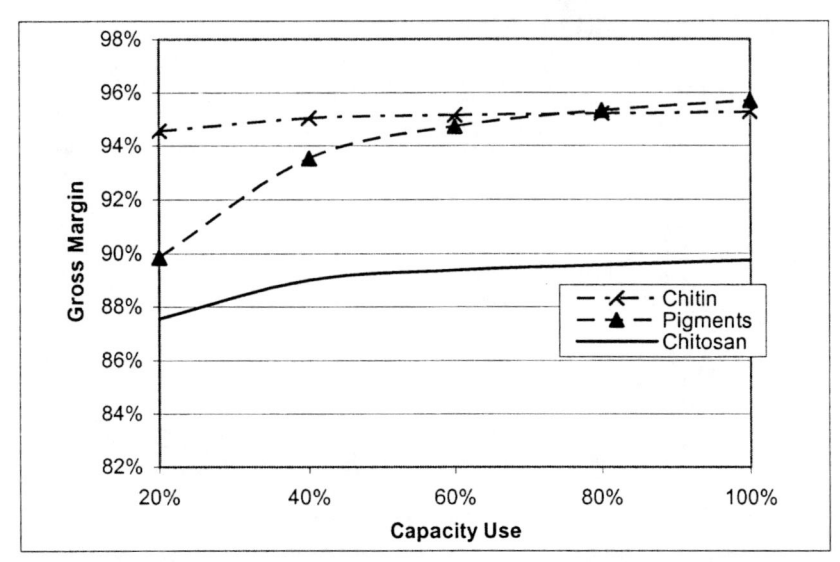

Figure 2. The Gross Margin of Producing the Value-added Products From Shrimp Shells at the Pilot Plant Level.

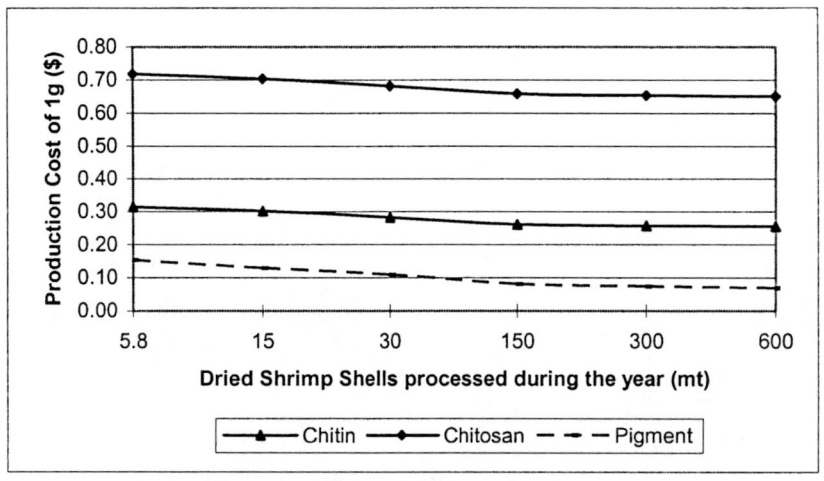

Figure 3. Estimated Production Costs of the Value-Added Products at the Industrial Level (cost of one gram).

Table 4. The Estimated Production Cost of One Gram of the Value-Added Products at the Industrial Level.

Direct production costs	5,850 kg	15,000 kg	Dried shrimp waste 30,000 kg	150,000 kg	300,000 kg	600,000 kg
Processing	$4,388	$11,250	$22,500	$112,500	$225,000	$450,000
Power	$640	$1,640	$3,281	$16,403	$32,805	$65,610
Heat	$28	$54	$87	$269	$437	$710
Labor	$1,487	$3,606	$3,150	$4,800	$5,446	$6,207
Supervision	$223	$472	$541	$720	$817	$931
Payroll charges	$910	$1,860	$2,466	$5,281	$7,714	$11,561

estimated production cost of one gram of the value added products at the industrial level.

Forecasting potential market demand for chitosan

Table 5 shows the estimated parameters of the Bass Model. Using the Nonlinear Least Squares Estimation (NLSE) procedure applied to 16 data points, convergence was achieved after 3 iterations. The parameters are statistically significant. The following values are obtained for the parameter estimates. The potential sales during the cycle of the product were approximately $37 million in 1999 constant dollar. The percentage of sales that can be made by innovative firms is estimated to be 5.4%.

Table 5. Regression Results of the Bass Model

Variable	Coefficient	Std. Error	t-Stat	P-Value
a = pm	1562.559007	344.6609	4.533612	0.001
b = q-p	0.119752	0.071950	1.664373	0.043
c = q/m	0.000006	0.000003	2.337184	0.036
$R^2 = 0.48$				

ANOVA					
Source	DF	SS	MS	F	P-Value
Regression	2	3014431	1507215	4.86	0.027
Error	13	4034956	310381		
Total	15	704938			

Table 6 is a summary of the potential sales of high-grade chitosan in Quebec and Canada. Accordingly, under this scenario the potential market for chitosan of all grades in Quebec is estimated to be about $37 million (cumulative for 20 years). The innovative firms will realize $2.03 million. The value in 1999 dollars was derived using the Industrial Product Price Indexes 111.7% "1986=1" (CANSIM, Matrix 5680) and 116.1% "1992=1" (CANSIM, MATRIX 1878)

CONCLUSIONS

Based on the stock assessment and future prospects from Fisheries and Oceans Canada, the total production of processed shrimp has been estimated with respect to total allowable catch and total shrimp landing in Quebec. This estimation was used to derive the quantity of shrimp waste in Quebec. The results indicate that 12,640MT of shrimp waste is expected to be generated annually in Quebec. It can be concluded from this that there is sufficiently large

Table 6. Chitosan Sales Forecasts: New Sales Each Year and Total (Cumulative) Sales - (thousand 1999 dollars)

Period	All Grades Quebec' (First time Purchase)	Cumulative: All Grades Quebec' (New & Repeat P.)	High Grade Quebec" (New & Repeat P.)	Cumulative: High Grade Canada* (New & Repeat P.)
1	2027	2027	1551	3232
2	2250	4277	3273	6819
3	2454	6731	5151	10731
4	2622	9353	7163	14922
5	2742	12095	9254	19280
6	2799	14894	11393	23735
7	2783	17677	13536	28199
8	2697	20374	15600	32499
9	2546	22900	17540	36553
10	2341	25261	19337	40285
11	2100	27361	20938	43620
12	1839	29230	22020	45876
13	1578	30800	23604	49175
14	1329	32137	24617	51285
15	1102	33239	25457	53036
16	901	34140	26145	54469
17	848	34988	26806	55847
18	773	35761	27393	57068
19	653	36414	27889	58102
20	556	36970	28312	58984

' These sales are derived from cellulose derivatives sales by applying the Bass model. The model estimates sales in 1986 dollars, which are then converted to 1999 dollar value.
" According to Technical Insight Study, 76.54% of Chitosan sales are expected to be of High Grade.
*It is assumed that the Quebec market for high-grade chitosan is equal to 48% of the Canadian market (the same proportion as for the Pharmaceutical & Medicine Market)

amount of raw material potentially available for further processing by biotechnology firms interested in producing chitin and other related industrial products.

The production cost of one gram of chitosan, chitin and carotenoprotein were $0.65, $0.26 and $0.07. The gross margins were estimated at 91% for chitosan, 96% for chitin and 98% for carotenoprotein. Based on these production costs and published catalogue prices of these three products it can be concluded that production at the pilot level and potentially at the industrial level is likely to be economically profitable.

Potential sales for chitosan in Quebec were estimated to be 2.03 million in 1999 constant dollars in the first year of launching and a cumulative market of about $37 million. This research shows that there is a great potential for chitin and chitosan industry in Quebec. However, these estimates are based on a theoretical approach of chitosan replacing an analogues cellulose product. Hence further market research to determine the actual market potential of these three by-products is needed.

REFERENCES

Bass, F.M. 1969. A New Product Growth for Model Consumer Durables. *Manage. Sci.* 15: 215 – 227.

Black, J.H. 1982. *Scaling Operating Costs by Cost-Capacity Factors.* A.A.C.E. Trans. pp. H.3.1-H.3.6.

Fisheries and Oceans Canada 1997. *The Quebec Seafood Industry Network.* DFO, Laurentian Region.

Georgianna, D.L. and Hogan, W.V. 1986. *Production Costs in Atlantic Fresh Fish Processing. Marine Resource Economics* 2: 275-292.

Hansen, M.E. and Illanes, A. 1994. Applications of Crustacean Wastes in Biotechnology. In *Fisheries Processing Biotechnological Applications.* A.M. Martin (ed.) Chapman & Hall, London, UK, 176 p.

Heeler, R.M. and Hustad, T.P. 1980. Problems in Predicting New Product Growth for Consumer Durables. *Management Science* 26 (October): 1007-1020.

Horton, F.W.H. Jr. 1994. *Analyzing Benefits and Costs: A Guide for Information Managers.* IDRC, Canada.

Lawrence, K.D. and Lawton, W.H. 1986. Applications of Diffusion Models: Some Empirical Results. In *Innovation Diffusion Models of New Product Acceptance,* 529-541. Volume 5 of Series in Econometrics and Management Sciences, Ballinger Publishing Company, Cambridge, UK.

Lovett, J. N. 1995. *Cost Estimating in Manufacturing*. In Cost Estimator's Reference Manual. Stewart et al. (ed.) John Wiley & Sons, Inc. N.Y

McGuigan, James R. and Moyer, C.R. 1975. *Managerial Economics: Private and Public Sector Decision Analysis*. The Dryden Press. Chicago, IL.

Pikulik, A. and Diaz, H.E. 1977. *Cost Estimating for Major Process Equipment*. In *Chemical Engineering, Modern Cost Engineering: Methods and Data*. McGraw-Hill, New York, NY, pp. 301-317.

Remer, D.S. and Idrovo, J.H. 1990. Cost Estimating Factors for Biopharmaceutical Process Equipment. *Biopharm.* 3: 36-42.

Simpson, B.K.; Gagne, N. and Simpson, M.V. 1994. *Bioprocessing of Chitin and Chitosan*. In *Fisheries Processing Biotechnological Applications*. A.M. Martin (ed.) Chapman & Hall, London, UK, p. 155.

Srinivasan, V. and Mason, C.H. 1986. *Nonlinear Least Squares Estimation of New Product Diffusion Models*. Marketing Science 5.

Statistics Canada. 1999. CANSIM- MATRIX # 1878.

Teftal, H 1999. Economics of Chitin and Chitosan.MSc Thesis, Department of Agricultural Economics, McGill University, Montreal QC.

Thomas, Robert J. 1985. Problems in Demand Estimation For New Technology. *J. Prod. Innov Manage.* 2: 145-157.

Williams, R. 1960. "Six -Tenth Factor" Aids in Approximating Costs. In Chilton. In *Cost Engineering in the Process Industry*. McGraw-Hill. New York, N.Y. pp 40-41.

Zugarramurdi, A. and Parin, M.A. 1995. *Economic Engineering Applied to the Fishery Industry*. FAO Fisheries Technical Paper #351. Rome, Italy.

CHAPTER 21

CHROMATOGRAPHIC METHOD FOR SIMULTANEOUS QUANTITATION OF N-ACETYLGLUCOSAMINE, GLUCOSAMINE AND ITS OLIGOMERS

Éric Demers[1], Diane Ouellet and Marie-Élise Carbonneau

[1]Centre Technologique des Produits Aquatiques
Ministère de l'Agriculture des Pêcheries et de l'Alimentation
Gouvernement du Québec, 96, montée Sandy Beach, bureau 205
Gaspé, Québec, G4X 2V6, Canada

The N-acetylglucosamine, glucosamine and its oligomers present in the reaction product resulting from acid hydrolysis of chitin are simultaneously assayed by means of a new high performance liquid chromatographic method (HPLC). The method uses a Shodex Asahipak NH_2P 50-4E column and acetronitrile/water eluent (80:20, v/v) as well as a refractometer detector. The method is reproduceable and affords mean variation coefficients of 4.7% for glucosamine and 10.2% for N-acetylglucosamine.

INTRODUCTION

A new high-performance liquid chromatographic (HPLC) method was developed to monitor the quality of industry-produced glucosamine from chitin extracted from crustacean wastes. The industrial process involves hydrolyzing and deacetylating the chitin by using a strong acid to form glucosamine (Novikov and Ivanov, 1997). An incomplete reaction leads to the formation of by-products such as N-acetylglucosamine and oligomers of glucosamine. HPLC analysis makes it possible to identify and quantify glucosamine and its oligomers, namely chitobiose, chitotriose and N-acetylglucosamine.

MATERIALS AND METHODS

Preparation of standards and samples

Chitobiose (# 400432), chitotriose (# 400433) and chitotétraose (# 400434) standards (Palace and Phoebe, 1997) were purchased from Seikagaku Corporation (Chuo-Ku, Tokyo, Japan). N-acetylglucosamine (# A 8625) and glucosamine (# G 4875) standards were procured from Sigma-Aldrich Company (Oakville, ON). The concentrations of standard N-acetylglucosamine and glucosamine solutions used varied from 0 to 60 mg/mL and from 0 to 160 mg/mL in the case of the glucosamine oligomers. The standards and samples were dissolved in Nanopure water (Barnstead; Dubuque, IA) and adjusted to pH 4.0 using a 0.1 N hydrochloric acid (# CR-0166, Laboratoire MAT; Beauport, QC, Canada) or 0.1 N sodium hydroxide (#B 30167, BDH Inc.; Toronto, ON, Canada). The solutions were then filtered through a 13-mm diameter nylon filter unit 0.45 µm thick (# 6513-06 Lida Manufacturing Corparation; Kenosha, WI) and kept at a temperature of at least 4°C. Samples were partially lyophilized to within the range of the proposed concentrations.

Chromatographic system

A chromatographic system manufactured by Waters Corporation (Milford, MA) was used. The system was equipped with a 717 Plus Autosampler, 600 Multisolvent Delivery System (flow rate 1 ml/min. with 20% helium sparge), 410 Differential Refractometer (temperature: 35°C, sensitivity: 4) and a column heater (temperature: 35°C). Data were recorded and processed using the following software: Waters Millenium 32, Version 3.2. The method uses a Shodex Asahipak NH_2P-50 4E (250 x 4.6 mm) column recommended by the manufacturer for the assay of mono and oligosaccharides (Showa Denko K.K.). The eluent was composed of acetonitrile (# A 998-4, Fisher Scientifique, Nepean, ON, Canada) and Nanopure water in proportions of 80:20 (v/v) was used. Before use, the eluent was filtered through a Millipore system (Bedford, MA), using a 0.5-µm filter (pore size) 47 mm in diameter (# FHUP 04700). An injection volume of 20 µl was used for standards and samples.

Results

Chromatography makes it possible to separate N-acetylglucosamine, glucosamine and its oligomers, namely chitobiose, chitotriose and chitotetraose in 45 min (Figure 1). It can be seen that the peaks became wider as retention

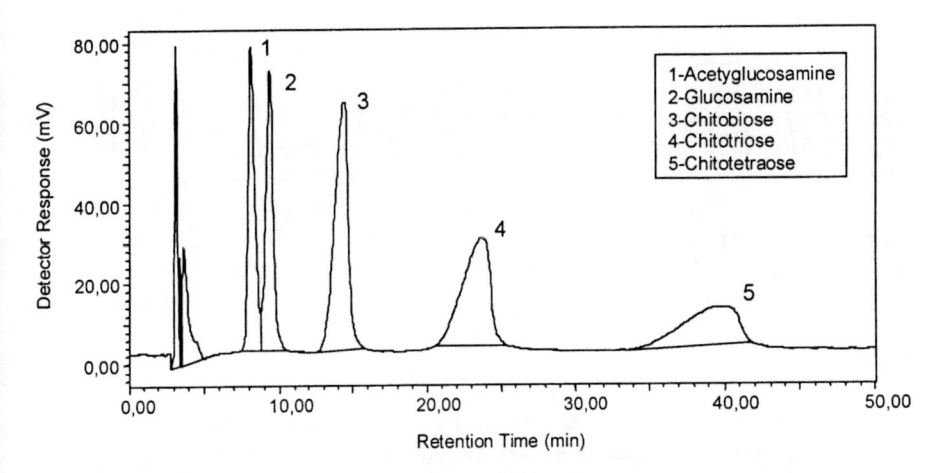

Figure 1. Chromatogram of N-acetylglucosamine, glucosamine and its oligomers.

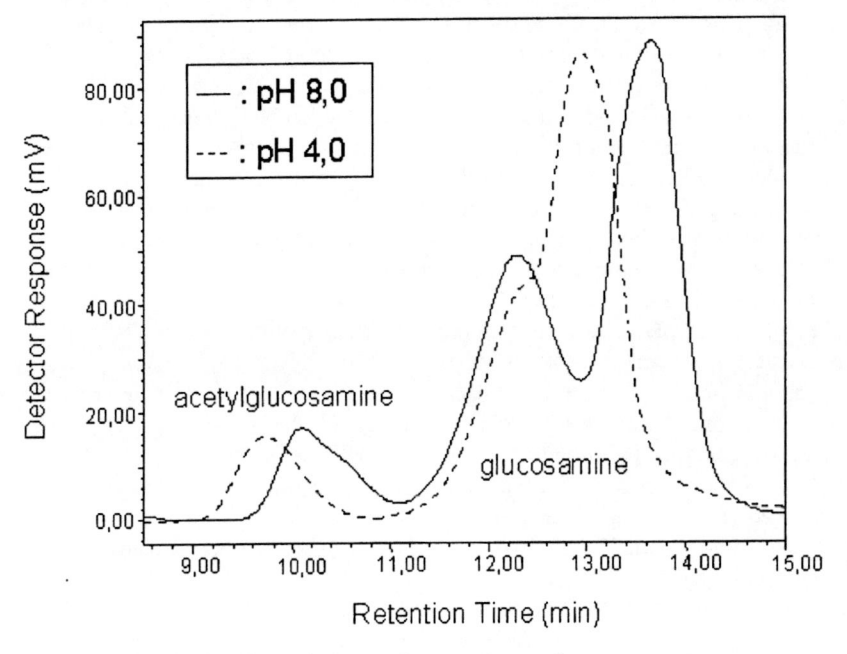

Figure 2. Split effect of pH on glucosamine peaks.

time increased, a phenomenon that becomes critical for chitotetraose. Quantification was done by means of a calibration curve; correlation coefficients were greater than 0.99. The chitotetraose was not assayed because an unacceptable correlation coefficient was obtained.

This chromatographic method was validated by examining reproducility (8 replicates) and recovery (5 replicates) of glucosamine and N-acetylglucosamine in commercial products. The quantification was reproduceable with mean variation coefficients of 4.7% for glucosamine and 10.2% for N-acetylglucosamine, the corresponding. Recovery values were 101.2% and 107.6%, respectively.

Discussion

The chromatographic method developed makes it possible to quantify N-acetylglucosamine, glucosamine as well as chitobiose and chitotriose. The low solubility of N-acetylglucosamine (100 mg/mL) and glucosamine (150 mg/mL) in water is greatly reduced in the presence of the acetonitrile/water (80:20, v/v) eluent. The concentration of the standards was therefore adjusted to obtain a maximum response between 0 and 80 mV from the detector. The use of high concentrations in case of glucosamine oligomers produced wider peaks that are attributable to an overloaded column. This effect was more pronounced for oligomers with long retention times. This overload was probably responsible for low correlation coefficients for the chitotetraose calibration curve.

Moreover, the number of theoretical plates (N) in the column is a critical factor influencing the resolution of chitotriose and chitotetraose peaks. Trials have shown that when there is a significant drop in the efficiency of the column (N < 5000) separation of these two components becomes impossible. Maximum efficiency occurred within the range of 7000 to 9000 theoretical plates.

The impact of pH was a key element in the resolution of peaks. For example, the glucosamine peak at a pH of 8.0 splited (Figure 2). This phenomenon is probably due to the presence of a mixture of α and β anomers (Schanzenbach and Peter, 1997) or a difference in the ionic form of the molecule. Adjusting the solution to a pH of 4.0 causes the amino function of the molecule to take on a cationic form, thus permitting quantification by means of the proposed method. The temperature of the column heater and the refractometer was set at 35°C to limit the influence of temperature fluctuations within the laboratory.

CONCLUSION

The HPLC analytical method described makes it possible to characterize the quality of chitin hydrolysis and deacetylation, and product purity in a single assay. The use of standards with lower concentrations should make it possible to quantify chitotetraose. Additional work should be conducted on the oligomers of N-acetylglucosamine in order to be able to fully monitor industrial processes.

REFERENCES

Novikov, V.Yu. and Ivanov, A.L. 1997. Synthesis of D(+)-glucosamine hydrochloride. *Russ. J. Appl. Chem.* 70: 1467-1470.

Palace, G.P. and Phoebe, C.H. Jr. 1997. Quantitative determination of amino acid levels in neutral and glucosamine-containing carbohydrate *Polymers. Anal. Biochem.* 244: 393-403.

Showa Denko, K.K. Shodex (Separations and HPLC) Group. Asahipak NH2P series. Instruction Manual.

Schanzenbach, D. and Peter, M.G. 1997 Chromatography of chito-oligosaccharides. In *Chitin Handbook*. R.A.A. Muzzarelli and M.G. Peter, (eds.) Atec, Grottammare, Italy, pp. 195-198.

CHAPTER 22

MODELLING THE EFFECT OF TEMPERATURE ON SHELF-LIFE AND ON THE INTERACTION BETWEEN THE SPOILAGE MICROFLORA AND *LISTERIA MONOCYTOGENES* IN COLD-SMOKED SALMON

Paw Dalgaard[1], Esther Murillo Iturrate[2] and Lasse Vigel Jørgensen[1]

[1] Danish Institute for Fisheries Research (DIFRES), Department of Seafood Research, Lyngby, Denmark
[2] Universidad de Zaragoza, Facultad de Veterinaria, Departamanto de Producción Animal y Ciencia de los Alimentos, Zaragoza, Spain

The effect of temperature (0-25°C) on shelf-life and microbial growth in sliced and vacuum packed cold-smoked salmon (SVP-CSS) was determined. Relative rate of spoilage (RRS) models were developed and successfully validated under dynamic temperature conditions as well as by comparison with shelf-life data from the literature. Simple models for growth of lactic acid bacteria (LAB), Enterobacteriaceae and Enterococcus were developed. Growth of LAB in SVP-CSS was 2.15 times slower than that predicted by a previously developed model for Lactobacillus curvatus. A new model to predict the effect of interaction between LAB and Listeria monocytogenes in SVP-CSS is presented. Interestingly, predicted growth of L. monocytogenes corresponded to the growth observed in naturally contaminated SVP-CSS and hence could be a valuable part of quantitative microbiological risk assessment.

INTRODUCTION

Cold-smoked salmon (CSS) is a lightly preserved and ready-to-eat seafood typically distributed as a sliced and vacuum-packed product (SVP-CSS). The global production of CSS is of approximately 70000 tons per year with Denmark, France and the UK being the main producing countries while Germany, Italy and France are the main importing countries (www.fao.org/fi/statist/FISOFT/FISHPLUS.asp). CSS is produced by salting, drying and cold-smoking at 18-27°C and this results in a product with 3-8 % water phase salt (WPS). These processes impart little bacteriocidal effect on spoilage microorganisms and they include no critical control point for *Listeria monocytogenes*. Consequently, growth of spoilage microorganisms and of *L. monocytogenes* is of major importance for shelf-life and safety of CSS (Huss *et al.*, 2000; Jørgensen, 2000).

Spoilage of SVP-CSS is primarily due to microbial activity (Truelstrup Hansen *et al.*, 1998; Jørgensen *et al.*, 2001; Stohr *et al.*, 2001) but individual measurements of microorganisms or microbial metabolites have not been useful as objective indices of spoilage (Hildebrandt and Erol, 1988; Truelstrup-Hansen *et al.*, 1995, 1998). Multiple compound quality indices (MCQI) that correlated reasonably closely with sensory acceptability or remaining shelf-life of SVP-CSS were recently suggested for products stored at 5°C (Jørgensen *et al.*, 2000a; Leroi *et al.*, 2001; Jørgensen *et al.*, 2001). The effect of storage temperature on shelf-life and on the validity of MCQI remains to be determined. This lack of information is somewhat surprising because SVP-CSS can be exposed to considerable temperatures abuses in the chill chain between production and consumption in various countries (James and Evans, 1990; O'Brien, 1997; Sergelidis *et al.*, 1997). It is also known that the effect of temperature can be more pronounced for shelf-life of lightly preserved seafood as compared to fresh seafoods (Dalgaard and Jørgensen, 2000).

The effect of temperature on growth of *L. monocytogenes* with respect to the safety of SVP-CSS is crucial. Prevalence of *L. monocytogenes* in CSS from different smokehouses can vary from close 0 % to near 100% (Jørgensen, 2000) and enormous efforts have been invested in reducing contamination during processing. Routes of contamination have been studied and both cleaning and the disinfection of processing equipment have been evaluated (Garland, 1995; Fonnesbech Vogel *et al.*, 2001). Progress has been made but low levels of *L. monocytogenes* can be present in SVP-CSS and it is important to evaluate growth as a function of product characteristics and storage conditions. Unfortunately, results from challenge tests with inoculated products, predictive models and storage trials with naturally contaminated products have been contradicting. In brief, challenge tests and predictive models have indicated *L. monocytogenes* typically grow to high levels above 10^6 cfu g^{-1} in SVP-CSS (Rørvik *et al.*, 1991; Peterson *et al.*, 1993; Dalgaard and Jøgensen, 1998;

Nilsson et al., 1999). In contrast, growth of L. monocytogenes to levels above 10^6 cfu g^{-1} have been extremely rare for naturally contaminated SVP-CSS (Cortesi et al., 1997; Jørgensen and Huss, 1998).

The objectives of this study were to model the effect of temperature on shelf-life of SVP-CSS and on the growth of L. monocytogenes in it. Firstly, the effect of temperature on shelf-life was modelled by empirical relative rate of spoilage (RRS) models. Secondly, the ability of available kinetic models to predict growth of the dominating spoilage microorganisms in SVP-CSS at 0-25°C was evaluated. Finally, the growth of L. monocytogenes was predicted at different temperatures by using the available models. Predicted growth rates were calibrated to include the inhibiting effect of smoke components in SVP-CSS. Furthermore, growth of L. monocytogenes was predicted by a model that take into account interactions with the dominating spoilage microflora i.e. the Jameson effect.

MATERIALS AND METHODS

Effect of temperature on shelf-life

Storage and sampling. Sliced and vacuum packed cold-smoked salmon (SVP-CSS) was obtained from a Danish processor. Packs, 200 g each were frozen and sent to DIFRES in Lyngby in the frozen form. At DIFRES packs were thawed at 5°C overnight. Two series of storage experiments were carried out: (i) storage of packs at constant temperatures of 0, 5, 10, 15, and 25 °C and (ii) storage under changing temperature conditions where one sub-batch was stored at 5°C during 8.9 days followed by transfer to 15°C and another sub-batch was kept at 15°C for 8.9 days and then transfer to 5°C. A third sub-batch was stored at a constant temperature of 10°C. Shelf-life of this sub-batch was used to calculate keeping quality under changing temperature conditions (Table 1). All packs were stored in cooling incubators and samples for sensory analyses were taken from the different sub-batches at regular intervals during storage. The temperature of each sub-batch of packs was recorded by using temperature data loggers (Tinytag-plus, Gemini Data Loggers, UK). In addition to these experiments shelf-life and microbial growth data from four different sub-batches of packs previously studies at a constant temperature of 5°C was analysed (Jørgensen et al., 2000a).

Sensory and chemical analyses. Packs of CSS were opened and left at room temperature for 15 min. The flavor of slices of the CSS was then evaluated by five to seven trained panellists. A simple three class scale with class III corresponding to rejection was used by each panellist as previously described (Jørgensen et al., 2000a). The panellists described the changes in appearance,

Table 1. Observed and predicted shelf-life of sliced vacuum packed cold-smoked salmon stored under changing temperature conditions.

Sub-batch	Storage temp. (°C) Phase I (8.9 days)	Storage temp. (°C) Phase II	Observed shelf life (d)	Predicted shelf life (days)[a] Eqn. 1b a = 0.0892	Predicted shelf life (days)[a] Eqn. 2b E_a = 60.5 kJ mol^{-1}	Predicted shelf life (days)[a] Eqn. 3b T_{min} = -10.2°C
A	5°C	15°C	20	20.0	20.1	20.5
B	15°C	5°C	25	24.1	24.3	26.0

a) Shelf life was predicted on the basis of (i) a temperature profile recorded during the entire storage period and (ii) a shelf-life of 23 days determined for the product stored at 10°C.

texture and off-flavors. The end of shelf-life of a sub-batch of CSS was determined as the time when 50% of the panellists judged the samples to be in class III. At the beginning of each series of storage experiments pH, water activity and the corresponding water phase salt content was determined for three different packs (Jørgensen *et al.*, 2000a).

Development and validation of relative rate of spoilage models. The rate of spoilage (RS, days^{-1}) for each sub-batch of SVP-CSS was calculated as the reciprocal of the shelf-life determined by sensory evaluation. Log-transformed RS-data were fitted to the exponential spoilage model (Eq. 1a) and to the the Arrhenius model (Eq. 2a) whereas square root transformed RS data were fitted to the square root spoilage model (Eq. 3a). Goodness-of-fit was compared for the three models by (i) estimating parameter values in eq. 1a, 2a and 3a from appropriately transformed data and (ii) calculating the sums of squared residual errors from non-transformed observed and predicted RS values.

$$Ln(RS) = k_1 + a \times T \qquad (1a)$$

$$Ln(RS) = k_2 - \frac{E_a}{R \times K} \qquad (2a)$$

$$\sqrt{RS} = k_3 \times (T - T_{min}) \qquad (3a)$$

In Eqs. 1, 2 and 3 T is temperature in °C, K is temperature in Kelvin, R is the gas constant (8.31 J mol^{-1} K^{-1}), k_1, k_2 and k_3 are constants and a, E_a and T_{min} are the temperature characteristics in the three models, respectively. After estimation of temperature characteristics using Eq. 1a, 2a and 3a relative rates of spoilage *i.e.* shelf-life at T (°C) divided by shelf-life at a reference temperature (T_{ref}) were predicted using Eq. 1b, 2b and 3b.

$$RSS = \frac{Shelf-life\ at\ T}{Shelf-life\ at\ T_{ref}} = Exp\left[a \times (T - T_{ref})\right] \qquad (1b)$$

$$RSS = \frac{Shelf-life\ at\ T}{Shelf-life\ at\ T_{ref}} = Exp\left[\frac{E_a}{R} \times (\frac{1}{(T+273)} - \frac{1}{T_{ref}+273})\right]$$
(2b)

$$RSS = \frac{Shelf-life\ at\ T}{Shelf-life\ at\ T_{ref}} = \left(\frac{T - T_{min}}{T_{ref} - T_{min}}\right)^2 \qquad (3b)$$

The three RRS-models developed from storage of products at constant temperatures were validated in two different ways. Firstly, observed and predicted shelf-life for the two sub-batches of SVP-CSS stored under dynamic temperature conditions were compared. Shelf-life under dynamic temperature storage conditions was predicted using a spreadsheet. For every 3 min time interval at a given temperature the corresponding time at 10°C was subtracted from the remaining shelf-life (RSL) at 10°C and this was continued until RSL reached zero. Secondly, RRS-values predicted by eqs. 1b, 2b and 3b were compared to RSS-values calculated from studies reported in the literature and where shelf-life of SVP-CSS have been determined at more than one storage temperature (RSS-observed). Observed and predicted RRS-values were compared and evaluated by calculation of bias- and accuracy factor values (eq. 4 and 5). These indices were previously suggested by Ross (1996) to evaluate the overall performance of microbial growth models.

$$\text{bias factor } (RRS) = 10^{(\Sigma \log(RRS-predicted / RRS-observed) / n)} \tag{4}$$

$$\text{accuracy factor } (RRS) = 10^{(\Sigma |\log(RRS-predicted / RRS-observed)| / n)} \tag{5}$$

Effect of temperature on growth of spoilage microorganisms

Microbiological changes. Data for growth of microorganisms at constant temperatures were obtained during the storage experiments described above. During storage of SVP-CSS at constant temperatures of 0, 5, 10, 15, and 25°C packs were removed at regular intervals and analysed for aerobic plate counts (APC), lactic acid bacteria (LAB), Enterobacteriaceae and *Enterococci*. APC was determined on spread plates of Long and Hammer agar (LH, 15°C, 7d), LAB was enumerated in pour plates of nitrite actidione polymyxin agar with pH 6.2 (NAP, 25°C, 3d). Enterobacteriaceae was enumerated in TSA/VRBG i.e. pour plates of 5 ml of tryptone soya agar (Oxoid, CM131), which after 2 h at 20-25°C (room temperature) were overlaid by 12-15 ml of violet red bile glucose agar (Oxoid, CM485). TSA/VRBG plates were incubated 48 h at 30°C. Enterococci was enumerated on Slanetz & Bartley agar (44°C, 48 h) (Oxiod CM377). Data for microbiological changes in the four batches of SVP-CSS stored at 5°C were obtained from Jørgensen *et al.* (2000a).

Modelling growth responses of potential spoilage microorganisms. The log-transformed Logistic model with three parameters was used as a primary growth model to estimate maximum specific growth rates (μ_{max}, h^{-1}) from growth curves of LAB, Enterobacteriaceae and *Enterococcus*. The simple square root growth model (Eq. 6, Ratkowsky *et al.*, 1982) was then used as a secondary growth model to describe the effect of temperature on the μ_{max}-values for the three groups of microorganisms.

$$\sqrt{\mu_{max}} = b \times (T - T_{min}) \qquad (6)$$

In eq. 6, *b* is a constant and T_{min} is the theoretical minimum growth temperature. Finally, μ_{max}-values determined for LAB in SVP-CSS were compared to values predicted by a mathematical model including the effect of temperature, pH and water activity on μ_{max} of *Lactobacillus curvatus* (Wijtzes *et al.*, 2001). Deviation between observed and predicted μ_{max}-values was quantified by calculation of bias- and accuracy factor values (Eqs. 4 and 5, Ross, 1996).

Effect of temperature on growth of *Listeria monocytogenes*

Modelling interaction between L. monocytogenes and LAB. The classical Logistic model (Eq. 7) describes how the volumetric growth rate of a culture (*dN/dt*) is dampened when the density of that culture approaches its maximum population density (N$_{max}$). In the present study growth of LAB and *L. monocytogenes* was predicted by expanded versions of the Logistic model (eq. 8 and 9). These models take into account the interaction between LAB and *L. monocytogenes*. As an example, Eq. 9 describes how high levels of LAB have the same inhibiting effect (1 – LAB(t)/ LAB$_{max}$) on μ_{max} of *L. monocytogenes* as the LAB cultures have on itself. The phenomenon that growth of all microorganisms on a food product ceases when the dominating microbial population achieves their MPD have been named the 'Jameson effect' (Ross *et al.*, 2000). Eq. 8 and 9 represent simple mathematical models that allow microbial interactions as expressed by the Jameson effect hypothesis to be taken into account when simultaneous growth of LAB and *L. monocytogenes* is predicted.

$$\frac{dN}{dt} = N(t) \times \mu_{max} \times \left(1 - \frac{N(t)}{N_{max}}\right) \qquad (7)$$

$$\frac{dLAB}{dt} = LAB(t) \times \mu_{max}^{LAB} \times \left(1 - \frac{LAB(t)}{LAB_{max}}\right) \times \left(1 - \frac{Lm(t)}{Lm_{max}}\right) \qquad (8)$$

$$\frac{dLm}{dt} = Lm(t) \times \mu_{max}^{Lm} \times \left(1 - \frac{Lm(t)}{Lm_{max}}\right) \times \left(1 - \frac{LAB(t)}{LAB_{max}}\right) \quad (9)$$

In Eq. 7, 8 and 9 *dN/dt*, *dLAB/dt* and *dLm/dt* represent volumetric growth rates, *N(t)*, *LAB(t)* and *Lm(t)* population densities (cfu g^{-1}) at the time t; N_{max}, LAB_{max} and Lm_{max} maximum population density (cfu g^1) and μ_{max}, μ_{max}^{LAB} and μ_{max}^{Lm} maximum specific growth rate (h^{-1}). *LAB* and *Lm* signifies lactic acid bacteria and *L. monocytogenes*, respectively.

Predicting growth of L. monocytogenes and LAB in SVP-CSS. In the present study μ_{max}^{Lm} was predicted at different temperatures by a mathematical model previously developed for *L. monocytogenes* Scott A growing in liquid media (Ross 1993). This model takes into account the effect of temperature (3 – 37°C), pH (5.6 – 6.2) and WPS (0.5 – 13%). In challenge tests with cured seafood, including cold-smoked salmon, the model was previously found to predict growth of *L. monocytogenes* more accurately than other available models (Dalgaard and Jørgensen, 1998). However, a μ_{max} bias-factor of 1.4 indicated predicted μ_{max}-values on average were higher than values observed in challenge studies. The faster predicted growth most likely reflect that this model does not include the inhibiting effect of smoke components in SVP-CSS. To correct for this incompleteness of the model μ_{max} value estimated by the Ross-model was divided by 1.4 before they were used in Eq. 9 to predict growth of *L. monocytogenes* in SVP-CSS. Growth was predicted both with and without a lag phase and in all cases a MPD of 1 x 10^8 cfu g^{-1} was assumed. Lag time values were obtained from the model included in the Pathogen Modeling Programme (Buchanan and Phillips, 1990).

Growth of LAB was predicted by Eq. 8 together with μ_{max} values obtained for *Lactobacillus curvatus* (Wijtzes et al., 2001). This model does not include the effect of lactate and smoke components and it has not previously been evaluated in validation studies with cold-smoked salmon. Therefore, the predicted growth rates were corrected by the bias-factor determined in the present study. Growth of LAB was predicted with different initial population density and a constant MPD of 1 x 10^8 cfu g^{-1}. A spreadsheet was used to predict growth of *L. monocytogenes* and LAB by Eq. 8 and 9. Predictions were made at pH 6.10 and for a water activity of 0.970 corresponding to 5.0 % WPS i.e. under conditions typical for SVP-CSS.

Comparison of observed and predicted growth

Predicted growth of *L. monocytogenes* was compared to the growth previously observed in naturally contaminated SVP-CSS at 5°C (Jørgensen and Huss, 1998; Dalgaard and Jørgensen, 1998). Growth of *L. monocytogenes* was predicted as described above (i) by including a lag phase and by taking into account interaction with LAB, (ii) without lag phase but still taking into account interaction with LAB, (iii) without lag phase and without interactions and finally (iv) by the commercially available Food MicroModel that include a lag phase but does not take into account interaction with other microorganisms (Anon, 1997).

RESULTS AND DISCUSSION

Effect of temperature on shelf-life

The shelf-life of SVP-CSS at 0, 5, 10, 15, and 25 °C was 52, 38±11 (n = 5), 23, 14-15, and 6 days, respectively. The pH, water activity and percentage WPS content were 6.12, 0.974 and 4.40%, respectively, for the batch of samples stored at different temperatures. The corresponding values for the additional four batches stored at 5°C were 6.01-6.14, 0.968-0.976 and 4.1-5.4%, respectively. Rates of spoilage (RS) calculated from these shelf-life data are shown in Figure 1. The temperature characteristics obtained by fitting of appropriately transformed RS data to Eq. 1a, 2a and 3a were 0.0892 °C^{-1} for a in eq. 1a; 60.5 kJ mol^{-1} for E_a in eq. 2a, and −10.2 °C for T_{min} in eqs. 3a. RS data fitted eq. 1a and 2a more closely than eq. 3a (Figure 1). The sums of squared residual errors were 2.0×10^{-5}, 2.4×10^{-5} and 7.4×10^{-5} for eqs. 1a, 2a and 3a, respectively. Although, RS-data did not fit the three models equally well all three models were included in validation studies.

Shelf-life under dynamic temperature storage conditions was predicted by using the three RRS-models developed above (Eq. 1b, 2b and 3b) together with a shelf-life of 23 (21-25) days determined at a constant storage temperature of 10°C. The three models all predicted shelf-life values close to those determined by sensory evaluation of products stored under variable temperature conditions (Table 1). Taken into account the precision of sensory shelf-life determination none of the three models could be singled out as being superior.

Bias-factor values between 0.75 and 1.25 has been suggested as a criterion of successful validation of models to predict the shelf-life of seafoods (Dalgaard 2000). Comparison of predicted RRS-values and RRS-values calculated from shelf-life data determined in previously published storage trials resulted in bias-factor

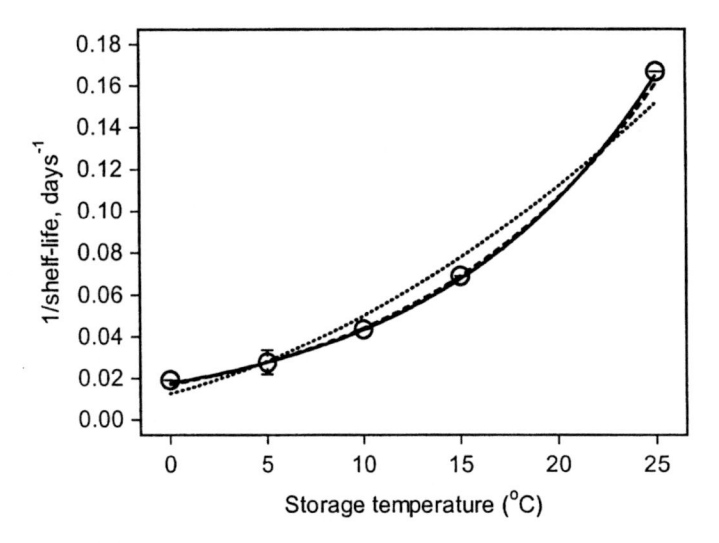

Figure 1. Rate of spoilage (RS) for sliced and vacuum packed cold-smoked salmon at different constant storage temperatures (open circles). Lines indicate how RS data fitted the exponential model (Eq. 1a, solid line), the Arrhenius model (Eq. 2a, dasked line), and the square root model (Eq. 3a, dotted line). Error bars indicate standard deviation for the average RS value determined for five batches of SVP-CSS.

values between 0.87 and 1.03 (Table 2). Thus, all three models predicted the effect of temperature on RRS of SVP-CSS in a way that allowed them to be used for shelf-life prediction.

The square root spoilage model (Eq. 3b) and a T_{min} value of –10.0°C have been used extensively to predict RRS values and shelf-life of fresh seafoods from temperate waters (Olley and Ratkowsky, 1973; Ratkowsky et al., 1982). The T_{min} value of –10.2°C determined in the present study for SVP-CSS indicated the effect of temperature to be practically the same for SVP-CSS and fresh seafoods from temperate waters. The simple square root model with a T_{min} value of –10.0°C i.e. RRS = $(1 + 0.1 * T°C)^2$ has been included in the seafood spoilage predictor (SSP). Thus, this software, which is available free of charge (www.dfu.min.dk/micro/ssp/)

can be used to predict shelf-life of SVP-CSS under constant and fluctuating temperatures.

In addition to the effect of temperature, shelf-life of SVP-CSS depends on the salt content of the water phase, concentration of smoke components, initial microbial contamination, permeability of the packaging film and possibly other factors (Jakobsen *et al.*, 1988; Truelstrup Hansen *et al.*, 1998; Leroi and Joffraud, 2000). Despite this complex situation the RSS models developed here can be used to predict the effect of temperature during distribution on shelf-life of products with relatively constant composition e.g. products from a particular smokehouse.

Eqs. 1b, 2b and 3b also allowed prediction of the effect of temperature on shelf-life of hot-smoked and aerobically stored eel, halibut and herring at 2-20°C. For these products bias- and accuracy-factor values calculated from 13 sets of shelf-life data were 0.91-0.95 and 1.15-1.18, respectively (Results not shown, Karnop 1980, Ristiniemi and Korkeala, 1998). Temperature, however, seems to influence shelf-life of hot-smoked and packed seafood to a much lesser extent (Cann *et al.*, 1984; Tinker *et al.*, 1985). For these products the bias-factor values calculated from eqs. 1b, 2b and 3b were 1.41, 1.43 and 1.75, respectively. Clearly, these RRS model cannot be used to predict the shelf-life of hot-smoked and packed seafoods and further studies are needed for the effect of temperature on shelf-life of hot-smoked and packed seafoods.

Effect of temperature on the growth of spoilage microorganisms

Microbiological changes. LAB dominated spoilage associations of SVP-CSS between 0 and 25°C. Irrespective of storage temperature LAB grew exponentially and without any lag phase until they reached their MPD of close to 10^8 cfu/g (Figure 2). At 0°C, however, MPD remained at about 10^6 cfu/g.

Previously, spoilage and shelf life of SVP-CSS at 2-10°C have been extensively evaluated. At these temperatures the spoilage microflora were dominated by (i) LAB in levels of 10^{7-8} cfu g^{-1} (ii) LAB together with psychrotolerant Enterobacteriaeceae or (iii) *Photobacterium phosphoreum* alone or in combination with LAB. In one study *Lactobacillus curvatus* and *P. phosphoreum* were identified as specific spoilage organisms but *Brochothrix thermospacta, Shewanella putrefaciens* and yeast have occasionally been found in high levels in this product (Jakobsen *et al.*, 1988; Leroi *et al.*, 1998; Paludan-Müller *et al.*, 1998; Truelstrup Hansen and Huss, 1998; Jørgensen *et al.*, 2000b). In the present study exponential growth of Enterobacteriaeceae at 5°C stopped at the time when LAB reached their MPD (Figure 2a). This growth pattern of Enterobacteriaeceae has been observed in several other studies with SVP-CSS at 5°C (Truelstrup Hansen *et al.*, 1995 1996, 1998; Paludan-Müller *et al.*, 1998; Leroi *et al.*, 2001). It is an example of the 'Jameson effect' and

Table 2. Observed and predicted relative rates of spoilage for sliced and vacuum-packed cold-smoked salmon.

Packaging	WPS	Shelf-life (d) and storage temp. (°C)	Shelf-life (d) and storage temp. (°C)	Observed relative rate of spoilage[a]	Predicted relative rate of spoilage Eq. 1b $b = 0.0892$	Predicted relative rate of spoilage Eq. 2b $E_a = 60.5$ kJ mol^{-1}	Predicted relative rate of spoilage Eq. 3b $T_{min} = -10.2$°C	Ref.
Vacuum	4.1	25, 0	21, 5	1.2	1.6	1.6	2.2	b
Vacuum	4.1	21, 5	14, 10	1.5	1.6	1.6	1.8	b
Vacuum	5.0	20 d, 6°C	6.5, 12	3.0	1.7	1.7	1.9	c
VP, Riloten	4.3	27 d, 5°C	13, 10	2.1	1.6	1.6	1.8	d
VP, Amilon	4.3	21.5 d, 5°C	18, 10	1.2	1.6	1.6	1.8	d
Vacuum	4.6	31.5 d, 5°C	10.5, 10	2.5	1.6	1.6	1.8	e
Vacuum	4.6	52.5 d, 5°C	38.5, 10	1.3	1.6	1.6	1.8	e
Vacuum	6.6	60 d, 2°C	20, 10	3.0	2.0	2.1	2.7	f
Bias factor					0.87	0.88	1.03	
Accuracy factor					1.40	1.39	1.40	

[a] Relative rates of spoilage was calculated as shelf-life at the lowest storage temperature divided by shelf-life at the highest storage temperature.
References are: b, Cann et al. 1984; c, Schneider and Hilderbrandt, 1984; d, Jakobsen et al., 1988; e, Trelstrays Hansen, 1995; and f, Cortesi et al. 1997.

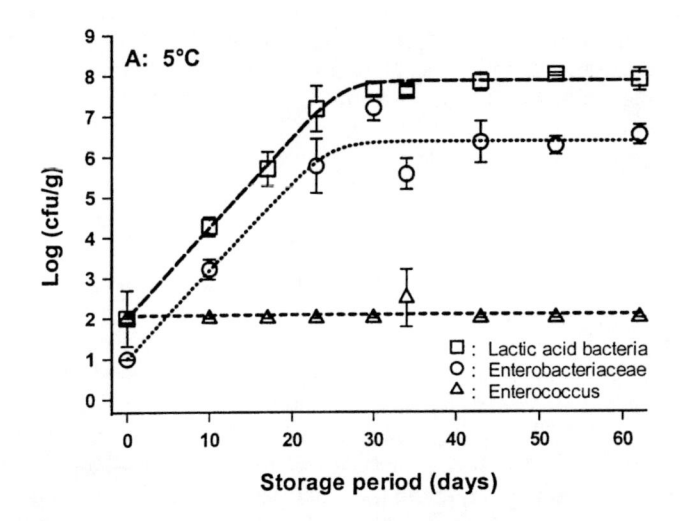

Figure 2. Growth of lactic acid bacteria, Enterobacteriaceae and *Enterococcus* in sliced vacuum packed cold-smoked salmon at 5°c (A) and at 25°C (B).

Enterobacteriaeceae probably does not influence spoilage SVP-CSS when inhibited by LAB.

At low storage temperatures the microbiological data in the present study concurred with numerous previous experiments. In addition, LAB can remain the dominant part of the spoilage microflora of SVP-CSS at 10, 15 and 25°C. Furthermore, Enterobacteriaeceae reached high levels of above 10^7 cfu g^{-1} in SVP-CSS when stored at 10, 15 and 25°C and at these temperatures they most likely contribute to product spoilage (Figure 2b). Seventeen strains of Enterobacteriaeceae were isolated from the spoilage microflora of SVP-CSS stored at 25°C. These isolates were identified as *Serratia liquefaciens* (53%), *Hafnia alvei* (41%) and *Citrobacter freundii* (6%) by simple biochemical tests and by using the API20E identification system (BioMérieux S.A., Marcy l'Etoile, France). Psychrotolerant Enterobacteriaceae previously isolated form the spoilage microflora of chilled SVP-CSS included *Enterobacter agglomerans, Hafnia alvei, Serratia liquefaciens, Serratia plymuthia* and *Rahnella aquatilis* (Truelstrup Hansen and Huss, 1998; Paludan-Müller *et al.*, 1998; Jørgensen *et al.*, 2000b). Compared to microbial changes in other seafoods it is somewhat surprising that *Serratia liquefaciens* and *Hafnia alvei* can be part of the spoilage microflora of SVP-CSS at both 5 and 25°C (Dalgaard and Jørgensen, 2000). Himelbloom *et al.* (1996) found *Citrobacter freundii*, *Micrococcus* sp. and *Staphylococcus* spp. as dominating parts of the spoilage microflora of Alaska native smoked salmon at 30°C. This different spoilage microflora could be due to aerobic storage conditions and a lower water activity of 0.95 for the product studied by Himelbloom *et al.* (1996).

As will be discussed below for *Listeria monocytogenes* the 'Jameson effect' hypothesis can be most valuable for evaluation of microbial developments in SVP-CSS. Therefore, it seems relevant to note here that *Enterococcus* at 25°C was growing exponentially in SVP-CSS although LAB had reached their MPD (Figure 2b). Clearly, the 'Jameson effect' hypothesis does not apply to the interaction between LAB and *Enterococcus* in SVP-CSS at 25°C.

Modelling growth responses of potential spoilage microorganisms. The Logistic model (Eq. 7) and the simple square root model (Eq. 6) seemed appropriate for growth of LAB, Enterobacteriaeceae and *Enterococcus* in SVP-CSS between 0 and 25°C (Fig. 2, Table 3). The models developed here (Table 3) can be used to predict the growth of the three groups of microorganism in SVP-CSS but only when their composition is similar to the products evaluated in the present study.

The predicted μ_{max} values were higher than those observed for LAB in SVP-CSS (Table 3). The Wijtzes-model did not include the effect of smoke components and lactate and this may at least partly explain the deviation between observed and predicted values. None of the models allowed prediction of the growth of Enterobacteriaeceae and *Enterococcus* in SVP-CSS. Growth of *Enterococcus* was only observed at 15 and 25°C and remained at below 10^6 cfu/g within the product shelf-life; thus it is unlikely to be important in the spoilage of SVP-CSS.

In some situations, models for growth of spoilage microorganisms can be used to predict the shelf-life of foods as the time they require for these organisms to grow from an initial level to a minimal spoilage level (MSL) e.g. 10^7 cfu g^{-1} (Dalgaard *et al.*, 2002). The models developed here for LAB and Enterobacteriaeceae cannot be used in this simple way to predict shelf-life of SVP-CSS. In fact, the dominating microflora of SVP-CSS may reach their MPD long before products are considered sensorially spoiled whereas in other cases the product spoiled at the time the dominating flora reached a level of 10^{6-7} cfu g^{-1} (Truelstrup Hansen *et al.*, 1995, 1996, 1998; Leroi *et al.*, 1998, 2001). Clearly, further research is needed before models for growth of spoilage microorganisms can be used to predict the shelf-life of SVP-CSS.

Table 3. Maximum specific growth rates (h^{-1}) of lactic acid bacteria, Enterobacteriaceae and *Enterococcus* in sliced vacuum packed cold-smoked salmon.

Temp. (°C)	Maximum specific growth rates (h^{-1})			
	Lactic acid bacteria		Enterobacteriaceae	Enterococcus
	Observed	Predicted[a]	Observed	Observed
0	0.009	0.008	NGD[c]	NGD
5	0.023±0.004[b]	0.051±0.003	0.021	NGD
10	0.037	0.114	0.041	NGD
15	0.123	0.213	0.093	0.154
25	0.275	0.503	0.269	0.233
b (h^{-1} °C^{-2})	0.018		0.020	0.014
T_{min} (°C)	-3.64		-0.7	+7.7
Bias factor		2.15	-	-
Accuracy factor		2.15	-	-

a) Values for *Lactobacillus curvatus* were predicted by the model of Wijtzes et al. (2001)
b) Average and standard deviation of data is indicated for five sub-batches of sliced and vacuum packed cold-smoked salmon studies at 5°C
c) NGD: No growth detected.

Effect of temperature on the growth of *Listeria monocytogenes*

In naturally contaminated SVP-CSS initial levels of LAB are typically between 10 and 10^4 cfu g^{-1}, whereas *L. monocytogenes* most often is absent or present at usually < 10 cfu g^{-1} (Jørgensen and Huss, 1998). As shown in Figure 3a, the higher initial level of LAB together with its relatively higher μ_{max}-value at 5°C resulted in predicted growth of *L. monocytogenes* being limited by the Jameson effect. At 5°C the level of *L. monocytogenes* was predicted to increase by a maximum of 3 or 4 log units depending on the presence or absence of a lag phase (Fig. 3a). This prediction is interesting as it explains the apparent contradiction that *L. monocytogenes* have been growing to high levels in challenge tests whereas this was never observed for naturally contaminated products. The model explains these growth responses by different ratios between *L. monocytogenes* and LAB in two situations (Table 4). At 10 and 15°C the predicted μ_{max} values were similar for *L. monocytogenes* and LAB (Table 4). As a result, the predicted maximum levels reached by *L. monocytogenes* increased by approximately one log-unit when the storage temperature was raised from 5 to 10°C and by *ca* two log-units with storage at 15°C as compared to 5°C (Fig.

Table 4. Predicted growth of *Listeria monocytogenes* and lactic acid bacteria (LAB) in sliced vacuum packed cold-smoked salmon[a].

Temp. (°C)	*Listeria monocytogenes* Population density, log cfu g^{-1} Highest	Initial	μ_{max} (h^{-1})	LAB Initial population density, log cfu g^{-1}	μ_{max} (h^{-1})
5	3.7	0	0.0142	2	0.0229
5	2.5	0	0.0142	5	0.0229
5	6.7	3	0.0142	2	0.0229
10	5.7	0	0.0538	2	0.0571
15	6.7	0	0.1187	2	0.1066
15	3.3	0	0.1187	5	0.1066

a) Growth of the two microorganisms were predicted at a pH value of 6.10 and for a water activity of 0.97 corresponding to 5.0 % water phase salt content i.e. under conditions typical for cold-smoked salmon. Eq. 8 and 9 were used as primary growth models and no lag phases were included. Maximum specific growth rate (μ_{max}) were obtained from the model of Ross (1993) for *Listeria monocytogenes* and from the model of Wijtzes et al. (2001) for lactic acid bacteria. μ_{max}-values predicted by the two models were divided by the respective bias-factor values to correct for the effect smoke components and other growth inhibiting parameters in cold-smoked salmon not yet included in these models.

3). These predictions indicate that *L. monocytogenes* can grow to high levels and thereby become an important safety risk in SVP-CSS if the product is temperature abused.

Validation studies showed that the growth of *L. monocytogenes* predicted by Eq. 9 at 5°C corresponds more closely with data from naturally contaminated SVP-CSS than growth predicted without taking into account the Jameson effect (Table 5). We were unable to evaluate the performance of the new model at higher storage temperatures as data for naturally contaminated SVP-CSS is lacking. Data to support or reject the hypothesis that the growth of *L. monocytogenes* exhibits a lag-phase in naturally contaminated SVP-CSS is also lacking. Therefore, the growth of *L. monocytogenes* has been predicted both with and without a lag-phase (Fig. 3, Table 5).

The use of Eq. 9 to model interactions between microbial populations and the correction of predicted growth rates by bias-factor values are both rather crude methods. Future studies could aim at a more detailed description of the actual growth kinetics in SVP-CSS e.g. by including the effect of smoke components and lactate in models to predict lag time and growth rates. Refinement of Eq. 9 can be obtained in various ways depending on the

Table 5. Observed and predicted growth of *L. monocytogenes* in naturally contaminated cold-smoked salmon stored at 5°

Lot	Storage time (days)	Observed[a], log cfu g^{-1} Average	Observed[a], log cfu g^{-1} Highest number in single packs	Predicted[b], log cfu g^{-1} With lag phase & Jameson effect	Predicted[b], log cfu g^{-1} No lag phase & with Jameson effect	Predicted[b], log cfu g^{-1} No lag phase & no Jameson effect	Predicted[c], log cfu g^{-1} Food MicroModel
A	8	0.8	2 - 3[d]	0.5	1.4	1.4	2.2
B	14	0.8	2.3[d]	2.2	3.0	3.0	4.5
C	14	0.8	2 - 3	2.0	2.8	2.8	4.3
D	16	1.1	2 - 3	0.9	1.7	2.4	1.7
E	16	0.5	1 - 2	0.5	1.2	2.7	3.7
F	20	2.1	2.4	1.1	2.3	3.9	4.7
G	21	0.4	1 - 2	0.8	1.6	3.6	5.7
H	21	0.3	1 - 2	0.6	1.5	3.5	3.8
I	21	0.9	3.2	0.7	1.6	4.1	5.0
J	21	0.0	<1	0.7	1.6	1.6	0.0
K	23	0.0	<1	1.1	2.9	4.3	6.2
L	43	0.0	<1	3.1	3.9	8.0	7.7
M	49	1.1	2.7	2.2	3.1	7.4	7.7

[a] See Dalgaard and Jørgensen (1998) for characteristics of the different lots of SVP-CSS,
[b] Growth predicted as described in the present study. See details in the text.
[c] Growth predicted by Food MicroModel (Anon, 1997), [d] Numbers indicated as a range e.g. 2-3 were determined by a semiquantitative method and numbers indicated with one decimal point were determined by a quantitative method (See Jørgensen and Huss, 1998).

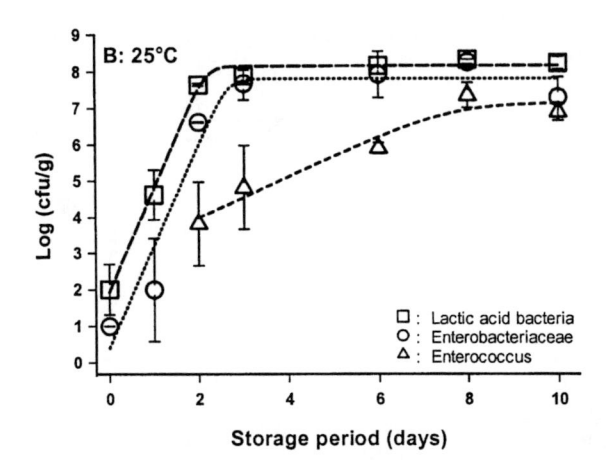

Figure 3. Predicted growth of *Listeria monocytogenes* (Lm) and lactic acid bacteria (LAB) at 5°C (A), 10°C (B) and 15°C (C). LAB (solid lines), Lm growing without a lag phase (dashed lines), Lm growing together with LAB and without a lag phase (dashed-dotted lines) and Lm growing together with LAB and with a lag phase (dotted lines).

mechanisms responsible for microbial interactions. Numerous studies have documented the ability LAB to inhibit growth of *L. monocytogenes* in seafoods (Huss *et al.*, 1995; Nilsson *et al.*, 1999; Duffes, 2000). In some cases bacteriocins can be involved in the interactions but the effects of other metabolic products and substrate competition remain poorly understood.

Only a few cases of listerioses have been related to cold-smoked salmon and trout (See Jørgensen, 2000). This is in agreement with the levels of *L. monocytogenes* detected in naturally contaminated SVP-CSS and predicted by the model suggested here but it is in sharp contrast to the high levels found in challenge tests and predicted by growth models that do not take into account interactions between *L. monocytogenes* and the remaining microflora in CSS. Although strictly empiricals the model suggested here can be valuable in quantitative microbiological risk assessments where exposure to *L. monocytogenes* from SVP-CSS is evaluated. In addition, the approach applied in this study, i.e. calibration of predicted growth rates by bias-factor values determined in validation studies with specific foods and application of Eq. 9 to take into account the effect of microbial interaction, will be useful in relation to several other foods.

ACKNOWLEDGEMENTS

This study was financially supported by the European Commission through the project 'Spoilage and Safety of Cold-Smoked Fish (FAIR CT95 1207). Support from the EU ERASMUS program allowed Esther Murillo Itturate to carry out a research visit at DIFRES.

REFERENCES

Anonymous 1997. *Food MicroModel - User Manual v. 2.5*. Food Micromodel Ltd., Surrey, UK.

Buchanan, R.L. and Phillips, J.G. 1990. Response surface model for predicting the effect of temperature, pH, sodium chloride content, sodium nitrite concentration and atmosphere on the growth of *Listeria monocytogenes*. *J. Food Protec.* 53: 370-376.

Cann, D.C.; Smith, G.L. and Houston, N.C. 1983. *Further Studies on Marine Fish Storage Under Modified Atmosphere Packaging*. Torry Research Station, MAFF, Aberdeen, UK.

Cann, D.C.; Houston, N.C.; Taylor, L.Y.; Smith, G.L.; Thomson, A.B. and Craig, A. 1984. *Studies of salmonids packed and stored under a modified atmosphere*. Torry Research Station, Ministry of Agriculture, Fisheries and Food, Aberdeen, Scotland.

Cortesi, M.L.; Sarli, T.; Santoro, A.; Murru, N. and Pepe, T. 1997. Distribution and behaviour of *Listeria monocytogenes* in three lots of naturally-contaminated vacuum-packed smoked salmon stored at 2 and 10°C. *Int. J. Food Microbiol.* 37: 209-214.

Dalgaard, P. and Jorgensen, L.V. 1998. Predicted and observed growth of *Listeria monocytogenes* in seafood challenge tests and in naturally contaminated cold smoked salmon. *Int. J. Food Microbiol.* 40: 105-115.

Dalgaard, P. 2000. Fresh and lightly preserved seafood. In *Shelf-Life Evaluation of Foods*. C.M.D. Man and A.A. Jones (eds.) Aspen Publishers, Inc. London, UK, pp. 110-139.

Dalgaard, P. and Jørgensen, L.V. 2000. Cooked and brined shrimps packed in a modified atmosphere have a shelf-life of >7 months at 0°C, but spoil at 4-6 days at 25°C. *Int. J. Food Sci. Technol.* 35: 431-442.

Dalgaard, P.; Buch, P. and Silberg, S. 2002. Seafood spoilage predictor - development and distribution of a product specific application software. *Int. J. Food Microbiol.* 73: 343-350.

Duffes, F. 1999. Improving the control of *Listeria monocytogenes* in cold smoked salmon. *Trends Food Sci. Technol.* 10: 211-216.

Fonnesbech Voge l,B.; Huss, H.H.; Ojeniyi, B.; Ahrens, P. and Gram, L. 2001. Elucidation of *Listeria monocytogenes* contamination routes in cold-smoked salmon processing plants detected by DNA-based typing methods. *Appl. Env. Microbiol.* 67: 2586-2695.

Garland, C.D. 1995. Microbiological quality of aquaculture products with special reference to *Listeria monocytogenes* in Atlantic salmon. *Food Australia,* 47: 559-563.

Hildebrandt, V.G. and Ero, I. 1988. Sensorische und mikrobiologosche untersuchung an vakuumverpacktem räucherlacks in scheiben. *Archiv für Lebensmittelhygiene* 39: 109-132.

Himelbloom, B.H.; Carpo, C.A. and Pfutzenreuter, R.C. 1996. Microbial quality of an Alask native smoked salmon process. *J. Food Protec.* 59: 56-58.

Huss, H.H.; Jeppesen, V.F.; Johansen, C. and Gram, L. 1995. Biopreservation of fish products - a review of recent approaches and results. *J. Aquatic Food Prod. Technol.* 4: 5-26.

Huss, H.H.; Jørgensen, L.V. and Fonnesbech Vogel, B. 2000. Control options for *Listeria monocytogenes* in seafoods. *Int. J. Food Microbiol.* 62: 267-274.

Jakobsen, M.; Benjaminsen, J.; Huss, H.H.; Jørgensen, B.R.; From, V. and Jakobsen, K. 1988. *Måling og styring af levnedsmidlers kvalitet.* (In Danish) Alfred Jørgensen Gæringfysiologisk Laboratorium A/S, Copenhagen, Denmark.

James, S. and Evans, J. 1990. Performance of domestic refrigerators and retail display cabinets for chilled products. *Refrig. Sci. Technol. Proc.* 4: 401-409.

Jørgensen, L.V. and Huss, H.H. 1998. Prevalence and growth of *Listeria monocytogenes* in naturally contaminated seafood. *Int. J. Food Microbiol.* 42: 127-131.

Jørgensen, L.V.; Dalgaard, P. and Huss, H.H. 2000a. Multible compound quality index for cold-smoked salmon (*Salmo salar*) developed by multivariate regression of biogenic amines and pH. *J. Agric. Food Chem.* 48: 2448-2453.

Jørgensen, L.V.; Huss, H.H. and Dalgaard, P. 2000b. The effect of biogenic amine production by single bacterial cultures and metabiosis on cold-smoked salmon. *J. Appl. Microbiol.* 89: 920-934.

Jørgensen, L.V. 2000. *Spoilage and Safety of Cold-Smoked Salmon*. Ph.D-thesis. Danish Institute for Fisheries Research, Lyngby, Denmark.

Jørgensen, L.V.; Huss, H.H. and Dalgaard, P. 2001. Significance of volatile compounds produced by spoilage bacteria in vacuum-packed cold-smoked salmon (*Salmo salar*) analysed by GC-MS and multivatiate regression. *J. Agric. Food Chem.* 49: 2381.

Karnop, G. 1980. Der einflus der lagertemperatur auf die halbarkeit von geräuchertem heilbutt, bückling un all. *Informationen fuer die Fischwirtschaft* 27: 125-128.

Leroi, F.; Joffraud, J.-J.; Chevalier, F. and Cardinal, M. 1998. Study of the microbial ecology of cold-smoked salmon during storage at 8°C. *Int. J. Food Microbiol.* 39: 111-121.

Leroi, F. and Joffraud, J.J. 2000. Salt and smoke simultaneously affect chemical and sensory quality of cold-smoked salmon during 5°C storage predicted using factorial design. *J. Food Protec.* 63: 1222-1227.

Leroi, F.; Joffraud, J.J.; Chevalier, F. and Cardinal, M. 2001. Research of quality indices for cold-smoked salmon using a stepwise multiple regression of microbiological counts and physico-chemical parameters. *J. Appl. Microbiol.* 90: 578-587.

Nilsson, L.; Gram, L. and Huss, H.H. 1999. Growth control of *Listeria monocytogenes* on cold-smoked salmon using a competitive lactic acid bacteria flora. *J. Food Protec.* 62: 336-342.

O'Brien, G.D. 1997. Domestic refrigerator air temperatures and the public's awareness of refrigerator use. *Int. J. Environ. Health Res.* 7: 141-148.

Olley, J. and Ratkowsky, D.A. 1973. Temperature function integration and its importance in the storage and distribution of flesh foods above the freezing point. *Food Technol. Australia* 25: 66-73.

Paludan-Müller, C.; Dalgaard, P.; Huss, H.H. and Gram, L. 1998. Evaluation of the role of *Carnobacterium psicola* in spoilage of vacuum- and modified atmosphere packed cold-smoked salmon stored at 5°C. *Int. J. Food Microbiol.* 39: 155-166.

Peterson, M.E.; Pelroy, G.A.; Paranjpye, R.N.; Poytsky, F.T.; Almond, J.S. and Eklund, M.W. 1993. Parameters for control of *Listeria monocytogenes* in smoked fishery products: Sodium chloride and packaging method. *J. Food Protec.* 56: 938-943.

Ratkowsky, D.A.; Olley, J.; McMeekin, T.A. and Ball, A. 1982. Relation between temperature and growth rate of bacterial cultures. *J. Bacteriol.* 149: 1-5.

Ristiniemi, M. and Korkeala, H. 1998. Shelf-life of smoked blatic herring stored at different temperatures. *Archiv für Lebensmittelhygiene* 49: 103-106.

Ross, T. 1993. A Philosophy for the Development of Kinetic Models in Predictive Microbiology. Ph.D-thesis. University of Tasmania, Hobart, Australia.

Ross, T. 1996. Indices for performance evaluation of predictive models in food microbiology. *J. Appl. Bacteriol.* 81: 501-508.

Ross, T.; Dalgaard, P. and Tienungoon, S. 2000. Predictive modelling of the growth and survival of *Listeria* in fishery products. *Int. J. Food Microbiol.* 62: 231-245.

Rørvik, L.M.; Yndestad, M. and Skjerve, E. 1991. Growth of *Listeria monocytogenes* in vacuum-packed, smoked salmon, during storage at 4°C. *Int. J. Food Microbiol.* 14: 111-118.

Schneider, V.W. and Hildebrandt, G. 1984. Untersuchungen zur lagerfähigkeit von vakuumverpacktem räucherlachs. *Arch. Lebensmittelhyg.* 35: 60-64.

Sergelidis, D.; Abrahim, A.; Sarimvei, A.; Panoulis, C.; Karaioannoglou, Pr. and Genigeorgis, C. 1997. Temperature distribution and prevalence of *Listeria* spp. in domestic, retail and industrial refrigerators in Greece. *Int. J. Food Microbiol.* 34: 171-177.

Stohr, V.; Joffraud, J.J.; Cardinal, M. and Leroi, F. 2001. Spoilage potential and sensory profile associated with bacteria isolated from cold-smoked salmon. *Food Res. Int.* 34: 797-806.

Tinker, B.L.; Slavin, J.W.; Learson, R.J. and Ampola, V.G. 1985. Evaluation of automated time-temperature monitoring system in measuring the freshness of chilled fish. *Refrig. Sci. Technol. Proc.* 4: 281-291.

Truelstrup Hansen, L.; Gill, T. and Huss, H.H. 1995. Effects of salt and storage temperature on chemical, microbiological and sensory changes in cold-smoked salmon. *Food Res. Int.* 28: 123-130.

Truelstrup Hansen, L.; Gill, T.; Røntved, S.D. and Huss, H.H. 1996. Importance of autolysis and microbiological activity on quality of cold-smoked salmon. *Food Res. Int.* 29: 181-188.

Truelstrup Hansen, L. and Huss, H.H. 1998. Comparison of the microflora isolated from spoiled cold-smoked salmon from three smokehouses. *Food Res. Int.* 31: 703-711.

Truelstrup Hansen, L.; Røntved, S.D. and Huss, H.H. 1998. Microbiological quality and shelf life of cold-smoked salmon from three different processing plants. *Food Microbiol.* 15: 137-150.

Wijtzes, T.; Rombouts, F.M.; Kant-Muermans, M.L.; Riet, K.V. and Zwietering, M. 2001. Development and validation of a combined temperature, water activity, pH model for bacterial growth rate of *Lactobacillus curvatus*. *Int. J. Food Microbiol.* 63: 57-64.

CHAPTER 23

AN OVERVIEW OF POSTHARVEST TREATMENTS TO REDUCE *VIBRIO* IN OYSTERS

Michael Morrissey, Hakan Calik, and Shiu-er Shiu

Department of Food Science and Technology, Oregon State University, Seafood Laboratory, Astoria, OR

Vibrio species are natural habitants of oysters and their numbers can increase rapidly if oysters are mishandled postharvest or during distribution. Several cases of severe gastroenteritis outbreaks in the U.S. have been linked to the consumption of raw oysters containing the Vibrio bacteria. It remains a challenge for the oyster industry to develop postharvest processes that would reduce or eliminate Vibrio while maintaining the flavor and texture of raw oysters. This chapter presents different methods such as depuration, chilling/freezing, acidic marinade, pasteurization, the Ameripure process, irradiation, and high pressure processing, which are known to reduce pathogens in oysters. The effects of these postharvest treatments on Vibrio and their effects on the sensory attributes of raw oysters are discussed.

INTRODUCTION

Several cases of severe illnesses and various forms of gastroenteritis associated with consumption of raw oysters have been reported in the U.S. (Brown and Dorn, 1977; Earampamoorty and Koff, 1975; Gerba and Goyal, 1978; Wood, 1976; Kaysner, 1998; Fyfe *et al.*, 1998). *Vibrio vulnificus* (Vv) and *Vibrio parahaemolyticus* (Vp) have been implicated in illness outbreaks in the Southeast and Pacific Northwest (Klontz *et al.*, 1993; Kaysner, 1998). The largest Vp outbreak with 110 confirmed cases was linked to the Galveston Bay area, Texas in 1998 (CDC, 2001). Vp infections associated with eating raw oysters were also found in Pacific Northwest both in 1997 and 1998 with the highest frequency in summer months with 209 confirmed cases of illness from

California, Oregon, Washington and British Columbia (CDC, 1998). CDC *Vibrio* surveillance reports (1999) show that Vv infections via oyster consumption cause several fatalities each year.

Oysters are filter feeders that are farmed in bays and estuaries and consequently take-up and accumulate the microflora prevalent in that particular water column. This flora eventually leads to spoilage of oysters and in some cases, if oysters are mishandled postharvest, growth of possible pathogenic bacteria could result as a threat to public health (Andrews *et al.*, 1975; Kaneko and Colwell, 1973; Vanderzant and Thompson, 1973). Pathogenic bacteria and viruses may accumulate inside the oyster's intestinal tracts resulting in high enough concentrations to cause disease (Buisson *et al.*, 1981). Furthermore, pathogenic organisms such as Vv and Vp are natural habitants in estuarine waters, and their numbers in harvested oysters depend on several factors such as season, location, environmental conditions, and possible temperature abuse postharvest (Cook, 2001a).

The U.S. Food and Drug Administration (FDA) has established a level of concern for total Vp numbers and will consider enforcement action against the sale of molluscan shellfish with levels equal to or greater than a most probable number (MPN) count of 10,000/g. The FDA's compliance program established a guidance level for Vv in ready-to-eat fish products as the "presence or absence of pathogenic organism showing mouse lethality" (Cook, 2001b). However, according to the Interstate Shellfish Sanitation Conference (ISSC) Vv Risk Management Plan for oysters, it is suggested that postharvest treatments should ensure Vv is less than 3 MPN/g when average monthly water temperatures exceed 75°F (ISSC, 2000). The ultimate goal of this risk management plan is to reduce the rate of etiologically confirmed shellfish-borne Vv septicemia illnesses from the consumption of commercially harvested raw or undercooked oysters by 40% at the end of 2005 and by 60% by the end of 2007 (ISSC, 2000). Due to serious public health concerns, there has been increased concentration in different treatment techniques to reduce pathogens in oysters to acceptable levels set by government authorities. The following is a review of several strategies to reduce pathogenic bacteria, especially *Vibrio* sp., in shellfish.

Depuration and Relaying

Oysters are filter feeders, trapping food particles in mucus on the gills, which is then drawn into the mouth; eventually, the particles are accumulated in the gut (Buisson *et al.*, 1981). When waters are contaminated by fecal material, pathogenic bacteria and viruses may also be accumulated. On the other hand, *Vibrios* are naturally occurring bacteria that grow in the marine environment and can be found in oysters from clean, non-polluted waters. As oysters are commonly eaten raw, they can present a potential public health risk (Buisson et al., 1981). Depuration is a dynamic process whereby the oysters are allowed to purge themselves of contaminants in tanks of sterilized or pathogen free seawater (Fleet, 1978). When placed in an environment free from pathogenic bacteria a contaminated oyster continues to feed, taking new material into the gut and expelling digested contaminated material as feces (Buisson et al., 1981). Depuration generally takes two or three days and is intended to remove coliform bacteria, which act as an indicator organism for presence of pathogens. Depuration techniques vary considerably as several factors, such as shellfish species, likely contaminant level, and water temperature will influence rates of microorganism reduction. Eyles and Davey (1984) studied the effectiveness of a commercial depuration plant, which purified oysters from a polluted estuary. Although depuration significantly reduced aerobic plate counts (APC) and counts of coliforms and *Escherichia coli*, studies revealed that it had no effect on naturally occurring Vp. Furthermore, a study by Tamplin and Capers (1992) showed that Vv can not be eliminated through depuration process. Although depuration is considered to be a successful means to reduce fecal coliform to acceptable levels complete elimination of *Vibrio* sp. is rarely achieved, indicating the need for better postharvest treatments to ensure shellfish safety.

Relaying is defined by FDA as the process of reducing pathogenic organisms that may be present in shellfish by transferring shellfish from a growing area classified as restricted to a growing area classified as approved ambient environments (CFAN, 2000). In a study by Jones (1994) oysters were relayed to high salinity waters in the state of Maine resulting in reduction of Vv levels after 7 to 28 days. Furthermore, Cook (2001b) reported that relaying Gulf Coast oysters to offshore waters reduced Vv densities to <10 MPN/g in 7 to 17 days. However, relaying is expensive and careful oversight, testing, and monitoring practices are necessary.

Pasteurization

Pasteurization, by definition, implies the use of heat to reduce or eliminate pathogenic and spoilage organisms in food products (Jay, 1996). Early work on pasteurization resulted in reduction of standard aerobic plate, coliform, and fecal coliform counts in oysters (Pringle, 1961). Seafood products are not easily sterilized by conventional heat processing methods due to unfavorable changes in taste and texture (Learson et al., 1969). However, according to sensory tests performed by Chai et al. (1984), there was little difference in the organoleptic characteristics of pasteurized and freshly shucked oysters. Additionally, Chai et al. (1991) found that a pasteurization process of 76 °C for 8 min produced oysters with good physical and sensory quality. In both studies, there was a 2 – 3

$\log_{10}$ reduction of bacteria in pasteurized oysters, yet no tests were done to investigate *Vibrio* sp. reduction.

Oyster texture and flavor are strongly affected by temperature and time of pasteurization (Chai *et al.*, 1991). Since many consumers prefer oysters raw (Jarosz *et al.*, 1989), it is a challenge for the oyster industry to choose and employ the best oyster postharvest treatment that would satisfy safety and organoleptic needs for the industry.

Heat Shock Treatment

Hesselman *et al.* (1999) showed that heat-shocking treatments are effective in reducing counts of Vv in oysters. Heat-shocking is a relatively simple method of heating the whole oysters in a heated water tank so that oyster meat reaches 50°C for 1 to 4 min. This method is also used to facilitate the shucking of oysters. The heat-shock method was shown to reduce Vv counts from 1 to 4 logs in the finished product. Reports by Green (2001) verified heat-shocking as a viable method for reducing Vv counts in oysters. Tank design and careful monitoring of the water temperature is important if the process is used commercially.

Ameripure Process

The Ameripure process is a mild thermal treatment of oysters in a warm water bath, followed by a rapid cool down in ice-water slurry. The length of the mild thermal treatment depends on the estimated contaminant level, the season, temperature, and amount of the oysters (Andrews, 1999). This FDA approved treatment raises the temperature of the oyster enough to destroy Vv but does not sterilize or cook the oyster. The effect of mild heat treatment on the survival of Vv in both pure culture and oysters was first studied by Cook and Ruple (1992). Rapid death of Vv in broth cultures was observed with a decimal reduction value of 78 sec at 47 °C. Furthermore, thermal inactivation experiments were performed with naturally-contaminated oysters, where a treatment of 50 °C for 10 min reduced Vv to non-detectable levels. Results showed that the mild heating process did not impart a noticeable cooked appearance or taste. Andrews *et al.* (2000), showed that low temperature pasteurization was effective in reducing *Vibrio* sp. from more than 1,000,000 CFU/g to non-detectable levels in less than 10 min of processing. The spoilage bacteria were reduced 2 to 3 fold resulting in increased shelf-life up to 7 days beyond live unprocessed oysters.

Cooling and Freezing Treatment

Cook (1997) showed that Vv counts increased in un-chilled oysters at temperatures above 18°C. These results strengthened the National Shellfish Sanitation Program (NSSP) recommendations of keeping oysters at temperatures below 8°C during storage to suppress growth of Vv. In general, the higher the temperature of harvest waters, the sooner refrigeration is necessary to prevent bacterial growth. Refrigeration and freezing have been studied as an alternative method to process oysters while preserving the integrity and organoleptic quality of raw oysters. Results by Oliver (1981) showed that Vv decreased rapidly in oyster broth or in whole shell oysters at 0.5°C to 4°C. However, studies by Kaspar and Tamplin (1993) showed that storage of naturally contaminated oysters at 2 and 4°C for 14 days caused only 1 $\log_{10}$ reduction and that storage at 30°C resulted in increases in Vv counts in oysters.

A freeze inactivation storage study revealed the durability of Vv to cold temperatures in that, it was possible to culture Vv from oysters after 12 weeks at –20 °C (Cook and Ruple, 1992). Even though there was 2 – 3 $\log_{10}$ reduction in numbers, complete elimination was not seen. A study by Parker *et al.* (1994) showed that after 70 days of –20 °C frozen storage, the organism was viable, further confirming the durability of Vv. A recent study by Schwarz (2000) demonstrated that Vv levels were reduced up to 99% following a rapid chilling treatment, whereas the conventionally cooled oysters required 4 days of cold room storage to achieve similar reduction. On the other hand, individual quick-frozen (IQF) process at –74°C (cryogenic freezing with carbon dioxide) eliminated almost all Vv in oysters while preserving quality and freshness (Berne, 1996). In general, *Vibrio* species show susceptibility to freezing, however decline in counts depend on freezing rate, method, and storage time (Cook, 2001a).

Acidic Marinade

The effect of low pH on microorganisms has been well studied and few pathogenic bacteria grow below pH 4.0 (Jay, 1996). Low pH affects the enzymatic functions and nutrient transportation eventually causing death of the cell. The treatment of soaking oysters in vinegar for several hours before consumption is based on the sterilizing effects of low pH. Since the pH of vinegar is close to 2.0, most foodborne bacteria cannot survive such acidic conditions. Shiu (1999) showed that an acidic marinade effectively inhibited all bacterial growth in oysters. However, the drop in pH will dramatically affect the sensory properties of the oyster. Acidic marinade is a popular style of eating raw oysters in Japan.

Irradiation Treatment

Food irradiation is a process for treatment of food products to increase quality, enhance shelf life, and improve microbial safety. The effects of this process are proportionally dependent on irradiation dose and treatment duration (Giddings, 1984). Generally, Cobalt-60 (^{60}Co) and Cesium-137 (^{137}Cs) are used for food preservation (Venugopal et al., 1999). Low gamma rays emitted by ^{60}Co have been proven effective in reducing spoilage and pathogenic microorganisms in a variety of seafood products. This technology has been recognized as a preservation method of foods since it can eliminate indigenous microflora in foods. The first use of irradiation on foods was approved by the FDA in 1963. The electron beam accelerator (electron beam) technology is a relatively new, flexible, and more effective source of food irradiation than the regular ^{60}Co source (Huang et al., 1997). The electron beam is easy to adapt to different radiation processes, such as operating at different beam energy levels. No radioactivity is present when the accelerator is off, and therefore, no radioactive waste accumulates, whereas up to 21 years are required to dispose ^{60}Co (Huang et al., 1997). To help combat *E. coli* 0157:H7 threat on public health, the FDA approved treating red meat products with a measured dose of irradiation (Henkel, 1998). In addition, in 1990 the FDA approved the use of this technology with poultry (dose range 1.5–3.0 kGy) to prevent frequent outbreaks of *Salmonellae* associated diseases (FDA, 1990). Currently, FDA does not allow irradiation treatment of seafood products and thus there are no approved irradiation dosages for seafood products (Hilderbrand, 2001).

Irradiation studies on the Eastern oyster, *Crassostrea virginica*, have demonstrated that *Vibrio* species are among the most radiation sensitive bacteria and that both *V. cholerae* and Vv can be eliminated with low doses of ^{60}Co-gamma radiation (Mallett et al., 1991). The investigators reported that doses less than 1 kGy were lethal to *V. cholerae*, Vv, and Vp. Based on these studies, it is clear that ionizing irradiation will deliver promising results in reducing *Vibrio* sp. in live oysters without killing the animals, which helps extend the shelf life of oysters. Novak et al. (1966) demonstrated that oysters treated with 2.0 kGy had a shelf life of 23 days under refrigeration, whereas, untreated oysters deteriorated within 7 days of storage. Despite the encouraging technical and scientific assessment of irradiation as a safe alternative for preserving foods, there are still a substantial fraction of consumers who perceive it to be unsafe and undesirable. The effect of irradiation on the sensory quality and consumer acceptance of such products is still the major concern for seafood industry (Chen et al., 1996).

High Pressure Processing

High pressure processing (HPP), is a novel food processing technology which subjects liquid and solid foods, with or without packaging, to pressures between 100 and 800 MPa at or above room temperature (Mertens and Deplace, 1993). The earliest work by Hite (1899) reported the reduction of microorganisms in milk and meat following pressure treatments. Renewed interest has been largely due to the successful development of commercialized high pressure processed jams, jellies, and beverages by the Japanese (Hoover, 1993; Farkas, 1993). A typical high-pressure system consists of a pressure vessel of cylindrical design, two end closures, a means for restraining the end closures, a low pressure pump, an intensifier which uses liquid from the low pressure pump to generate high pressure process fluid for system compression, and necessary system controls and instrumentation. The pressure-transferring medium is usually water.

Pressures used in the process of foods appear to have little effect on covalent bonds (Tauscher, 1998); thus, foods subjected to the treatment at or near room temperature do not undergo significant chemical changes. Only non-covalent bonds such as hydrogen, ionic, and hydrophobic bonds are destroyed or formed during the pressure treatment; this enables the foods to maintain their natural vitamin content, flavor and taste even after the treatment (Hayashi *et al.*, 1989). Knorr (1993) reported that HPP treatment could be used not only to cause microbial inactivation but also for enzyme inactivation or activation, gel formation, and textural modification in foods.

HPP causes changes in morphology, biochemical reactions, generic mechanisms, and cell membrane/wall of microorganisms. The high pressurization process distorts bacterial cell membranes and denatures enzymes and nucleic materials necessary to carry on biological activities and reproduction, eventually leading to bacterial cell death. The cell wall completely covers and maintains the shape of the cell; however pressures of 40 MPa can cause the cell to lyse due to mechanical disruption of the stressed cell wall (Berger, 1959). Gram-negative bacteria, such as *Vibrio* species, appear to be more susceptible to the effects of pressure and/or heat than gram-positive bacteria, solely due to relatively brittle structure of gram-negative cell membrane. Carlez *et al.* (1994), confirmed this phenomenon in a study on bacterial growth during chilled storage of pressure treated minced meat (lean beef muscle).

In 1974, Schwarz and Colwell examined the effect of hydrostatic pressure of 20 to 100 MPa on the growth and viability of three strains of Vp. No sustained growth was observed at pressure of 40 MPa and above at 25 °C and with 20 MPa at 15 °C. The response of Vp to HPP was further studied by Styles *et al.* (1991). A 10^6 colony forming unit (CFU)/mL population of Vp was eliminated by 172 MPa for 10 min and 30 min in clam juice and phosphate buffer, respectively. Yukizaki *et al.* (1992) found Vp, *V. mimicus*, and *V. cholerae* to be destroyed by 193, 294, and 486 MPa for 10 min in liquid buffer at 0 °C, respectively. Work by Berlin *et al.* (1999) has shown that pathogenic *Vibrio* species are very susceptible to pressure treatments at levels between 200–310 MPa. A recent study with Vv by Kilgen (2000) demonstrated that the pathogen was reduced by 5 $\log_{10}$ cycles to non-detectable levels after pressure treatments between 200 and 345 MPa. Another recent study by Calik *et al.* (2001) confirmed that Vp strains *in vitro* or in artificially inoculated oysters were very susceptible to high pressure and were reduced below detection levels after 1 min at 345 MPa. Results showed Vp inactivation by HPP in both pure culture and whole oysters was dependent on treatment time and pressure. Optimum conditions for reducing Vp in pure culture and oysters from 10^9 to 10^1 CFU/mL were achieved at 345 MPa for 30 and 90 s, respectively. Resistance variations were detected between Vp in pure culture and in oysters. A process time of 5 min at 310 MPa was necessary for complete elimination of clinical 03:K6 Vp strain (isolate from the 1998 Texas outbreak), making the isolate the most baro-resistant. In this study, HPP proved to be an efficient means of reducing Vp in oysters.

CONCLUSIONS

Due to important economic and public health constraints, methods to control pathogens in raw oysters have received considerable attention from the industry and public. It remains a challenge for the oyster industry to develop postharvest processes that would reduce or eliminate pathogens while maintaining the flavor and texture of raw oysters. Depuration is considered to be a successful means to meet these standards; however, complete elimination of *Vibrio* sp. is rarely achieved. Studies show relaying process could decrease Vv counts however results are variable and transportation and other costs present certain drawbacks. Although, several investigators report that pasteurization treatments of oysters achieve elimination of Vv without any significant changes in organoleptic qualities, consumer acceptance of pasteurized oysters remains to be seen. Irradiation remains a promising approach to inactivate *Vibrio* sp. in raw oysters, however practical application and consumer acceptance of this control strategy remains unknown. While the proposed approaches for producing safe, high quality oysters vary, HPP is a promising solution. Recent studies and industry applications confirm that HPP is a viable means of non-thermal treatment of raw oysters for reduction of pathogenic and spoilage bacterial loads without causing significant changes in appearance, flavor, texture and nutritional qualities. The ultimate aim of oyster postharvest treatments is to reduce processing costs, increase yield, eliminate pathogens and extend shelf life. Ultimately, the effects of postharvest treatment on oyster demand will depend on consumer acceptance of postharvest treated oysters.

REFERENCES

Andrews, L. Sugar Processing Research Institute, New Orleans, LA. Dec 13, 1999. Personal communication with Mary Muth, Research Triangle Institute.

Andrews, L.; Park, D.L. and Chen, Y.P. 2000. Low temperature pasteurization to reduce the risk of Vibrio infections from raw shellstock oysters. Presented at the 25th Annual Meeting of the Seafood Science & Technology Society Meeting of the Americas, Oct. 9-11, Longboat Key, FL.

Andrews, W.H.; Diggs, C.D.; Presnell, M.W.; Miescier, J.J.; Wilson, C.R.; Goodwin, C.P.; Adams, W.N.; Turfari, S.A. and Musslman, J.F. 1975. Comparative validity of members of the total coliform and fecal coliform group for indicating the presence of *Salmonella* in the eastern oyster, *Crassostrea virginica. J. Milk Food Technol.* 38: 453-458.

Berlin, D.L.; Herson, D.S.; Hicks, D.T. and Hoover, D.G. 1999. Response of pathogenic *Vibrio* species to high hydrostatic pressure. *Appl. Environ. Microbiol.* 65: 2776-2780.

Berger, L.R. 1959. The effect of hydrostatic pressure on cell wall formation. *Bacteriol. Proc.* 59: 129.

Berne, S. 1996. Cryogenic freezing gives oysters six-month shelf life. *Prepared Foods* 165: 16.

Brown, L.D. and Dorn, C.R. 1977. Fish, shellfish and human health. *J. Food Protec.* 40: 712-717.

Buisson, D.H.; Fletcher, G.C. and Begg, C.W. 1981. Bacterial depuration of the pacific oyster (Crassostrea gigas) in New Zealand. *New Zealand J. Sci.* 24: 253-262.

Calik, H.; Morrissey, M.T.; Reno, P.W. and An, H. 2002. Effect of high pressure processing on survival of *Vibrio parahaemolyticus* in pure culture and Pacific oysters. *J. Food Sci.* 67: 1506-1510.

Carlez, A.; Rosec, J.P.; Richard, A. and Cheftel, J.C. 1994. Bacterial growth during chilled storage of pressure-treated minced meat. *Lebensm-Wiss. U-Technol.* 27: 48-54.

Center for Disease Control and Prevention (CDC). 1998. Outbreaks of Vibrio parahaemolyticus infections associated with eating raw oysters–Pacific Northwest, 1997. *MMWR.* 47: 457-462.

Center for Disease Control and Prevention (CDC). 1999. Outbreaks of Vibrio parahaemolyticus infection associated with eating raw oysters and clams harvested from Long Island Sound –Connecticut, New Jersey, and New York, 1998. *MMWR*. 48: 48-51.

Center for Disease Control and Prevention (CDC). 2001. *Vibrio* Surveillance System Summary Data, 1997-1998. Accessed in 2001. Available from: http://www.cdc.gov/ncidod/dbmd/diseaseinfo/cstevib99.pdf

Center for Food Safety and Applied Nutrition Food and Drug Administration (CFSAN). 2000. Draft: Risk Assessment on the Public Health Impact of *Vibrio parahaemolyticus* in Raw Molluscan Shellfish. *Vibrio parahaemolyticus* Risk Assessment Task Force. Accessed in 2001. Available at: http://vm.cfsan.fda.gov/~acrobat/vprisk.pdf

Chai, T.; Pace, J. and Cossaboom, T. 1984. Extension of shelf life of oysters by pasteurization in flexible pouches. *J. Food Sci*. 49: 331-333.

Chai, T.; Liang, K.T.; Pace, J. and Schlimme, D.V. 1991. Effect of heat processing on quality of pasteurized oysters. *J. Food Sci*. 56: 1292-1294.

Chen, Y.P.; Park, D.L.; Andrews, L.S. and Gridner, R.M. 1996. Reduction of *Vibrio vulnificus* in Gulf Coast oysters (*Crassostrea virginica*) by mild treatment. Presented at the annual meeting of the Institute of Food Technologists, June 22-26, New Orleans, LA.

Cook, D.W. and Ruple, A.D. 1992. Cold storage and mild heat treatment as processing aids to reduce the numbers of *Vibrio vulnificus* in raw oysters. *J. Food Protec*. 55: 985-989.

Cook, D.W. 1997. Refrigeration of oyster shellstock: Conditions which minimize the outgrowth of *Vibrio vulnificus*. *J. Food Protec*. 60: 349-352.

Cook, D.W. 2001a. Personal communications regarding *Vibrio vulnificus* in shellfish.

Cook, D.W. 2001b. An overview of postharvest treatment of oysters to eliminate *Vibrios*. Presentation at Gulf and South Atlantic States Shellfish Conference April 24-27, Biloxi, MS.

Earampamoorthy, S. and Koff, R.S. 1975. Health hazards of bivalve mollusk ingestion. *Ann. Int. Med*. 83: 107-110.

Eyles, M.J. and Davey, G.R. 1984. Microbiology of commercial depuration of the Sydney rock oyster, *Crassostrea commercialis*. *J. Food Protec*. 47: 703-706.

Farkas, D.F. 1993. 'Heatless' sterilization. Food Processing. Aug 72-75.

Fleet, G.H. 1978. Oyster depuration–a review. *Food Technol. Australia* 30: 444-454.

Food and Drug Administration (FDA). 1990. Irradiation in the production processing and handling of food. *Fed. Regist*. 55: 18538-18544.

Fyfe, M.; Kelly, M.T.; Yeung, S.T.; Daly, P.; Schallie, K.; Buchanan, S.; Waller, P.; Kobayashi, J.; Therien, N.; Guichard, M.; Lankford, S.; Stehrgreen, P.; Harsch, R.; Debess, E.; Cassidy, M.; Mcgivern, T.; Mauvais, S.; Fleming, D.; Lippmann, M.; Pong, L.; Mckay, R.W.; Cannon, D.E.; Werner, S.B.; Abbott, S. and Hernandez, M. 1998. Outbreak of *Vibrio parahaemolyticus* infections associated with eating raw oysters—Pacific Northwest, 1997. *Morbid. Mortal. Weekly Rep.* 47: 457–462.

Gerba, C.P. and Goyal, S.M. 1978. Detection and occurrence of enteric viruses in shellfish: a review. *J. Food Protec.* 41: 743-754.

Giddings, G.G. 1984. Radiation processing of fishery products. *Food Technol.* 38: 61-96, 118.

Green, D. 2001. Personal communication regarding heat-shock treatments. North Carolina State University Seafood Laboratory, Moorehead City, NC.

Hayashi, R.; Kawamura, Y.; Nakasa, T. and Okinaka, O. 1989. Application of high pressure to food processing: Pressurization of egg white and yolk and properties of gels formed. *Agric. Biol. Chem.* 53: 2935.

Henkel, J. 1998. Irradiation. *FDA Consumer.* 32: 12-17.

Hesselman, D.M.; Motes, M.L. and Lewis, J.P. 1999. Effects of a commercial heat-shock process on *Vibrio vulnificus* in the American oyster, *Crassostrea virginica* harvested from the Gulf Coast. *J. Food Protec.* 62: 1266-1269.

Hite, B.H. 1899. The effect of pressure in the preservation of milk. *Bull. W. Va. Agr. Exp. Sta.*, Morgantown. 58: 15-35.

Hilderbrand, K. 2001. Personal communication regarding irradiation of seafoods.

Hoover, D.G. 1993. Pressure effects on biological systems. *Food Technol.* 47: 150-155.

Huang, S.; Herald, T.J. and Mueller, D.D. 1997. Effect of electron beam irradiation on physical, physicochemical, and functional properties of liquid egg yolk during frozen storage. *Poultry Sci.* 76: 1607-1615.

Interstate Shellfish Sanitation Conference (ISSC). 2000. A *Vibrio vulnificus* Risk Management Plan for Oysters. Accessed 2001. Available at: http://www.issc.org/News/FL-Vv-Risk-Reduction-Plan/pdf/ISSC-Issue-00-201.PDF

Jay, J.J. 1996. Modern food microbiology, 5th ed. New York, NY: Chapman & Hall. 127p.

Jarosz, L.A.; Timmer, J. and Rach, E.C. 1989. Restaurateur reaction to irradiated shellfish. *J. Food Safety* 9: 283-290.

Jones, S.H. 1994. Relaying: Oyster Relay and Depuration Experience in New Hampshire. Proceedings of the 1994 *Vibrio vulnificus* Workshop. FDA-CFSAN. Accessed in 2001. Available at: http://vm.cfsan.fda.gov/~ear/VV.html

Kaneko, T. and Colwell, R.R. 1973. Ecology of *Vibrio parahaemolyticus* in Chesapeake Bay. *J. Bacteriol.* 113: 24-32.

Kaspar, C.W. and Tamplin, M.L. 1993. Effects of temperature and salinity on the survival of *Vibrio vulnificus* in seawater and shellfish. *Appl. Environ. Microbiol.* 59: 2425-2529.

Kaysner, C.A. 1998. *Vibrio parahaemolyticus*: a new pathogen for the Pacific oyster. Presented at the PacRim Shellfish Sanitation Conference, April 7-9, Portland, OR.

Kilgen, B.M. 2000. Processing controls for *Vibrio vulnificus* in raw oysters – commercial hydrostatic high pressure. Presented at the 25th Annual Meeting of the Seafood Science & Technology Society Meeting of the Americas, Oct. 9-11, Longboat Key, FL.

Klontz, K.C.; Williams, L.; Baldy, L.M. and Casmpos, M.J. 1993. Raw oyster associated *Vibrio* infections; linking epidemiological data with laboratory testing of oysters obtained from a retail outlet. *J. Food Protec.* 56: 977-979.

Knorr, D. 1993. Effects of high-hydrostatic pressure processes on food safety and quality. *Food Technol.* 47(6): 156-161.

Learson, R.J.; Ronsivalli, L.J. and Spracklin, B.W. 1969. Process criteria for producing radiation sterilized fish products. *Food Technol.* 23: 85-91.

Mallett, J.C.; Beghian, L.E.; Metcalf, T.G. and Kaylor, J.D. 1991. Potential of radiation technology for improved shellfish sanitation. *J. Food Safety* 11: 231-245.

Mertens, B. and Deplace, G. 1993. Engineering aspects of high-pressure technology in the food industry. *Food Technol.* 47: 164-169.

Novak, A.F.; Liuzzo, J.A.; Grodner, R..M. and Lovell, R.T. 1966. Radiation pasteurization of Gulf coast oysters. *Food Technol.* 20: 201-202.

Oliver, J.D. 1981. Lethal cold stress of *Vibrio vulnificus* in oysters. *Appl. Environ. Microbiol.* 41: 710-717.

Parker, R.W.; Maurer, E.M.; Childers, A.B. and Lewis, D.H. 1994. Effects of frozen storage and vacuum-packing on survival of *Vibrio vulnificus* in Gulf coast oysters (*Crassostrea virginica*). *J. Food Protec.* 57: 604-606.

Pringle, S.B. 1961. The South Carolina Report on the "hot dip" and "shock" method of opening oysters. In *Proceedings-Shellfish Sanitation Workshop*. US Department of Health, Education, and Welfare, US Public Health Service, Washington DC, pp. 62-74.

Schwarz, J.R. and Colwell, R.R. 1974. Effect of hydrostatic pressure on growth and viability of Vibrio parahaemolyticus. *Appl. Microbiol.* 28: 977-981.

Schwarz, J.R. 2000. Rapid Chilling of Oyster Shellstock: A Postharvest Process to Reduce *Vibrio vulnificus*. Presented at the 25th Annual Meeting of the Seafood Science & Technology Society Meeting of the Americas, Oct. 9-11, Longboat Key, FL

Shiu, S. 1999. The Effect of High Hydrostatic Pressure (HHP) on the Bacterial Count and Quality of Shucked Oysters. M.S. Thesis, Oregon State University, Corvallis, OR.

Styles, M.F.; Hoover, D.G. and Farkas, D.F. 1991. Response of *Listeria monocytogenes* and *Vibrio parahaemolyticus* to high hydrostatic pressure. *J. Food Sci.* 56: 1404-1407.

Tauscher, B.K. 1998. Effect of high pressure treatment to nutritive substances and natural pigments. Fresh Novel Foods by High Pressure. VTT Symposium 186. Technical Research Centre of Finland. Helsinki, Finland.

Tamplin, M.L. and Capers, G.M. 1992. Persistence of *Vibrio vulnificus* in tissues of Gulf Coast oysters, Crassostrea virginica, exposed to seawater disinfected with UV light. *Appl. Environ. Microbiol.* 58: 1506-1510.

Vanderzant, C. and Thompson, C.A. Jr. 1973. Microbiological flora and level of *V. parahaemolyticus* of oyster (*Crassostrea virginica*), water sediment from Galveston Bay. *J. Milk Food Technol.* 30: 447-451.

Venugopal, V.; Doke, S.N. and Thomas, P. 1999. Radiation Processing to improve the quality of fishery products. *Crit. Rev. Food Sci. Nutr.* 39: 391-440.

Wood, P.C. 1976. Guide to shellfish hygiene. WHO Offset publication no 31. World Health Organization, Geneva.

Yukizaki, C.; Kawano, M. and Tsumagari, H. 1992. The sterilization of sea urchin eggs by high hydrostatic pressure. In *High Pressure Bioscience and Food Science*, R. Hayaski (ed.). San-ei Pub. Co., Kyoto, Japan, pp. 225-228.

CHAPTER 24

MICROBIAL QUALITY OF CULTURED NEWFOUNDLAND MUSSELS AND SCALLOPS

Muhammad Ahmad Khan[1,2] Christopher C. Parrish[2] and Fereidoon Shahidi[1,2]

[1]Department of Biochemistry, Memorial University of Newfoundland, St. John's, NL, A1B 3X9, Canada and [2]Ocean Sciences Centre, Memorial University Newfoundland, St. John's, NL, A1C 5S7, Canada

Microbial quality of various samples of wild and cultured blue mussels and scallops were examined using marine and plate count agars. Microbial quality of cultured blue mussels stored at three temperatures, -12, 2 and 9°C were examined using aerobic and psychotropic counts. The effects of season of harvest and aquaculture site on the microbial quality of blue mussels were evaluated. Marine agar was more efficient in enumerating bacteria from bivalves compared to plate count agar. Storage temperature affected the microbial shelf life of mussels. The microbial quality of mussels harvested from different aquaculture sites, different seasons and different years was within the acceptable limits.

INTRODUCTION

The microbial quality of cultured fish and shellfish may be affected by various factors such as initial microbial load, aquaculture site, temperature during storage and distribution, season of harvest, manual or mechanical washing, sorting, packaging and processing. Aquatic species are cultured throughout Newfoundland in sheltered cold ocean waters to protect the animals from icebergs and violent waves due to storms. The culturing sites may be close to the shore with slow water and nutrient movements or open to the ocean with more vigorous mixing of marine algae and microorganisms.

Blue mussels (*Mytilus edulis*) are the single largest shellfish aquaculture product of Newfoundland (Anonymous, 2000). Mussels are cultured by using the rope technique. The ropes are anchored at both sides and supported by floating buoys at certain intervals. Mussel spat (approximately one year old mussels) are loaded into the mesh tubing or socks and tied at specific intervals on the long ropes and suspended approximately 4 m below the surface of water. The mussels reach market size after one or two years in the socks with a bluish-black color of the shell and white, pink or dark blue color of the meat. The cultured mussels (2-3 years old) are harvested by pulling the socks out of water using powerful winches, loading in small boats and transporting to the packaging plants. Upon arrival, the mussels are stripped from the socks and washed thoroughly. Filtered sea water (5-6°C) is used in all washing steps and the temperature in the processing plant is usually kept at 8-10°C. Clumps of mussels are separated, sorted and washed again before final packaging in bags with wide mesh. The bags are placed in insulated containers or special boxes and kept at 2-4°C. Scallops (*Placopecten magellanicus*) are also cultured in the cold water of Gilbert Bay, Labrador, using a similar technique. The cultured scallops are kept inside lantern nets at a depth of 8 m and the market size scallops are usually 48 months old.

Microbial quality of seafoods is an important criterion and this is generally determined by the conventional agar method (Harrigan, 1998). Total aerobic plate count represents the total number of bacteria that are capable of forming visible colonies (Busta *et al.*, 1984). Aerobic plate counts (APC) or total heterotrophic bacterial (THB) count in filter feeders such as mussels and scallops indicate the general microbial quality of the seafood and surrounding water as well as hygiene during handling, processing and storage (Hunt *et al.*, 1984). Therefore, high APC counts ($>10^5$ (or $\log_{10}$ 5 CFU/g) may indicate the potential health hazard for consumers (Neufeld, 1984). Meanwhile, psychotropic bacterial counts (PPC) are used as the spoilage indicator of seafoods at refrigeration temperatures (Gilliland *et al.* 1984). Therefore, APC and PPC can be used to evaluate the microbial quality of both wild and cultured products.

The nutritional requirements of heterotrophic bacteria vary significantly. Therefore, it is not possible to use an agar medium capable of supporting the growth of all microorganisms. However, any medium must include water, peptides, amino acids, carbohydrate as an energy source, and accessory growth factors such as vitamins (Harrigan, 1998). Therefore, there is a constant search for an agar medium that gives the highest and most consistent recovery of microorganisms.

The objectives of this study were to 1) evaluate the microbial quality of various samples of wild and cultured blue mussels and scallops using different types of agar; 2) examine microbial quality of cultured blue mussels stored at three temperatures, -12, 2 and 9°C using APC and PPC on marine (MA); and 3) examine the effects of season of harvest and aquaculture site on the microbial quality of blue mussels.

MATERIALS AND METHODS

Experiment I

Cultured blue mussels (*Mytilus edulis*) were obtained from a mussel packaging plant in Charles Arm, Notre Dame Bay, Newfoundland, Canada. Wild blue mussels as well as cultured and wild scallops (*Placopecten magellanicus*) were obtained from Gilbert Bay, Labrador, Canada. Microbial quality of the samples were analyzed using APC and PPC on plate count agar and marine agar.

Experiment II

To examine the effects of storage temperature on the microbial quality of cultured mussels, thirty five months old cultured blue mussels (*Mytilus edulis*) were obtained from a mussel packaging plant in Fortune Harbor, Notre Dame Bay, Newfoundland, Canada. Mussels from three batches were stored for 10 days at three different temperatures (-12±2°C), (2±2°C) and (9±2°C) in plastic bags with holes to drain the discharges. Samples at 2 °C were stored in a Foster refrigerator (Drummond Ville, Quebec), while samples at 9 and –12°C were stored in a household Cold Spot refrigerator (9°C) with upper freezer compartment (-12°C). The samples were analyzed for microbial quality on days 0, 3, 7 and 10 using APC and PPC on marine agar.

Experiment III

To examine the effects of season of harvest and aquaculture site on microbial quality of cultured mussels, two commercial aquaculture sites were selected with two stations at each site, in June 2000. In each station or sampling place, mussel spats (approximately one year old mussels) were loaded into the mesh tubing or socks and tied at specific intervals on long ropes and suspended approximately 4 m below the surface of water. Mussels were removed from the socks and examined for microbial quality at specific intervals, which represent different seasons. Mussels from three socks per station were examined for microbial quality using APC on MA in June, August, September, October, 2000 and April, May, June, August and October 2001.

Examination of microbial quality

Mussels and scallops (3-6 animals for experiment I and II, and 10-15 mussels for experiment III) were opened and the shell liquor and meat were collected in a sterile beaker and homogenized using a commercial Waring blender (Dynamic Corporation of America, New Hartford, CT) for 60 s. Ten grams of the homogenate were mixed with 90 mL of 0.1% peptone water for another 30 s. Serial decimal dilutions from this homogenate were prepared in 9 mL of 0.1% peptone water. Amounts ranging from 0.1 to 0.5 mL of appropriate dilutions were spread with a sterile glass spreader on pre-poured PCA and/or MA plates (Difco laboratories, Detroit, IL). Both media were incubated at 4°C for 10 days to determine PPC counts, and at 30°C for 48 h to determine APC counts (Hunt et al., 1984). Bacterial colonies on PCA and MA were counted using a Quebec counter and recorded as colony forming unit CFU/g of the meat.

Statistical analysis

All experiments were replicated three times except for cultured and wild scallops because of limited numbers of samples. Mean values ± standard deviations were reported. Prior to statistical analysis, results of microbiological analysis were transformed to $\log_{10}$ values. Analyses of variance (ANOVA) and Tukeys studentized test were performed at a level of p<0.05 to evaluate the significance of differences between mean values.

RESULTS AND DISCUSSION

Effect of agar type on examination of microbial quality of wild and cultured mussels and scallops

Bacterial counts (Log CFU/g) on PCA and MA of wild and cultured mussels and scallops are shown in Table 1. Results indicate that bacterial counts on MA were approximately 1-3 log CFU/g higher than their corresponding counts on PCA for all samples examined. This can perhaps be explained by the type of growing bacteria and the agar used. Marine agar is composed of the following components (g/L distilled water), agar (15.0), yeast extract (1.0), glucose (-), pancreatic digest of casein (-), peptone (5.0), beef extract (-), NaCl (19.45), $MgCl_2$ (8.8), $MgSO_4$ (-), Na_2SO_3 (3.24), $CaCl_2$ (1.8), KCl (0.55), $NaHCO_3$ (0.16), Ferric citrate (0.10), KBr (0.08), $SrCl_2$ (0.03), H_3BO_3 (0.03), Na_2HPO_4 (8.0), Na_2SiO_3 (4.0), NaF (2.4) and NH_4NO_3 (1.6), (Atlas, 1995). Meanwhile, plate count agar is composed of (g/L distilled water) agar (9.0), yeast extract

(2.5), glucose (1.0) and pancreatic digest of casein (5.0), (Atlas, 1995). Therefore, many of the elements that support the growth of marine bacteria, and are predominant in seafoods, are not present in plate count agar.

Table 1. Effects of agar type on the microbial quality[1] of wild and cultured mussels and scallops during storage

Site	Aerobic plate counts		Psychotropic plate counts	
	PCA	MA	PCA	MA
Charles Arm cultured mussels	3.33 ± 0.06^b	4.32 ± 0.08^d	2.87 ± 0.05^a	3.38 ± 0.05^c
Gilbert Bay wild mussels	3.21 ± 0.16^b	3.53 ± 0.01^e	3.11 ± 0.07^{ab}	3.36 ± 0.02^c
Gilbert Bay cultured scallops[2]	ND	2.79	ND	2.83
Gilbert Bay wild mussels[2]	ND	2.83	ND	2.93

[1]Based on aerobic counts and psychotropic plate count (Log CFU/g) on plate count (PCA) and marine agar (MA), [2]Average of duplicate determinations. Values are means of three determinations ± SD. Values within each row with different superscripts letters are different (p<0.05) from one another. ND; no bacteria detected at the lowest dilution (10^{-1}).

Although it is possible bivalves are exposed to non-marine bacteria during transportation, distribution and storage, this exposure is limited because of the normal closure of the shells and accumulation of marine bacteria within the bivalve tissues during the feeding and filtration processes. Furthermore, numerous studies have indicated that various marine bacteria are the main contributor to overall microbial quality of bivalves (Colby et al., 1995; Hoi 1998; Ray, 1996).

Results from this experiment also indicate to close Log CFU/g counts of APC and PPC on MA with some exceptions. This can be explained, in part, by the incubation temperatures of 30°C used in this study to obtain APC, which may have allowed the growth of psychotrophic bacteria. This explanation can be supported when examining the results from microbial evaluation of refrigerated and frozen seafoods such as calms, oysters, crabmeat, shrimp and frozen raw lobster tails, in which PPC on agar incubated at 30°C were higher than those incubated at 35°C (Swatzentruber *et al.*, 1980; Jay, 1996).

Effects of storage temperature on the microbial quality of mussels

Microbial analysis of seafoods at different storage temperatures is performed to estimate their microbial shelf life and to comply with the regulations of food safety and protection agencies (Colby *et al.*, 1995; Harrigan, 1998; and Park, 1991). Blue mussels were stored at three temperatures: -12±2°C which considered to be the frozen storage temperature of seafoods, 2±2°C which simulate the normal temperature in household refrigerators, and 9±2°C is considered to be an abuse refrigeration temperature.

Changes in APC and PPC counts (Log CFU/g) on MA of Fortune Harbour mussels stored at three different temperatures are presented in Tables 2 and 3, respectively. Results indicate bacterial counts of mussels were affected by storage temperature. APC and PPC on MA of samples stored at 9°C were significantly higher ($p<0.05$) than samples stored at 2 and −12°C. After 3 days of storage at 9°C, APC counts on MA were 6.06±0.10 Log CFU/g, but only 4.07±0.06 Log CFU/g and 3.39±0.04 Log CFU/g for samples stored at 2 and −12°C.

Microbial spoilage of seafoods stored at refrigerated temperatures is shown to be primarily due to the presence of psychrotrophic Gram-negative bacteria such as *Pseudomonas, Vibrio, Falvobacterium* and *Moraxella* (Cook 1997; Hoi *et al.*, 1998; Jay, 1996). Nonetheless, microflora of many spoiled seafoods, including bivalves, at refrigeration temperatures have been reported to be composed mainly (90-100%) of *Pseudomonas* species (Ray, 1996). Therefore, it is expected that many of the bacteria enumerated on MA belong to the *Pseudomonas* species. However, it is difficult to reach firm conclusions regarding the bacteria that contributed to the spoilage of blue mussels without further identification. At very high bacterial counts, > log 7 CFU/g, microbial activity may produce off-odors because of production of undesirable primary metabolites such as trimethylamine from trimethylamine oxide, other amines and ammonia, hydrogen sulfide and carbonyl compounds from degradation of lipids (Jay, 1996). Therefore, microbial examination of Fortune Harbor mussels stored at 9°C was terminated after 7 days of storage.

Table 2. Effects of storage temperature on the microbial quality (based based on aerobic plate counts [Log CFU/g] on marine agar) of Fortune Harbor mussels stored at three different temperatures

Storage time (days)	Storage temperature (°C)		
	-12	2	9
0	3.51 ± 0.01^a	3.51 ± 0.01^a	3.51 ± 0.01^a
3	3.39 ± 0.04^a	4.07 ± 0.06^b	6.06 ± 0.10^c
7	3.83 ± 0.58^a	7.06 ± 0.06^b	8.01 ± 0.09^c
10	3.39 ± 0.01^a	8.9 ± 0.13^b	ND

Values are means of three determinations ± SD. Values within each row with different superscripts are different ($p<0.05$) from one another. ND; not determined.

Table 3. Effects of storage temperature on the microbial quality (based on psychotropic plate counts [Log CFU/g] on marine agar) of Fortune Harbor mussels stored at three different temperatures

Storage time (days)	Storage temperature (°C)		
	-12	2	9
0	3.24 ± 0.06^a	3.24 ± 0.06^a	3.24 ± 0.06^a
3	3.39 ± 0.04^a	4.43 ± 0.03^b	6.40 ± 0.04^c
7	3.2 ± 0.07^a	6.22 ± 0.04^b	6.84 ± 0.24^c
10	3.25 ± 0.10^a	7.17 ± 0.03^b	ND

Values are means of three determinations ± SD. Values within each row with different superscripts are different ($p<0.05$) from one another. ND; not determined.

Effect of season of harvest and aquaculture site on the microbial quality of mussels

Changes in APC of mussels harvested from different aquaculture sites and from different seasons are shown in Table 4. Statistical analysis using ANOVA indicated that neither the aquaculture site/station nor the season of harvest affected the microbial quality of mussels, with few exceptions. Furthermore, various maximum acceptable levels of APC (10^5 to 10^7) in bivalves have been set in different countries (Colby *et al.*, 1995; Harrigan, and Park, 1991). However, it is generally recommended that APC should not exceed 10^5 or log 5 CFU/g. Therefore, the microbial quality of mussels harvested from different sites/stations, at different seasons and years were within the acceptable levels.

This perhaps can be explained, in part, by the fact that the mussels in Newfoundland and Labrador are cultured in cold ocean water and this limits the growth of harmful and spoilage bacteria.

Table 4. Effects of season of harvest and aquaculture site on the microbial quality[1] of cultured mussels from Fortune Harbour and Charles Arm

Month	Year	Charles Arm Station I	Charles Arm Station II	Fortune Harbor Station I	Fortune Harbor Station II
June	2000	3.40	3.39	3.32	3.32
Aug.	2000	3.37	3.45	3.47	3.37
Sept.	2000	3.31	3.36	3.36	3.31
Oct.	2000	3.26	3.28	3.28	3.23
April	2001	3.36	ND	3.27	ND
May	2001	3.37	3.39	3.16	3.17
June	2001	3.43	3.55	3.24	3.25
Aug.	2001	3.66	3.69	3.53	3.58
Oct.	2001	3.21	3.19	3.17	3.18

[1] based on aerobic plate counts (Log CFU/g) on marine agar. Values are means of three determinations. Standard deviation ranged from 0.01 to 0.04 for all samples. ND; not determined.

REFERENCES

Anonymous. 2000. Canadian aquaculture production statistics. Statistics Canada Catalogue. 23-603-UPE.

Atlas, R.M. 1995. *Handbook of media for environmental microbiology*. CRC Press, Boca Raton, FL, pp. 265-365.

Busta, F.F.; Peterson, E.H.; Adams, D.M. and Johnson, M.G. 1984. Colony count method. In *Compendium of methods for the microbiological examination of foods*. M. Speck (ed.) American Public Health Association 2nd. Inc., Washington, DC, pp. 62-111.

Colby, J.-W.; Enriquez-Ibara, L.G. and Flick, G.J. 1995. Shelf life of fish and shellfish. In *Shelf life studies of foods and beverages, chemical, biological, physical and nutritional*. G. Charalambous (ed.) Aspects, Elsevier Applied Science, New York, NY, pp. 85-143.

Cook, D.W. 1997. Refrigeration of oysters shellstock: conditions which minimize the outgrowth of *vibrio vulnificus*. *J. Food Protec.* 349-352.

Gilliland, S.E.; Michener, H.D. and Kraft, A.A. 1984. Psychrotrophic microorganisms. In *Compendium of methods for the microbiological examination of foods*. M. Speck (ed.) 2nd. Ed, American Public Health Association. Inc. Washington, DC, pp. 135-141.

Harrigan, W.F. 1998. Composition of cultured media. In *Laboratory methods in food microbiology*. 3rd. ed. Academic Press, London, UK, pp 79-88.

Harrigan, W.F. and Park, R.W.A. 1991. Methods for assessing microbiological quality. In *Making safe food: A management guide for microbiological quality*. Academic Press, London, UK, pp 99-112.

Hoi, L.; Larsen, J.L.; Dalsgaard, I. and Dalsgaard, A. 1998. Occurrence of *vibrio-vulnificus* biotypes in Danish marine environments. *Appl. Environ. Microbiol.* 64: 7-13.

Hunt, D.A.; Miescier, J.; Redman, J.; Salinger, A. and Lucas, J.P. 1984. Molluscan shellfish, fresh or fresh frozen oysters, mussels, or clams. In *Compendium of methods for the microbiological examination of foods*. M. Speck (ed.) 2nd. Ed., American Public Health Association. Inc. Washington, DC, pp. 590-610.

Jay, J.M. 1996. Seafoods. In *Modern food microbiology*, 5th ed., Chapman and Hall. New York, NY, pp. 118-130.

Neufeld, N. 1984. Procedures for the bacteriological examination of sea water and shellfish. In *Laboratory procedures for the examination of sea water and shellfish*. A.E. Greenberg and D. A. Hunt (eds.) 5th. ed., American Public Health Association. Inc. Washington, DC, pp. 37-63.

Ray, B. 1996. Spoilage of specific food groups. In *Fundamentals of food microbiology*. CRC Press, New York, NY, pp. 213-232.

Sakata, T. 1989. Microflora of healthy animals. In *Methods for the microbiological examination of fish and shellfish*. B. Austin and D. A. Austin (eds.) Halstead Press, London, UK, pp. 141-163.

Swartzentruber, A.; Schwab, A.H.; Duran, A.P.; Wentz, R.B.; Reed, Jr. R.B. 1980. Microbiological quality of frozen shrimp and lobster tail in retail market. *Appl. Environ. Microbiol.* 40 : 765-769.

CHAPTER 25

HIGH PRESSURE DESTRUCTION KINETICS OF MICROORGANISMS IN TROUT AND SHRIMP

N. Houjaij[1], H.S. Ramaswamy[1] and E. Idziak[2]

[1]Department of Food Science and Agricultural Chemistry
[2]Department of Natural Resource Sciences
Macdonald Campus of McGill University, Ste-Anne-de-Bellevue, Quebec, H9X 3V9, Canada

High pressure (HP) kinetics of the microbial destruction in both trout and shrimp was studied. Fresh samples of trout fillets and shrimp containing indigenous microflora of approximately 10^6 CFU/mL were vacuum sealed in pouches and subjected to HP treatment (200-400 MPa) with holding times of 0-90 min. The kinetic parameter, D-value, was evaluated based on first order rate of destruction following a small reduction in load ascribed to an instantaneous pressure kill (IPK) effect. Pressure sensitivity of kinetic parameters were evaluated based on pressure death time. The study showed a 5 and 4 log-cycle microbial reduction was achieved in trout and shrimp, respectively, by pressure treatment at 350 MPa for 60 min. On the other hand, a 20 min treatment at 400 MPa reduced the viable count to zero in trout while causing only a 2 log-cycle reduction in shrimp. The pressure sensitivity of microbial destruction was therefore higher in trout than in shrimp.

INTRODUCTION

Given the perishable nature of fish, development of satisfactory methods for processing and shelf life extension has for long preoccupied the attention of food technologists. Traditional methods of fish preservation include various forms of chilling (chilling in melting ice, chilling at subzero temperatures, super-chilling, ice slurries, refrigerated sea water), chilling coupled with barriers (chemical treatments, modified atmospheres, low-dose irradiation, enzyme/protein inhibitors), freezing, canning, curing, drying and fermentation. The principal agents of decay are endogenous enzymes and bacteria, and their effects are

© 2004 ScienceTech Publishing

apparent at different times. They produce two overlapping phases: autolysis, which disrupts the balance of the fish's metabolic functions, and bacterial activity, which produces more pronounced effects of spoilage.

The organisms of significance in the spoilage of aerobically refrigerated fish include (1) endogenous microorganisms which are aerobic psychrotrophic spoilage microorganisms e.g. *Pseudomonas, Alteromonas,* and to a lesser extent, *Flavobacterium, Achromabacter, Acinetobacter* and *Vibrio,* and (2) external sources of contamination such as handling, processing equipment, etc. (Kruk and Lee, 1982; Jaenicke, 1987).

Increasing consumer demand for minimally processed, additive-free, shelf-stable products has prompted search for novel physical treatments in the field of food processing. High pressure (HP) treatment to preserve foods has been known for a century (Hite, 1899). Progress though, initially, had been relatively slow since it was not commercially feasible to subject foods to pressures necessary to either preserve them or to makedly modify their functional qualities. However, more recently there have been significant developments in the engineering aspects of high-pressure equipment to suggest that the process is both economically and technically feasible (Farr, 1990). Success in the effectiveness of HP processing has led to the use of this technology as a viable means to reduce spoilage caused by microbial and enzymatic activities while minimizing the destruction of sensory and nutritional qualities of foods (Smelt, 1998).

The applications of HP in the food industry are extensive and several commercial products are being marketed today in Japan, Europe, and the United States of America. Extensive research work has been carried out on HP processing and its benefits, as well as its disadvantages in many different food products, but little is known about its effect on aquatic freshwater and marine organisms such as trout and shrimp. Further, not much is known about the effect of high-pressure processing on the microbiological quality of fresh fish. The practical use of HP processing requires careful adjustment of operating parameters like pressure, processing time and processing temperature. Optimal operating parameters would be those that offer enhanced quality and shelf-life extension without any detrimental effects. The main objective of this study was to evaluate the pressure destruction kinetics of microbial destruction of naturally occurring microflora in trout and shrimp.

MATERIALS AND METHODS

Sample preparation and treatment

Fresh samples of trout fillets and shrimp were obtained from a local fish market. Test samples were placed in an incubator for 5-6 h at 30°C in order to boost the indigenous microbial population to approximately 10^6 CFU/g. Samples were then vacuum sealed in polyethylene pouches, cooled and kept at 4°C prior to pressure treatment.

High pressure equipment

A cold isostatic press (ABB Cold Isostatic Press Model #CIP42260 – Autoclave Engineers, Subsidiary of ABB Autoclave System, Columbus, OH) with a cylindrical pressure chamber (55 cm height x 10 cm diameter) was used for pressure treatments. The equipment was rated for operation up to a maximum pressure of 414 MPa. Distilled water was used as the pressure transfer medium. The pressure come-up time ranged from 30 s to 3 min depending on the pressure level, and depressurization time was ~10 s. All treatments were given at 20°C, by quick dissipation of the generated heat. The details of pressure treatments given are detailed in Tables 1 and 2 for trout and shrimp, respectively. All samples, treated and control, were stored at 4°C, prior to enumeration.

Microbial Analysis

A standard plate count was used for the enumeration of microorganisms by using the pour plate method. An enriched medium, tryptic soy agar (TSA) (Fisher Scientific, Ottawa, ON) was used. A known weight of test sample (about 1 g) was transferred aseptically from test pouches into a stomacher bag to which 99 mL of 0.1% sterile peptone water was added and thoroughly mixed for 1 min. Samples were then transferred and plated onto TSA. The plated samples were then incubated at 25°C for 48 h and enumerated. The residual microbial count after a pressure treatment was expressed as a fraction based on the initial count (control).

Data Analysis

The pressure destruction of microorganisms was analyzed based on the dual effect of pressure inactivation behavior previously reported by Basak and Ramaswamy (1996) in studies carried out on pectin methyl esterase (PME) in orange juice. It was reported that an instantaneous change in the PME activity (IPK) resulted from a pressure pulse (pressurization to the desired level followed immediately by a quick de-pressurization, thus no holding time involved), after which the inactivation followed the conventional first order model with pressure holding time. As part of our experimental setup, a zero holding time was included in the experimental plan from which the IPK values could be calculated. The pressure inactivation kinetics of microorganisms during the pressure-hold time phase was analyzed assuming a first-order reaction indicating a logarithmic order of death, and expressed as:

$$Ln\ (N/No) = -k\ (t) \tag{1}$$

Where N = number of surviving microorganisms following pressure treatment at time t, No = initial number of microorganisms with no pressure treatment, k = reaction rate constant, The k-value was obtained as the negative slope from Eq. (1). The decimal reduction time, D value (time for 90% destruction), was obtained from the k-value:

$$D = 2.303/k \tag{2}$$

The pressure dependence of the kinetic parameters was analyzed by the pressure death time (PDT) model. In this model, logarithm of D values were plotted against the pressure and the pressure z-value (z_p) was obtained as the negative reciprocal of the slope of regression:

$$z_p = (P_2-P_1) / [\log(D_1/D_2)] \tag{3}$$

where D_1 and D_2 represent the D-values at pressures, P_1 and P_2, respectively, and z_p represents the pressure change that results in a 10-fold change in D-values.

RESULTS AND DISCUSSION

IPK and D values

Microbial survivor curves expressed in pressure-treated samples of trout fillets and shrimp are presented in Figure 1 as a semi-logarithmic relationship of survivor fraction ($x10^6$) with pressure treatment time. The associated IPK values are summarized in Table 2. From the figures, it can be seen that the destruction of microorganisms in the trout and shrimp is somewhat bi-phasic with an instantaneous drop in survivors due to a pressure pulse (pressure treatment with zero holding time) followed by conventional first order reaction kinetics. IPK was clearly noticeable at pressures higher than 250 MPa. As indicated in Table 2, IPK increased with pressure. Instantaneous pressure kill can occur due to rapid pressure pulses. Pressurization causes water compression, the magnitude of which is proportional to the amount of pressure applied. A sudden depressurization of a compressed system leads to an adiabatic expansion of water, which, in turn, will increase the impact on the cell membranes of the microbes. Hayakawl et al. (1998) reported that this force upon rapid depressurization is

Table 1. Experimental setup for microbiological evaluation of trout

Pressure (MPa)	Holding Time (min)			
250	0	45	90	-
300	0	30	60	-
350	0	15	30	60
400	0	20	40	-

Table 2. Experimental setup for microbiological evaluation shrimp

Pressure (MPa)	Holding Time (min)				
200	0	40	80	-	-
250	0	30	45	60	90
300	0	20	45	90	-
350	0	30	60	-	-
400	0	20	40	-	-

much stronger than that initially caused by pressurization. With bacterial spores, elevated temperatures around 70°C were needed in conjunction with pressure treatments to produce effective IPK's in order to soften the spore coats (Hayakawa et al., 1994). Ludwig and Shreck (1997) reported that the pressure destruction of *P. aeruginosa* changed considerably from first order kinetics to more complex ones when treatment temperatures were below 10°C.

The survivor curves of natural microflora in trout and shrimp as a function of pressure-treatment times (Figure 1) generally indicated that the pressure destruction during the holding period followed a first order model with $R^2 > 0.92$ (lowest at 250 MPa) for trout and R^2 values > 0.75 (lowest at 200 MPa) for shrimp (Table 3). Higher processing pressures and longer holding times clearly favored a larger destruction of microorganisms. Therefore, at higher operating pressures, the corresponding D values were lower for trout and shrimp. A comparison between the D values associated with pressure destruction of microorganisms in trout and shrimp revealed some interesting observations. At the lower pressure level of 250 MPa, the D value in trout was much higher than that in shrimp, but the trend got reversed after 300 MPa. In fact, at 400 MPa, it was possible to completely reduce the viable counts to zero in trout after a 20 min pressure treatment while it only resulted in a 2-log cycle reduction of survivors in shrimp.

Pressure sensitivity

The pressure sensitiveity of D-values is normally expressed as a pressure resistance curve with log D-values plotted against pressure as described in

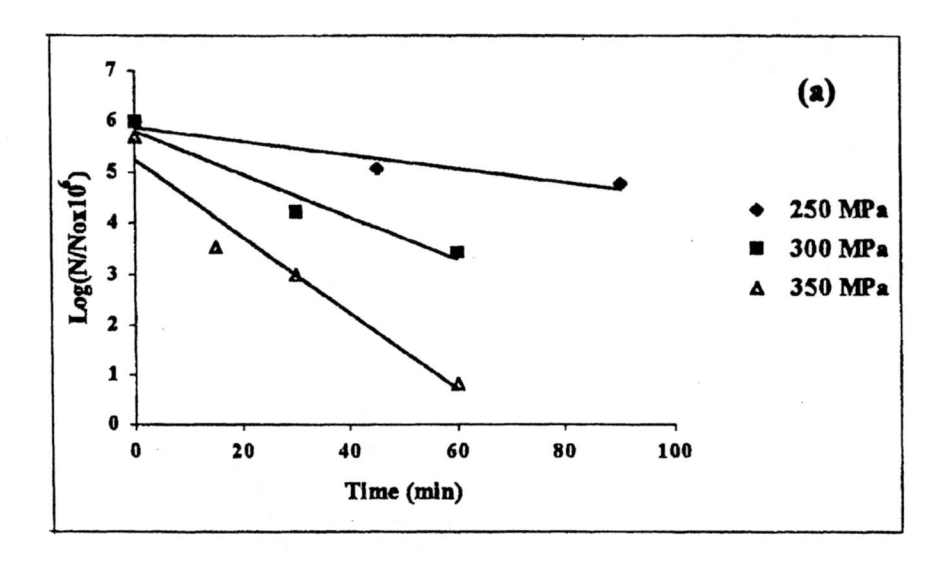

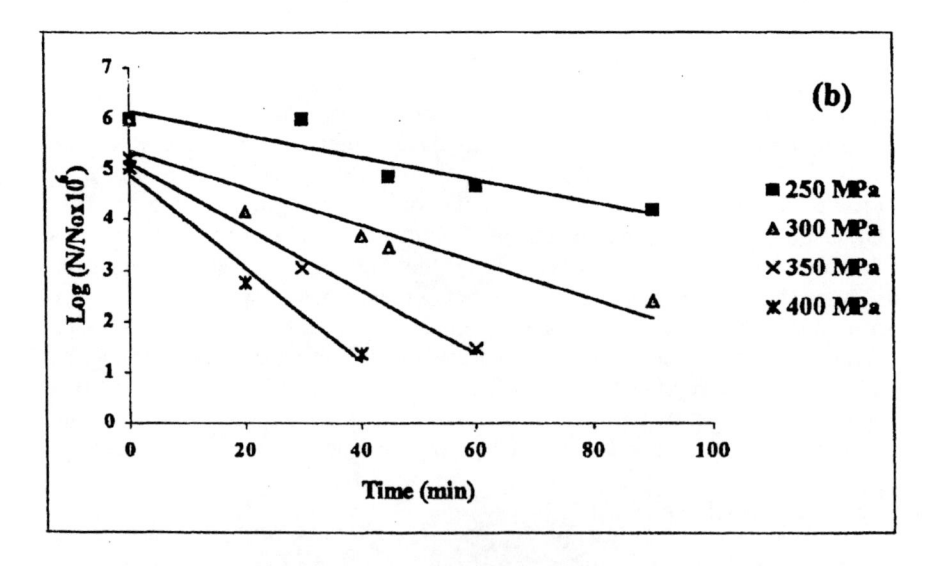

Figure 1. Survivor curves of microorganisms in (a) trout and (b) shrimp at different pressures.

Figures 2 (a) and (b) for trout and shrimp, respectively. The negative reciprocal slope of the curve is the z_p value. Alternately, these curves can also be represented by an Arrhenius model with ln(k) plotted against pressure (Mussa, 1998). Lower values of z_p indicate higher sensitivity of D values to pressure. The computed z values and the associated R^2 are also shown in Table 3. Trout had about 10% lower z value than shrimp. Hence, microbial destruction D values in shrimp are more pressure sensitive than their counterparts in shrimp.

HP treatment for microbial control

It has been recognized that HP treatment, especially at pressures higher than 300 MPa and longer holding times, can be responsible for the alteration of certain quality attributes in fish. Indeed, such changes were observed in previous studies (Houjaij, 2000) depending on the combination of selected process variables. Optimized process variables for trout and shrimp qualitative attributes gave pressure in the range of 175-200 MPa, a holding time of approximately 30 min and a temperature of 10°C. Unfortunately, such HP treatments are inadequate from the point of view of microbial destruction sought in these products (trout and shrimp) for shelf-life extension.

Like their thermal processing counterparts, a more severe pressure treatment as necessary for microbial destruction generally imparts a cooked appearance and results in some changes in physicochemical characteristics of fish. Such changes may not be undesirable if the finished product were to be cooked prior to consumption. The present study indicated that approximately a 5 log-cycle reduction in microbial count was achievable in trout at 350 MPa and for 60 min. At 400 MPa, there was no count after 20 min. While these times are still fairly long for commercial applications, the lower D at higher pressure concept would mean one could reduce the treatment time significantly at higher pressures. This is even more so with shrimp because even after a 60 min treatment at 350 MPa, only a 4 log-cycle reduction in viable counts were achieved. Even so, since the associated z value was 165 MPa, it can be shown that the time could be reduced to under 10 min at 500 MPa for a similar destruction, and that a treatment less than a minute at 700 MPa would result in a nil count. In general, pressures between 300 MPa and 600 MPa have been shown to inactivate yeasts, molds and most vegetative bacteria including most infectious food-borne pathogens. Bacterial spores on the other hand, can only be killed by very high pressures >1000 MPa in combination with high temperatures.

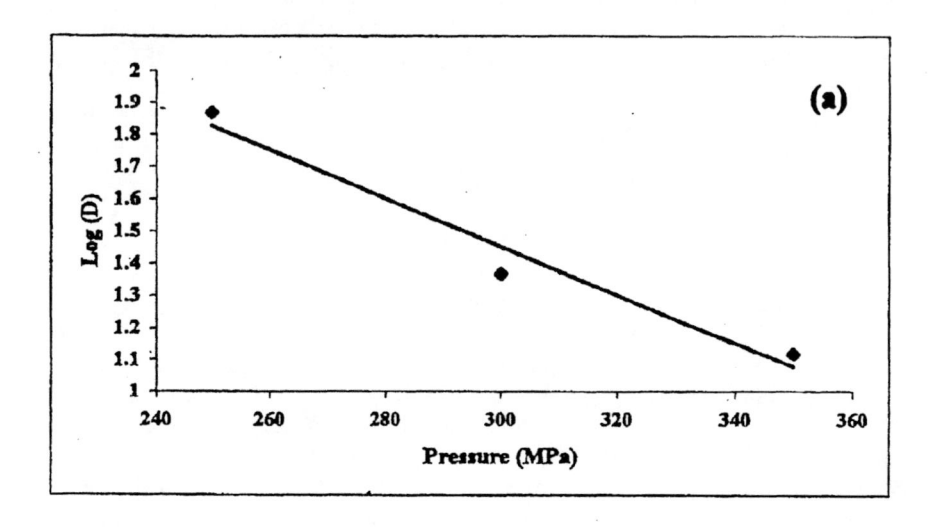

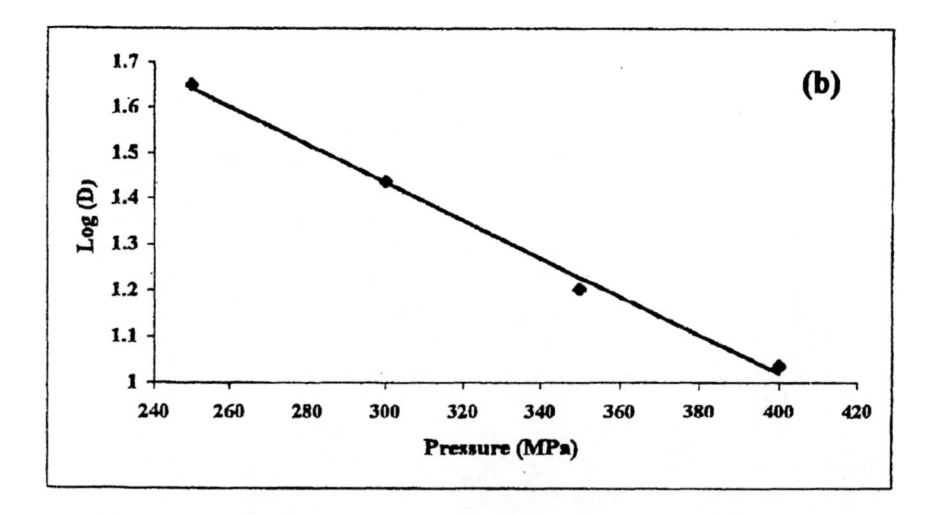

Figure 2. Decimal reduction time curve (D) for microorganisms in pressure treated (a) trout and (b) shrimp.

Table 3. Calculated IPK values for pressure-treated trout and shrimp

Pressure (MPa)	IPK (log) Trout	IPK (log) Shrimp
200	-	0.0015
250	0.0248	0.0015
300	0.1249	0.1326
350	0.3010	0.8985
400	1.2253	1.0746

Table 4. k and D values for the native microorganisms in trout and shrimp samples treated at different pressure levels.

Pressure (MPa)	Trout D value (min)	R^2	Shrimp D value (min)	R^2
250	70.27	0.91	43.03	0.83
300	23.38	0.94	27.34	0.88
350	12.69	0.91	15.88	0.99
400	-		10.81	0.99
	Z_p = 135 MPa		Z_p = 165 MPa	
	R^2 = 0.97		R^2 = 0.90	

CONCLUSIONS

Fresh samples of trout fillets and shrimp containing natural microbial populations of approximately 10^6 CFU/g were subjected to HP treatment at room temperature (20°C). The microbial destruction studied showed that the pressure destruction followed a dual destruction kinetics characterized by an instantaneous pressure kill (IPK) dependent only on the pressure level and a rate constant (decimal reduction time, D), also dependent on pressure level. The associated IPK values were slightly higher in trout. As expected, higher pressures had lower D values. The pressure z-value (z_p) in trout samples was lower than in shrimp indicating the D values in trout to be more pressure sensitive than in shrimp.

ACKNOWLEDGEMENTS

Financial support for this project from the Ministry of Agriculture and Fisheries of Quebec (MAPAQ) is greatly appreciated.

REFERENCES

Basak, S. and Ramaswamy, H.S. 1996. Ultra high pressure treatment of orange juice A kinetic study on inactivation of pectin methyl esterase. *Food Res. Int.* 29: 601-607.

Farr, D. 1990. High pressure technology in the food industry. *Trends Food Sci. Technol.* 1: 15-16.

Hayakawa, I.; Kanno, T.; Tomita, M. and Fujio, Y. 1994. Application of high pressure for spore inactivation and protein denaturation. *J. Food Sci.* 59: 159-163.

Hayakawa, I.; Furakawa, S.; Midzunaga, A.; Horiuchi, H.; Nakashima, T.; Fuji, Y.; Yano, Y.; Ishikura, T. and Sasaki, K. 1998. Mechanism of inactivation of heat-tolerant spores of *Bacillus stearothermophilus* IFO 12550 by rapid decompression. *J. Food Sci.* 63: 371-374.

Hite, B.H. 1899. The effect of pressure on the preservation of milk. West Virginia *University Agricultural and Experimental Station Bulletin* 58: 15-35.

Houjaij, N. 2000. The application of high pressure treatment and its effect on the quality attributes of trout and shrimp. M.Sc. Thesis. Department of Food Science and Agricultural Chemistry, Macdonald Campus, McGill University, Montreal, QC.

Jaenicke, R. 1987. Cellular components under extremes of pressure and temperature: Structure-function relationship of enzymes under pressure. In *Current Perspectives in High Pressure Biology*. H.W. Jannasch, R.E. Marquis and A.M. Zimmerman (eds.) Academic Press, London, UK, pp. 252-272.

Kruk, M. and Lee, J.S. 1982. Inhibition of *Escherichia coli* trimethylamine-N-oxide reductase by food preservatives. *Food Process.* 45: 241-243.

Ludwig, H. and Schreck, C.H. 1997. The inactivation of vegetative bacteria by pressure. In *High pressure research in the biosciences and biotechnology: Proceedings of the XXXIVth Meeting of the European High Pressure Research Group*. K. Heremans (ed.). Leuven University Press, Leuven, Belgium, 534 p.

Mussa, D.M. 1998. High pressure processing of milk and muscle foods: evaluation of process kinetics, safety and quality changes. Ph.D. Thesis. Department of Food Science and Agricultural Chemistry, Macdonald Campus, McGill University, Montreal, Quebec.

Smelt, J.J.P.P.M. 1998. Recent advances in the microbiology of high pressure processing. *Trends Food Sci. Technol.* 9: 152-158.

CHAPTER 26

IDENTIFICATION OF SPOILAGE MICRO-ORGANISMS AND RESEARCH ON QUALITY INDICES FOR COLD-SMOKED SALMON

Françoise Leroi[1], Jean Jacques Joffraud[1], Frédérique Chevalier[1], Mireille Cardinal[1], Valérie Stohr[1], Jean-Louis Berdague[2] and Jean-Luc Vallet[1]

[1]IFREMER, Laboratoire de Génie Alimentaire, BP 21105, 44 311 Nantes Cedex 3, France
[2]INRA DE THEIX, Station de Recherches sur la Viande, 63122 Saint Genès Champanelle, France

> Five isolates of 9 bacterial groups were inoculated in sterile cold-smoked salmon blocks and microbial, chemical and sensory changes were analyzed during chilled vacuum storage. The principle component analysis of the sensory data allowed association of some specific descriptors to the volatile compounds. Lactobacillus sakei, L. farciminis and Brochothrix thermosphacta were the more active spoilers, whereas Shewanella putrefaciens and L. alimentarius were neutral. Photobacterium phosphoreum, Aeromonas spp., Carnobacterium piscicola and Serratia liquefaciens produced weak or strong off-odors depending of the strains. Quality indices for cold-smoked salmon were researched in thirteen batches representative of the French production. Shelf-life was very variable (1->6 weeks) and was related to the initial Enterobacteriaceae load, depending of hygienic conditions in the smokehouse. A high correlation existed between the remaining-shelf-life (RSL) and lactobacilli and yeasts counts and total volatile basic-nitrogen (TVBN) concentration.

INTRODUCTION

Vacuum-packed sliced cold-smoked salmon is a highly perishable product, because of light preservative treatments (salt ranging from 2.5 to 3.5% (w/w) and phenol generally less than 0.5 mg 100 g^{-1}) and no other additives such as nitrate or nitrite allowed in France. Many authors (Truelstrup Hansen *et al.*, 1996; Joffraud *et al.*, 1998) have demonstrated that sensory deterioration of cold-smoked salmon was mainly caused by microorganisms, with off-odors and off-flavors described as sour, acid, pungent and occasionally fecal (Truelstrup Hansen, 1995; Leroi *et al.*, 1998). However, spoilage mechanisms are not well understood. Different studies indicate that various bacterial groups, including lactic acid bacteria, marine Vibrionaceae, Enterobacteriaceae and *Brochothrix thermosphacta* dominate the cold-smoked salmon microflora (Truelstrup Hansen *et al.*, 1995, 1998; Leroi *et al.*, 1998; Lyhs *et al.*, 1998; Paludan Müller *et al.*, 1998; Truelstrup Hansen and Huss, 1998; Leroi *et al.*, 2000; Jorgensen *et al.*, 2000), but only some species might participate in spoilage (Gram and Huss, 1996; Leroi *et al.*, 1998) and until now the literature has not yet established a clear link between the aroma and specific microbial species. The first purpose of this study was to investigate the off-odor production and the associated volatile compounds of bacteria belonging to nine different taxonomic groups inoculated on sterile cold-smoked salmon blocks. The second purpose was to develop a multiple compound quality index for cold-smoked salmon, as many authors have shown that neither microbial nor chemical parameters could be used alone as a quality indicator for cold-smoked salmon (von Rakow, 1977; Cann *et al.*, 1984; von Hildebrandt and Herol, 1988; Dodds *et al.*, 1992; Truelstrup Hansen *et al.*, 1995, 1996, 1998; Truelstrup Hansen and Huss, 1998).

MATERIALS AND METHODS

Identification of spoilage microorganisms

Five isolates of 9 bacterial groups, previously isolated from cold-smoked salmon by Leroi *et al.* (1998), were studied. The bacterial groups examined were *Aeromonas* spp, *Brochothrix thermosphacta*, *Carnobacterium piscicola*, *Lactobacillus alimentarius, sakei and farciminis*, *Photobacterium phosphoreum*, *Serratia liquefaciens* and *Shewanella putrefaciens*. Sterile cold-smoked salmon blocks were prepared as described by Joffraud *et al.* (1998). For each strain, one bag of 360 g sterile blocks was inoculated with 2% (v/w) of the appropriate diluted culture in order to obtain an initial cell concentration of 10^4-10^5 cfu g^{-1}. A control assay was conducted in the same way by inoculation of 2% of sterile physiological water. Blocks were then allotted in 17 different bags required for all kinds of analyses, vacuum-packed and stored at 6°C for 5 weeks. Directly after inoculation and each week until week 5, one 15 g -bag of each assay sample was opened for microbial analysis as described by Stohr *et al.* (2001).

Remaining 20 g-packages were used after 5 weeks for sensory and chemical analysis *i.e.* pH, total volatile basic nitrogen (TVBN), trimethylamine (TMA) and volatile compounds.

Research of quality indices

Thirteen batches of sliced vacuum-packed cold-smoked salmon (Atlantic *Salmo salar*) representative of the French traditional production *i.e.* dry salted and traditionally smoked at temperature ranging from 20 to 26°C, were collected just after processing in 5 French smokehouses and transported to the laboratory in frozen conditions. Three to five batches studied in a work session, including 35 100-200-g bags for each batch, were thawed overnight and stored at 5°C for 5-6 weeks. Each week from week 0 until one week after sensory spoilage was evident, microbial, chemical and sensory analysis were performed.

Microbial analysis

Total psychrotrophic microorganisms (TPC), lactic acid bacteria (LAB), lactobacilli, *B. thermosphacta*, Enterobacteriaceae and yeast were enumerated under conditions described by Leroi *et al.* (2001).

Chemical analysis

TVBN and TMA were measured in duplicate by the Conway micro-diffusion method (Conway and Byrne, 1933). The pH value was measured in the five-fold-diluted flesh with a pH-meter (Mettler Delta 320, AES, Combourg, France). Lipids, dry matter, sodium chloride and total phenols were quantified by methods described by Leroi *et al.* (2000). The volatile components, desorbing from salmon, were extracted at room temperature (20°C) by a stream of helium and separated and identified by gas chromatography-mass spectrometry (GC-MS) according to the procedure described by Joffraud *et al.* (2001).

Sensory analysis

For the inoculated samples, twelve trained panellists were asked to perform a profiling test using 14 attributes previously chosen as good spoilage indicators for cold-smoked salmon (Stohr *et al.*, 2001). During each session, six 20 g-inoculated samples and one control were presented in a randomised order to panellists, who smelled and marked each attribute on a non-structured 80 mm line scale, anchored at each end. At the end of the profiling test, panelists had to classify each sample depending on their spoilage level (NS: non-spoiled, LS: lightly spoiled; SS: strongly spoiled).

For commercial samples, two to three bags per lot were opened and divided in 20-g portions in aluminium foil in order to keep the odors intact. Fourteen trained panellists smelled the samples and classified them in the 3 classes described previously. The sensory rejection time (SRT) was determined when at least 7 judges estimated that the product was in the SS class. The remaining shelf-life (RSL) of a sample was the difference between the SRT in week (known at *posteriori*) and the week of analysis.

Statistical analysis

Production of volatile compounds in the inoculated samples and results of the profiling tests were treated by principal component analysis (PCA). Results of the classification in 3 spoilage levels were analysed by correspondence factorial analysis (CFA) to discriminate isolates according to their spoilage potential. For quality indices, one-way variance analysis (ANOVA, Statgraphics Plus, version 4, Sigma Plus) was used to test differences between groups of samples having the same RSL using successively each microbiological or chemical index. Means were compared by the least significance difference (LSD) test at a 0.05 level of probability. For the multi-factorial approach, a polynomial fitting the RSL to the microbiological and chemical data was calculated using the stepwise forward multiple regression method (Statgraphics Plus). This method is preferred to classical multiple regression when correlation between factors is suspected.

RESULTS

Identification of spoilage microorganisms

All species inoculated on sterile cold-smoked salmon blocks contaminated and multiplied well (data not shown). With the exception of *Sh. putrefaciens* and *P. phosphoreum*, all groups reached their maximum levels after two or three weeks of storage and their final counts varied generally between 10^8 and 10^9 cfu g^{-1}.

The results of the profiling test were treated by principal component analysis. The projection of the individuals on the plane formed by the first two components of the PCA which restored 75% of inertia (Figure 1) showed that the bacterial group used to inoculate the product had a real effect on the off odors produced. In most cases, specific sensory descriptors may be correlated to

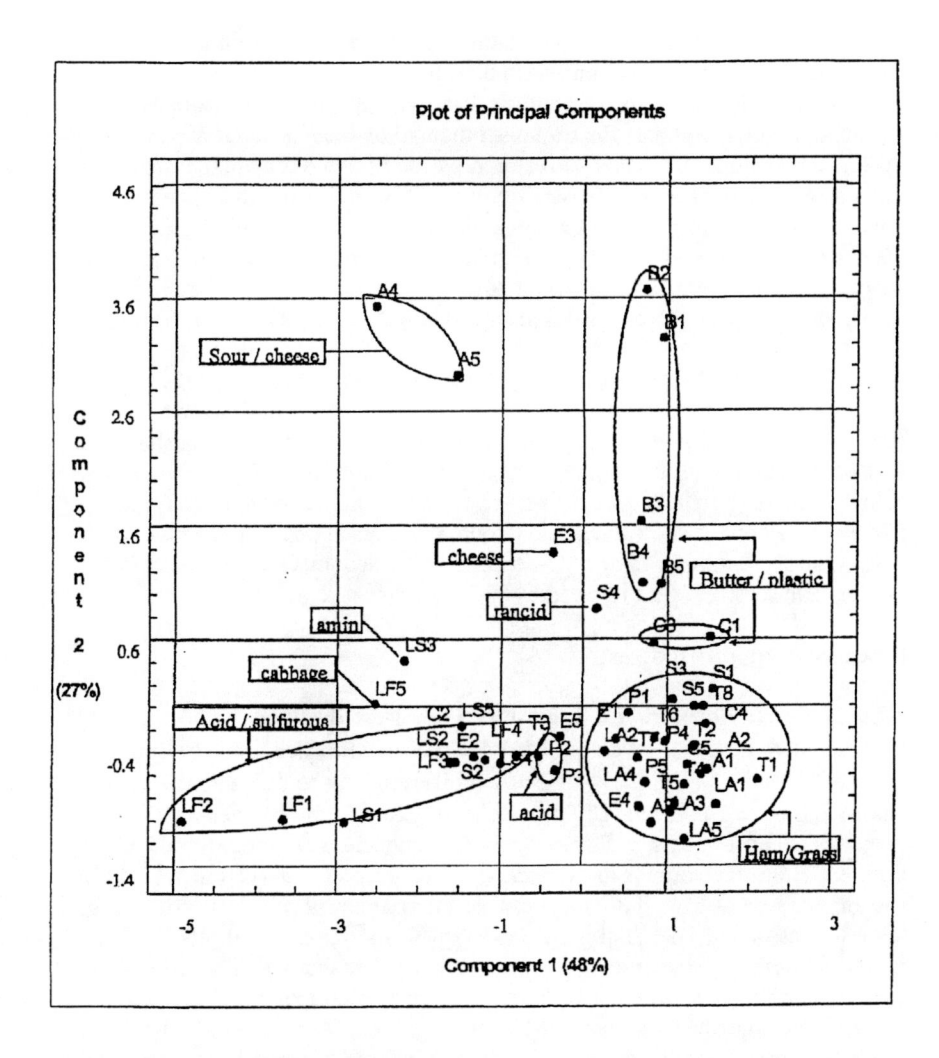

Figure 1. Correlation between bacterial strains and sensory descriptors according to plane 1-2 of principal component analysis. Component 1: 48% of variance; Component 2: 27% of variance. A: *Aeromonas* spp; B: *Brochothrix thermosphacta;* C: *Carnobacterium piscicola*; LA: *Lactobacillus alimentarius*; LS: *Lb sakei*; LF: *Lb farciminis*; P: *Photobacterium.*

a given bacterial species. The same mathematical treatments were performed for the volatile compounds as summarised in Table 1.

Correspondence factorial analysis realised for different spoilage levels allowed characterisation of the spoilage potential of each bacterial group (Figure 2). The total information may be restored by the first two principal components. Therefore the first axis condensed by itself 80% of inertia and appeared as a "spoilage axis" ascending from a non-spoiled to a strongly spoiled pole. On the other hand, the second axis, representing 20% of inertia, discriminated mildly spoiled samples from a non-spoiled to a lightly spoiled pole. Three spoilage ability behaviors could be distinguished. Group 1 was composed of samples inoculated with *Sh. putrefaciens*, *Lb. alimentarius* and the controls and were considered as not spoiled by the panel. Group 2 included species that spoiled moderately the product. It included *P. phosphoreum*, *Aeromonas* spp, *C. piscicola* and *S. liquefaciens*. The behavior of these bacterial groups was variable depending on the isolates considered. Isolates of the same species were able to spoil the product strongly while others should not be particularly able to affect sensory quality detrimentally. Group 3 was made up of *Lb. farciminis*, *Lb. sakei* and *B. thermosphacta*. All the isolates tested in this case were strong spoilers and samples were rejected by the sensory panel.

Research of quality indices

Characterisation of samples. The thirteen lots were relatively homogeneous in their initial composition, with 14.0 ± 1.4% (95% confidence limit) of lipid, 60.5 ± 1.8% of water, 3.1 ± 0.3% of NaCl (corresponding to 5.2 ± 0.5% in water phase) and 0.55 ± 0.15 mg 100 g^{-1} of phenols. Shelf-life observed by the panel ranged between 1 and more than 6 weeks (data not shown). Initial flora was very different from one sample to another, ranging from 10^2 to 10^6 cfu g^{-1}. During vacuum storage at 5°C, TPC increased to a maximum level of 10^7-10^9 cfu g^{-1}, more or less quickly depending of the samples, and remained at this level until spoilage sometimes several weeks later (data not shown). Variation in the composition of microflora between lots was very important and three scenarios could be distinguished. In scenario 1 (lots 3, 4, 6 and 10, Figure 3a), lactobacilli were the dominating flora. *B. thermosphacta*, yeasts and Enterobacteriaceae were in a minority, never exceeding 1% of TPC. In scenario 2 (lots 1, 2, 5 and 9, Figure 3b), the spoilage microflora was mainly represented by lactobacilli and Enterobacteriaceae and to a lesser extent by yeasts. *B. thermosphacta* counts were generally below the detection threshold. In scenario 3 (lots 7 and 8, Figure 3c) TPC was dominated by total LAB and *B. thermosphacta*. LAB probably belonged to the *Carnobacterium* genus because the lactobacilli count on Rogosa

Table 1. Odours and volatile compounds released by bacterial strains

Strains	smells	volatile compounds
Lactobacillus alimentarius	neutral green	acetic acid
Lactobacillus sakei	sulphurous weakly acid	
Lactobacillus farciminis		
Carnobacterium piscicola	butter/plastic rubber neutral	2,3-butanedione
		2,3-pentanedione
Brochothrix thermosphacta	butter/plastic rancid	2-heptanone
		2-hexanone
Photobacterium phosphoreum	acid, neutral	
Aeromonas sp.	cheese/sour	TMA
Shewanella putrefaciens	neutral	dimethyldisulfide
Serratia liquefaciens		2,3-butanol
		2-pentanol

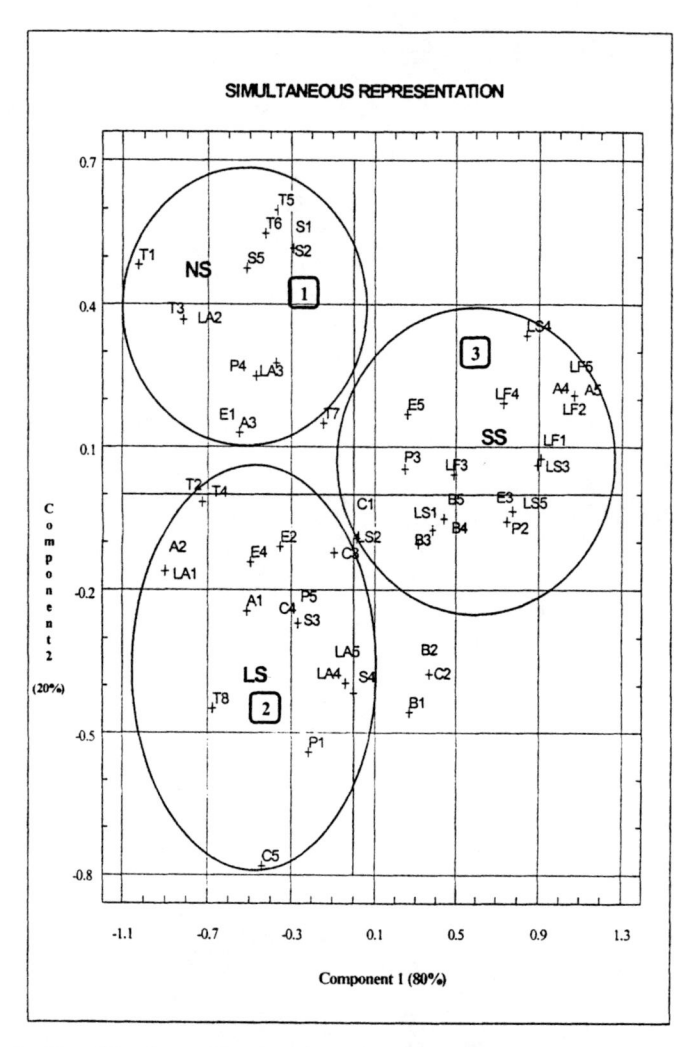

Figure 2. Classification of bacterial strains depending on their spoilage abilities (NS: not spoiled; LS: lightly spoiled; SS: strongly spoiled) according to plane 1-2 of the correspondence factorial analysis. A: *Aeromonas* spp; B: *Brochothrix thermosphacta;* C: *Carnobacterium piscicola*; LA: *Lactobacillus alimentarius*; LS: *Lb sakei*; LF: *Lb farciminis*; P: *Photobacterium*.

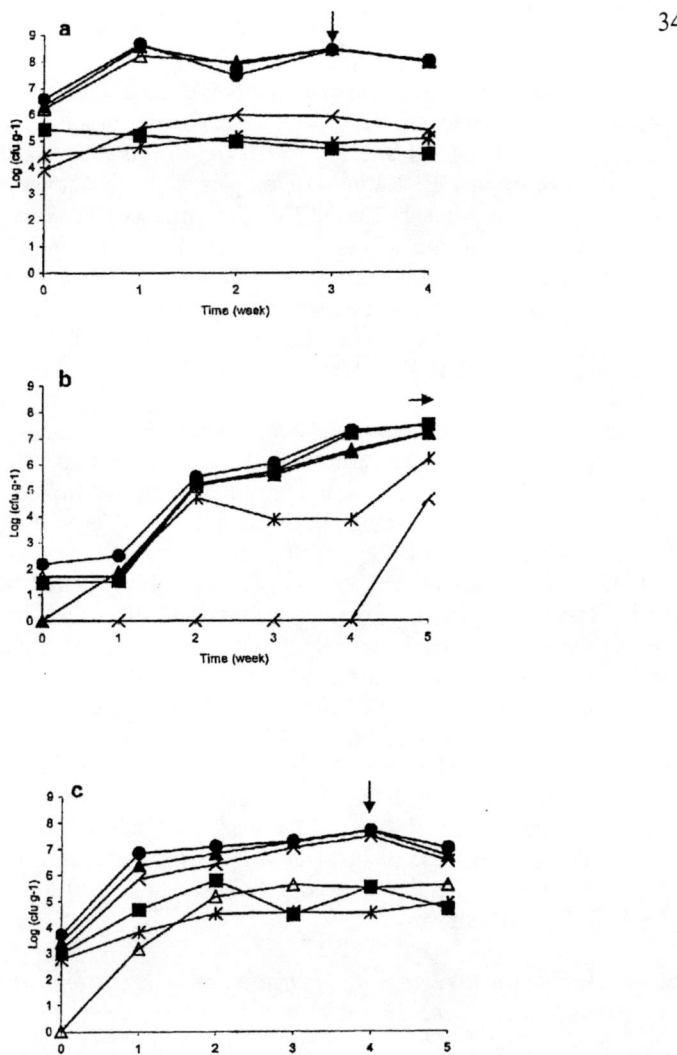

Figure 3. Micro-flora evolution in a) sample 3, b) sample 9 and c) sample 7 during the vacuum storage at 5°C. ●: total psychrotrophic count; ▲: total lactic acid bacteria; Δ: lactobacilli; ■: Enterobacteriaceae (VRBG total count); x: *Brochothrix thermosphacta*; *: yeast. Arrow indicates the sensory rejection time.

agar was always 2 log lower than count on NAP. Just after the smoking process, TMA and TVBN concentrations were rather constant in all the lots, with average values of 1.6 ± 0.1 and 16.1 ± 0.3 mg-N 100 g^{-1}, respectively. During the vacuum storage at 5°C, 2 groups of samples could be distinguished: Group 1 (lots 2, 5, 7, 8, 9) in which TMA and TVBN never exceeded 6 and 30 mg-N 100 g^{-1}, respectively, and group 2 (lots 1, 3, 4, 6, 10, 11, 12, 13) for which TMA and TVBN reached always concentrations higher than 11 and 37 mg-N 100 g^{-1} respectively. The pH, initially equal to 6.20 ± 0.04, was rather constant during the storage of most of the samples except for samples 3, 6 and 10 in which a significant acidification to pH 5.9 was observed (data not shown).

Relationship between shelf-life and initial composition. The results of fitting a multiple linear regression model to describe the relationship between shelf-life and initial pH, lipid, water, NaCl and phenol contents confirmed that there was no statistically significant correlation between the variables at the 90% or higher confidence level (data not shown). The relationships between shelf-life and initial microbiological load of the samples were investigated. Results of the stepwise forward multiple regression showed that the shelf-life was mostly linked to the initial Enterobacteriaceae count ($P<0.05$), the higher initial total count on VRBG agar, the shorter the shelf-life. However, the low R-squared statistic (0.69) indicated that this measure could not be used alone to precisely predict the shelf-life. The initial level of Enterobacteriaceae seemed to be related to the smokehouse rather than to the raw material quality. Samples coming from plants C and D, which had the shorter shelf-life (1-3 weeks), had an initial Enterobacteriaceae load always higher than $10^{4.6}$ cfu g^{-1}, whatever the raw material processed in these plants, and samples from plants A, B and E, which had the longer shelf-life (4->6 weeks), had an initial load always lower than $10^{3.4}$ cfu g^{-1} (data not shown).

Relationship between remaining shelf-life and microbiological and chemical data. When observing the microbial growth curves for the thirteen samples, it appeared difficult to find a single rule for prediction of shelf-life. In some cases, the product was rejected several weeks after all the enumerated microorganisms had reached their maximum levels and in other cases very early during the exponential growth phase of the microorganisms. Also different microorganisms *i.e.* lactobacilli, lactobacilli/Enterobacteriaceae and carnobacteria/*B. thermosphacta* dominated the spoilage microflora at the SRT. Results of the one-way ANOVA confirmed that there was no statistical difference between groups of samples that had reached their lifetime and samples that were not yet rejected by the panelists for any of the microbiological responses measured. As an example, Figure 4a shows the means plot and 95% LSD intervals for TPC. Although the average TPC was lower at the beginning of the storage period (RSL of 3 to 5 weeks), no significant difference could be observed between samples with RSL ranging between 2 and –2 weeks. On the opposite, TVBN index could be used to discriminate samples with a RSL of 1, 0, -1 and –2 weeks but not samples with a higher RSL (Figure 4b). TMA was less discriminant and no statistical difference in the pH means was noticed. Although TVBN concentration in the flesh seemed to be of most value for estimation of cold-smoked salmon quality, it could not be used alone to precisely predict the shelf-life. A multi-factorial approach was developed using the forward stepwise multiple regression method. Results showed that there was a statistically significant relationship at the 99% confidence level between the RSL and lactobacilli count and TVBN concentration, and at the 95% confidence level for yeast count. The equation of the fitted polynomial model was: $RSL_{(week)} = 5.65 - 0.31 \times Log\ (OGA\ count)_{cfu\ g^{-1}} - 0.25 \times Log\ (ROG\ count)_{cfu\ g^{-1}} - 0.06 \times (TVBN)_{mg-N\ 100\ g^{-1}}$. The model was successfully validated with 3 left out samples (data not shown). R^2 indicated that the model explained 80% of the variability in the RSL. R^2 was not significantly increased by adding the other microbial and chemical descriptors indicating they were either not good quality indices for smoked salmon, or highly correlated with the 3 selected factors. Figure 5 represents the RSL as a function of lactobacilli count and TVBN concentration for a yeast count fixed at 10^4 cfu g^{-1}. Assuming that lactobacilli could not exceed 10^9 cfu g^{-1}, a minimum of 36 mg-N 100 g^{-1} was necessary for a product to be rejected (RSL = 0). With lower values such as 10^7, 10^4 or 10^2 cfu g^{-1}, products were rejected for TVBN concentrations reaching 44, 57 and 65 mg-N 100 g^{-1}, respectively.

DISCUSSION

Among the nine bacterial groups investigated, three Gram-positive bacteria could be considered as strong specific spoilage organisms (*B. thermosphacta, Lb. sakei* and *Lb. farciminis*) and two Gram-negative organisms as weak spoilers, depending on the strain (*S. liquefaciens* and *P. phosphoreum*). These results are very different from those observed on other fresh or vacuum-packed fish products where *Sh. putrefaciens* and *P. phosphoreum* are the main spoilage organisms (Gram *et al.*, 1987; Dalgaard, 1995; Gram and Huss, 1996). In cold-smoked salmon, *Sh. putrefaciens* did not really spoil the product and did not release strong off-odors. This is in accordance with the moderate growth on the product. However, this species was one of the highest producers of TMA (data not shown), and this tended to prove that TMA can not be considered as a good

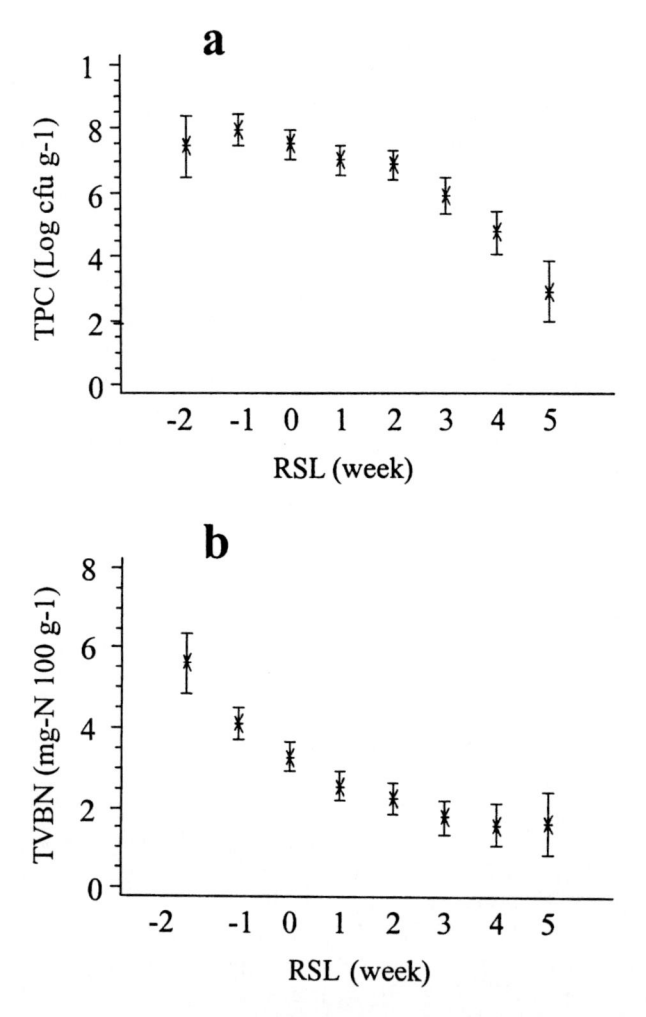

Figure 4. Means plot and 95% intervals for (a): total psychrotrophic count and (b): total volatile basic nitrogen versus remaining shelf-life.

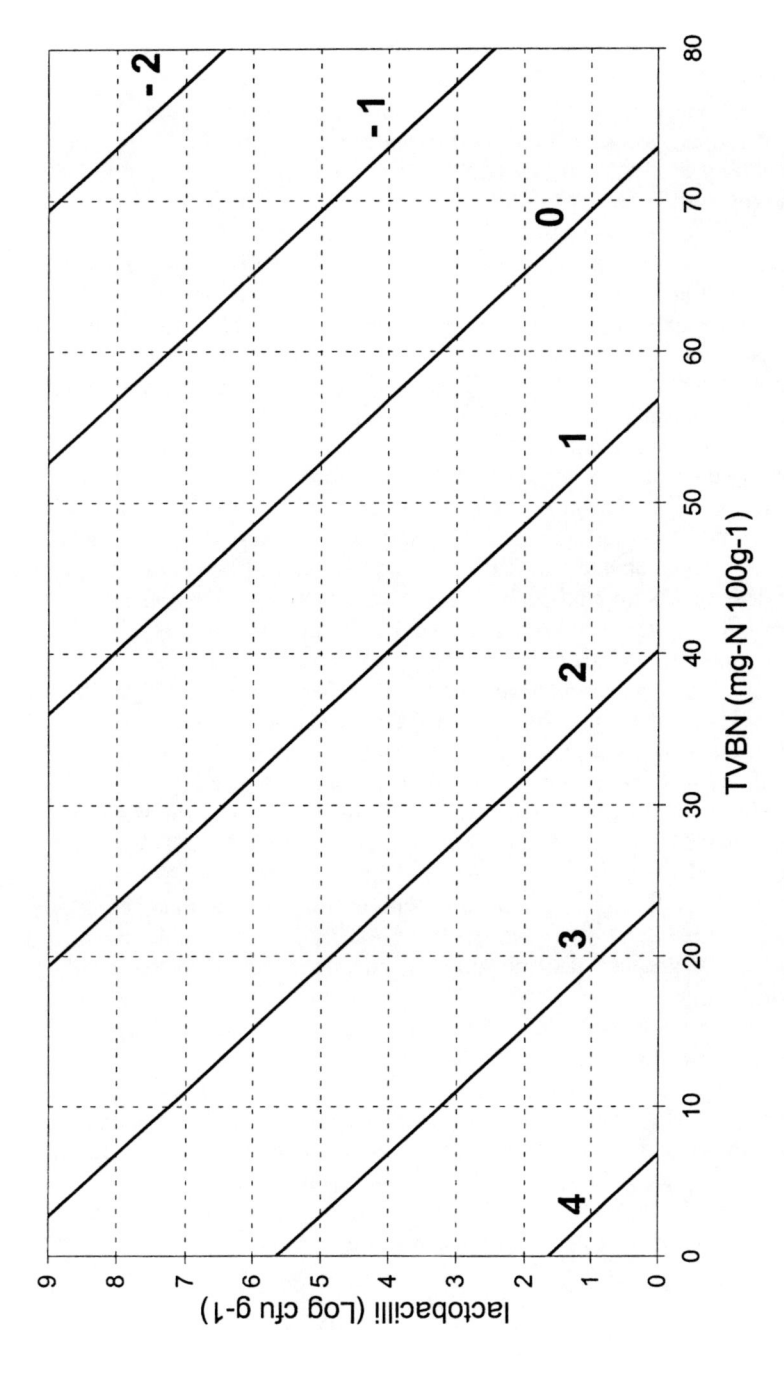

Figure 5. Isoresponse curves for cold-smoked salmon remaining shelf-life (week) versus total volatile basic nitrogen concentration and lactobacilli count on Rogosa agar (yeast count fixed to 10^4 cfu g^{-1}).

spoilage indicator for cold-smoked salmon. *S. liquefaciens* has been shown to be one of the major Gram-negative bacteria responsible for spoilage even though no particular off-odors could be associated with this organism. *S. liquefaciens* was the main representative species of *Enterobacteriaceae* isolated from cold-smoked salmon by Truelstrup-Hansen (1995) and in our laboratory (unpublished data). Hydrogen sulfide and methyl disulfide production has been linked to the presence of *Enterobacteriaceae* in fresh meat by Dainty and Makey (1992). The role of *P. phosphoreum* in spoilage was quite similar to that of *S. liquefaciens*. In most cases no sensory descriptor could be associated with the sensory quality damages but two isolates were characterized by acidic odors. The production of these types of aromas has already been revealed by Truelstrup Hansen (1995) who underlined the production of acetic acid from amino acids and carbohydrates. The greater cold-smoked salmon spoilers were undeniably encountered among the Gram-positive bacteria, for which homogeneity of behavior among the isolates was important. *B. thermosphacta, Lb. sakei* and *Lb. farciminis* appeared as the main organisms involved in spoilage. For *B. thermosphacta*, the strong off odors detected were associated with butter/plastic or rancid descriptors, linked to a high desorption of ketones (2-heptanone and 2-hexanone). The production of such off-odors by this bacterial species has already been noticed on spoiled fresh red meat by Dainty and Mackey (1992). Moreover, Talon *et al.* (1994) indicated that *B. thermosphacta* produced acids during meat spoilage which conferred rancid and a cheesy odor to the product. *Lb. sakei* and *Lb. farciminis* isolates released strong sulfurous, and to a lesser extent acid odors which could be linked to acetic acid revealed by GC-MS analysis. Sulfurous odors have not been directly linked to hydrogen sulfide desorption which was not well detected by this method. On vacuum-packed meat, Dainty and Mackey (1992) and Borsch *et al.* (1996) indicated that LAB produced lactic and acetic acids from sugars associated with strong acid odors. Huss *et al.* (1995) noticed the same phenomenon on fresh vacuum-packed cod. Lastly, Truelstrup Hansen (1995) showed that LAB isolated from cold-smoked salmon were able to produce acid and sulfurous odors on salmon blocks. Within the *Lactobacillus* species, *Lb. alimentarius* isolates did not spoil the product. This species produced TVBN in quantities two fold less than the other species, although the acidification was identical (data not shown). This result seems to indicate that acid components are not sufficient to provoke sensory rejection of the product by the panel whereas volatile amines are more directly involved. *C. piscicola* isolates were not considered as spoilage organisms by the panel. Leroi *et al.* (1996), Paludan-Muller *et al.* (1998) and Duffes *et al.* (1999) had previously suggested that Carnobacteria were not involved in cold-smoked salmon spoilage. Butter/caramel and plastic odors noticed for half of the tested isolates and may be linked to volatile compounds such as 2,3-butanedione and 2,3-pentanedione. Further works is necessary to investigate the incidence of bacteria in mixed cultures, as competition between the species may change the global metabolism and associations of volatile compounds may modify the perception of off-odors.

The large variation in the initial contamination of cold-smoked salmon coming from different smokehouses (10^2 to 10^6 cfu g^{-1}) and the differences in the quantitative and qualitative microbiological composition at the SRT had already been observed by Truelstrup Hansen and Huss (1998), Truelstrup Hansen *et al.* (1998) and Jorgensen *et al.* (2000). Two of the 3 scenarios proposed, *i.e.* domination of lactobacilli or a mixture of lactobacilli and Enterobacteriaceae have also been found by those authors whereas the last one, *i.e.* domination of Carnobacteria and *B. thermosphacta* was less current. As shown by Truelstrup Hansen *et al.* (1998) in 3 different Danish processing plants, the shelf-life of cold-smoked salmon was highly related to hygienic conditions in the smokehouse rather than to the raw material quality or to the processing parameters. Some authors have established that the shelf-life of cold-smoked salmon was extended with increasing salt and/or phenol concentration in the flesh (Shimasaki *et al.*, 1994; Truelstrup Hansen *et al.*, 1995; Leroi and Joffraud, 2000), but those results were obtained with samples processed under otherwise identical conditions, and with higher differences in salt and phenol levels. As demonstrated by many studies, no single chemical compound nor microbiological count could be used as an index of quality for vacuum-packed cold-smoked salmon. However, a combination of the three parameters TVBN, lactobacilli and yeast counts could be used to successfully predict the RSL. The quality of the fitted model ($R^2 = 0.80$) was identical to the model developed by Jorgensen *et al.* (2000) relating sensory data with biogenic amines and pH ($R^2 = 0.79$). According to Stohr *et al.* (2001), the higher producers of TVBN in cold-smoked salmon were Enterobacteriaceae, *Photobacterium* spp. and *Lactobacillus* spp. Thus, the model here integrated most of the potential spoilage organisms identified by other authors. Although *B. thermosphacta* was found to be a strong spoiler organism (Stohr *et al.*, 2001), it was not included in the model used here. In a set of experiments, *B. thermosphacta* never reached levels higher than 10^{6-7} cfu g^{-1}, probably explaining that this organism was not considered important in estimation of RSL. With the intention of lowering the number of routine analysis, for development of a standard for example, simplification of the model by eliminating the yeasts count could be proposed without losing too much precision ($R^2 = 0.77$). The simplified model was: RSL$_{(week)}$ = 4.78 - 0.34 Log (ROG count)$_{cfu\ g}^{-1}$ – 0.06 x (TVBN)$_{mg-N\ 100\ g}^{-1}$. With this model, a product was rejected with counts on ROG agar of 10^9, 10^7, 10^4 or

10^2 cfu g^{-1} associated with TVBN concentrations of 30, 40, 57 and 68 mg-N 100 g^{-1}, respectively.

ACKNOWLEDGEMENTS

The authors thank L. Campello for the chemical analysis and J. Cornet for sensory analysis. This study was part of a EU-FAIR project PL95-1207.

REFERENCES

Borsch, E.; Kant-Muermans, M-L and Blixt, Y. 1996. Bacterial spoilage of meat and cured meat products. *Int. J. Food Microbiol.* 33: 103-120.

Cann, D.C.; Houston, N.C.; Taylor, L.Y.; Smith, G.L.; Smith, A.B. and Craig, A. 1984. Studies of salmonids packed and stored under a modified atmosphere. Report - Torry Research Station, Aberdeen, Scotland.

Dainty, R.H. and Mackey, B.M. 1992. The relationship between the phenotypic properties of bacteria from chill-stored meat and spoilage processes. *J. Appl. Bacteriol. Symposium Suppl.* 73: 103S-114S.

Dalgaard, P. 1995. Qualitative and quantitative characterization of spoilage bacteria from packed fish. *Int. J. Food Microbiol.* 26: 319-333.

Dodds, K.L.; Brodsky, M.H. and Warburton, D.W. 1992. A retail survey of smoked ready-to-eat fish to determine their microbiological quality. *J. Food Protec.* 55: 208-210.

Duffes, F.; Corre, C.; Leroi, F.; Dousset, X. and Boyaval, P. 1999. Inhibition of *Listeria monocytogenes* by *in situ*-produced and semi-purified bacteriocins of *Carnobacterium* spp. on vacuum-packed refrigerated cold-smoked salmon. *J. Food Protec.* 62: 1394-1403.

Gram, L. and Huss, H.H. 1996. Microbiological spoilage of fish and fish products. *Int. J. Food Microbiol.* 33: 121-137.

Gram, L. and Huss, H.H. 1996. Microbiological spoilage of fish and fish products. *Int. J. Food Microbiol.* 33: 121-137.

Gram, L.; Trolle, G. and Huss, H.H. 1987. Detection of specific spoilage bacteria from fish stored at low (0°C) and high (20°C) temperatures. *Int. J. Food Microbiol.* 4: 65-72.

Huss, H.H.; Jeppesen, V.F.; Johansen, C. and Gram, L. 1995. Biopreservation of fish products; A review of recent approaches and results. *J. Aquatic Food Prod. Technol.* 4: 5-26.

Joffraud, J.J.; Leroi, F. and Chevalier, F. 1998. Development of a sterile cold-smoked fish model. *J. Appl. Microbiol.* 85: 991-998.

Joffraud, J.J.; Leroi, F.; Roy, C. and Berdagué, J.L. 2001 Characterization of volatile compounds produced by bacteria isolated from cold-smoked salmon flora. *Int. J. Food Microbiol.* 66: 175-184.

Jorgensen, L.V.; Dalgaard, P. and Huss, H.H. 2000. Multiple compound quality index for cold-smoked salmon (*Salmo salar*) developed by multivariate regression of biogenic amines and pH. *J. Agric. Food Chem.* 48: 2448-2453.

Leroi, F. and Joffraud, J.J. 2000. Salt and smoke simultaneously affect chemical and sensory quality of cold-smoked salmon during 5°C storage predicted using factorial design. *J. Food Protec.* 63: 1222-1227.

Leroi, F.; Joffraud, J.J. and Chevalier, F. 2000. Effect of salt and smoke on the microbiological quality of cold-smoked salmon during storage at 5°C as estimated by the factorial design method. *J. Food Protec.* 63: 502-508.

Leroi, F.; Joffraud, J.J.; Chevalier, F. and Cardinal, M. 1998. Study of the microbial ecology of cold smocked salmon during storage at 8°C. *Int. J. Food Microbiol.* 39: 111-121.

Leroi, F.; Joffraud, J.J.; Chevalier, F. and Cardinal, M. 2001. Research of quality indices for cold-smoked salmon using a stepwise multiple regression of microbiological counts and physico-chemical parameters. *J. Appl. Microbiol.* 90: 578-587.

Leroi, F.; Arbey, N.; Joffraud, J.J. and Chevalier, F. 1996. Effect of inoculation with lactic acid bacteria on extending the shelf-life of vacuum-packed cold-smoked salmon. *Int. J. Food Sci. Technol.* 31: 497-504.

Lyhs, U.; Björkroth, J.; Hyytiä, E. and Korkeala, H. 1998. The spoilage flora of vacuum-packaged, sodium nitrite or potassium nitrate treated, cold smoked rainbow trout stored at 4°C ou 8°C. *Int. J. Food Microbiol.* 45: 135-142.

Paludan-Müller, C.; Dalgaard, P.; Huss, H.H. and Gram, L. 1998. Evaluation of the role of *Carnobacterium piscicola* in spoilage of vacuum and modified atmosphere-packed-smoked salmon stored at 5°C. *Int. J. Food Microbiol.* 39: 155-166.

Shimasaki, T.; Miake, K.; Tsukamasa, Y.; Sugiyama, M.A.; Minegishi, Y. and Shimano, H. 1994. Effect of water activity and storage temperature on the quality and microflora of smoked salmon. *Nippon Suisan Gakkaishi* 60: 569-576.

Stohr, V.; Joffraud, J.J.; Cardinal, M. and Leroi, F. 2001. Spoilage potential and sensory profile associated with bacteria isolated from cold-smoked salmon. *Food Res. Int.* 34: 797-806.

Talon, R.; Montel, M.C.; Labadie, J.C.; Larpent, J.P. and Fournaud, J. 1994. Altération des viandes par les bactéries lactiques. In Bactéries Lactiques:

Aspects fondamentaux et technologiques. (2nd vol. Lorica éd.) Uriage, France, pp. 573-580.

Truelstrup Hansen, L. 1995. Quality of chilled vacuum-packed cold-smoked salmon. *Ph.D. Thesis,* Danish Institute for Fisheries Research and The Royal Veterinary and Agricultural University of Copenhagen, Denmark.

Truelstrup Hansen, L. and Huss, H.H. 1998. Comparison of the micro-flora isolated from spoiled cold-smoked salmon from three smokehouses. *Food Res. Int.* 31: 703-711.

Truelstrup Hansen, L.; Drewes Ronved, S. and Huss, H.H. 1998 Microbiological quality and shelf life of cold-smoked salmon from three different processing plants. *Food Microbiol.* 15: 137-150.

Truelstrup Hansen, L.; Gill, T. and Huss, H.H. 1995. Effects of salt and storage temperature on chemical, microbiological and sensory changes in cold-smoked salmon. *Food Res. Int.* 28: 123-130.

Truelstrup Hansen, L.; Gill, T.; Drewes Rontved, S. and Huss, H.H. 1996. Importance of autolysis and microbiological activity on quality of cold-smocked salmon. *Food Res. Int.* 29: 181-188.

Von Hildebrandt, G. and Erol, I. 1988 Sensorische und mikrobiologische Untersuchung an vakuumverpackten Räucherlachs in Scheiben (Sensory and microbiological analysis of vacuum packed sliced smoked salmon. *Arch. Lebensmittelhyg.* 39: 120-123.

Von Rakow, D. 1977. Bemerkungen zum Keimgehaltvon im Handel befindlichen Räucherfischen. *Arch. Lebensmittelhyg.* 28: 192-195.

CHAPTER 27

ISOLATION OF BACTERIOCIN-PRODUCING LACTIC ACID BACTERIA FROM REFRIGERATED SMOKED SALMON, MUSSELS AND SHRIMP

Michel Desbiens, Sharon Thibault, Geneviève Imbeault

Ministère de l'Agriculture, Pêcheries et Alimentation du Québec, Direction de l'Innovation et des Technologies, Gaspé, Quebec, G4X 2V6, Canada

Isolation of acclimatized lactic acid bacteria (LAB) strains from refrigerated ready-to-eat marine food products, showing inhibition against the pathogen Listeria monocytogenes that sporadically contaminate marine food products was intended. From hundreds of strains isolated, two strains having strongest anti-listeria activity were selected, as tested by deferred antagonism assay, and tentatively identified as Carnobacterium divergens and C. piscicola. Inhibition activity was observed with cultures on solid media, but also with sterile supernatant at 5° and 30°C. The growth rate of LAB strains was very good at 5°C. Inactivation of the antimicrobial properties occurred when proteolytic enzymes were added to the cultures, confirming the peptidic composition of the substance produced. Pure cultures of these strains were inoculated on smoked salmon slices and kept for 21 days at 5°C. Preliminary sensory evaluation showed that a moderate inocula level did not spoil the product.

INTRODUCTION

The occurrence of bacterial pathogens in ready-to-eat seafoods constitutes an important problem for the industry. Because no lethal treatments are applied to lightly preserved refrigerated products before consumption, the growth of pathogens should be controled during storage. Traditional hurdles can be used to limit the growth of microorganisms, but addition of sodium chloride and other chemical preservatives are not preerred by the consumers who wish a low salt content. So, new ways to limit the proliferation of pathogens should be explored. For years, bacteriocins produced by lactic acid bacteria (LAB) have been characterized and several experiments aimed to introduce bacteriocins in food as a microbiological barrier. Bacterocins are antimicrobial peptides usually active against a limited range of species closely related to the bacteriocin-producing bacteria (Tagg et al., 1976). Nisin is the only commercially available bacteriocin, and its use is permitted for certain applications in foods in more than 50 countries (Turtell and Delves-Broughton, 1998). The objective of this study was to isolate acclimated bacteriocin-producing LAB from refrigerated cold-smoked salmon, smoked mussels and brined shrimp, that could inhibit the growth of *Listeria monocytogenes*, a bacterial pathogen frequently detected in lightly preserved seafoods. The final aim was to elaborate a method to inoculate selected LAB strains in lightly preserved foods as protective cultures that could inhibit the pathogen efficiently.

MATERIALS AND METHODS

Selection of LAB strains

Commercial packages of frozen cold-smoked salmon, smoked mussels and lightly brined shrimps were thawed at 5 °C and kept at this temperature for 2-8 weeks. After maturation, samples were homogenized in stomacher with refrigerated 0.1 % peptone water (1:9). Spread plates were made on MRS agar incubated at 30°C. An average of 100 well defined colonies were picked from these plates and inoculated in 3 mL of APT broth, then incubated at 30°C. These test cultures were evaluated by deferred antagonism assay, adapted from the method of Tagg et al. (1976). A sample of 2 µL of each test culture was deposited on a plate of APT agar supplemented with yeast extract, and kept at room temperature for 30 min to dry. Ten marked spots of different cultures were placed on each plates. They were incubated for 18-24 h at 30°C anaerobically to avoid production of H_2O_2. Then, an overlay of BHI with 0.75 % agar with catalase (500 UI/mL) and 2 % glycerophosphate, seeded with *Listeria monocytogenes* ATCC 19115 and 19112 strains at a concentration of 10^5-10^6 cells/mL was added to the APT-YE plates. After incubation at 30°C for 24 h, the plates were examined for clear zones surrounding the inoculated spots, indicating an inhibitory activity against the target organism. LAB strains showing a zone wider than 1 mm were selected for further examination. To determine if the inhibition was attributed to bacteriocins, the antagonism assay was repeated with the selected strains, on plates with proteolytic enzymes. APT-YE plates on which 2 µL spots of enzymes (proteinase K, α-chymotrypsin, protease, trypsin; Sigma) at a concentration of 10 mg/mL in PBS 0.01M pH 7.5 were placed at about 2 mm close to the LAB spots previously incubated at 30°C. Absence of inhibition at these precise points was recorded.

Characterization of the LAB strains

The characterization tests of the strains included Gram staining, cell morphology of a 24 h culture from APT broth; catalase (30 % H_2O_2); oxidase test strips (Oxoid); gas production from glucose on APT broth was monitored for up to 10 days as described by Dicks and van Vuuren (1987); growth on Rogosa agar at 30°C; arginine degradation on Moeller decarboxylase agar with 0.5 and 2.0 % glucose; rhamnose, D-xylose, mannitol, ribose and inuline fermentation (API 50 CHL wells); growth at 5 and 45°C, growth at 10 % NaCl. Lactic acid configuration was determined enzymatically with a Bohringer commercial kit. The presence of mesodiaminopimelic acid in the cell wall was tested using the method of Bousfield et al. (1985).

Antilisterial activity of the strains

The selected LAB strains were transfered in APT-YE broth and incubated for 18-24 h at 30°C. The cultures were centrifugated at 3800 g, and the supernatants filtered on a Millipore Durapore HVLP 0,45 µm membrane. Sterile supernatants were added at a concentration of 20 % in series of screw cap tubes containing APT-YE broth, at pH 5.2, 6.0 and 6.6. Series of tubes were inoculated with *Listeria monocytogenes* 19112 and 19115 strains at a concentration of 10^2 cells/mL. Parallel series of tubes were incubated at 5° and 30°C. Absorbance readings at 600 nm were made using a Biochrom Novaspec ll spectrophotometer.

Other tubes of LAB cultures were placed in a boiling water bath for 10 min to kill the cells. The content of the tubes was filtered on Millipore filtration membranes, then the membranes were scraped with a glass rod and rinsed with 2-3 mL of 0.85 % saline to remove the cellular fragments. One milliliter of the rinse liquid was added to 8 mL tubes of APT-YE broth pH 6.6 seeded with 10^3 cells/mL *L. monocytogenes* ATCC 19112; the tubes were incubated at 30°C in a water bath with orbital shaking. Subsamples were removed after 0, 5 and 30 h to perform spread plate counts on TSA-YE agar at for 30°C 24-96 h.

Challenge studies on cold-smoked salmon

Cold-smoked Atlantic salmon was bought from a supplier; the salmon had been processed 3 days before the experiment and kept frozen. The thin slices of salmon from the same animal were thawed at 5-7°C. The flesh had a pH of 6.2, an Aw of 0.97 and contained 1.4-1.5 % total NaCl (w/w). Cultures of M-35, CS-74 LAB strains and *L. monocytogenes* were initially grown in TSB without glucose for 20 h at 30°C. Two culture blends were prepared, containing 0.1 % peptone water with 10^3 cells/mL of *L. monocytogenes* and 10^4 cells/mL of CS-74 or M-35 strains. A 0.5 mL portion of the blends was added separately on each slice and spread on the entire surface with a sterile glass rod, in duplicate. Controls with only *L. monocytogenes* were included. The slices were individually vacuum packaged with a Winpak Vak 4 R film (oxygen transmission rate 40 mL/m^2/24h) and incubated at 5 and 10°C. Subsamples were taken at 0, 7, 14, 21 and 28 days. Bacterial counts were performed on NAP agar (72 h at 30°C) and PALCAM agar (48 h at 35°C).

Other series of slices were inoculated only with the selected LAB (0.5 mL of 10^4 and 10^7 cell/mL) grown on APT-YE and kept at 5°C for sensory evaluation. A panel of 5 trained persons evaluated color (this parameter was also measured with a CS-300 Colorimeter, Minolta), appearance and odor of the inoculated salmon slices (but not the taste), and compared them to non-inoculated refrigerated and frozen controls, over a period of 21 days.

RESULTS

From several hundreds of bacterial strains isolated from smoked salmon, mussels and shrimps, two strains produced large inhibition zones (> 10 mm) on the test plates, although ten produced smaller zones. These two LAB strains (M-35 and CS-74) were active against both *L. monocytogenes* strains. The strains were tentatively identified as *Carnobacterium divergens* and *C. piscicola* respectively (Table 1), mainly based on the identification scheme of Mauguin and Novel (1994). The cell morphology varied between small rods and coccoid rods. Very low amounts of gas were observed from M-35, but no detectable gas from glucose was detected from CS-74, in spite of the use of Durham tubes and agar plugs covering the broth. Both strains degraded arginine at glucose concentration of 0.5 % but not at 2.0 %. Mannitol was fermented only by CS-74. No growth was observed on Rogosa agar, probably because of the presence of acetate in the media. Growth at 5°C was rapid (Figure 1), and no growth was observed at 45°C. Both strains were able to grow in 10 % NaCl.

The inhibitory activity of the strains was suppressed during the antagonism assay by the incorporation of proteolytic enzymes. Trypsin had no effect, while protease, α-chymotrypsin and proteinase K stopped the activity, indicating that the inhibition was caused by a proteinaceous subtance.

Table 1. Biochemical and growth profiles of M-35 and CS-74 strains

	Strain M-35	Strain CS-74
Catalase	-	-
Oxidase	-	-
Growth on Rogosa Agar at 30 C	-	-
Gas in APT broth	+	-
Arginine degradation - glucose 0.5 %	+	+
Arginine degradation - glucose 2 %	-	-
Growth at 45 C	-	-
Growth 10 % NaCl	+	+
Rhamnose fermentation	-	-
D-Xylose fermentation	-	-
Mannitol fermentation	-	+
Ribose fermentation	+	+
Inuline fermentation	-	-
D-Lactic acid	-	-
L-Lactic acid	+	+
DL-Lactic acid	-	-
mDAP (diaminopimelic acid)	+	+

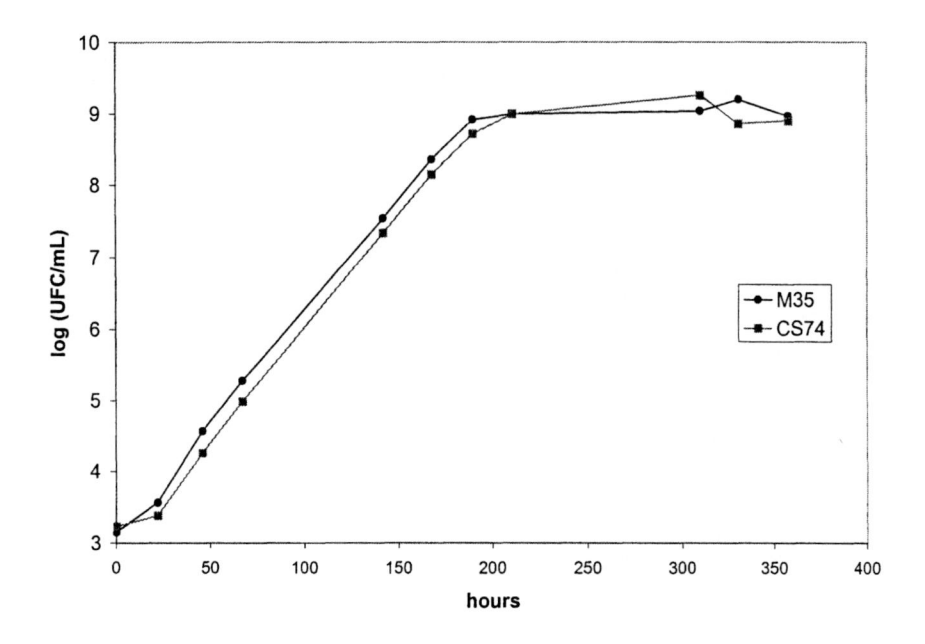

Figure 1. Growth of M-35 and CS-74 strains at 5°C in APT-YE broth.

The addition of sterile supernatant from the LAB cultures negatively affected the growth of *L. monocytogenes* in liquid media at 5°C and pH 6.6, 6.0 and 5.2, as measured by absorbance (Figure 2), in comparison with control. At pH 5.2, there was no observed growth of *L. monocytogenes* with CS-74 and M-35 supernatants during the experiment at refrigeration temperature. The inhibition was also evident at 30°C at the same pH values. Figure 3 shows the activity of CS-74 at 30°C; M-35 had approximately the same behavior. The killed LAB cells resuspended and added to the test cultures also limited the growth of *L. monocytogenes* (Table 2). The growth increased of only 1 log in the presence of these cell fragments at 30°C, as compared to the control cultures of *Listeria* (7.1 log increase).

Table 2. Effect of killed LAB cells on Listeria monocytogenes growth in APT-YE broth at 30°C and pH 6.6

	Listeria cell counts (log UFC/mL) after:		
	0	5 hours	30 hours
L. monocytogenes (control)	2.8	3.6	9.9
L. monocytogenes + killed LAB cells	2.3	2.2	3.3

The challenge on cold-smoked salmon demonstrated that both M-35 and CS-74 LAB strains can interfere with the growth of *L. monocytogenes* at 10°C, an abuse temperature frequently encountered inside consumers' refrigerators. The antagonistic activity against *L. monocytogenes* is not noticeable at day 7, as indicated by the cell counts that increased from 2 to 5.5 logs on PALCAM during this period (Figure 4). However, from day 14, the growth of *L. monocytogenes* was almost stopped in slices containing inoculated LAB. After 28 days, the *Listeria* counts in samples challenged with M-35 and CS-74 were respectively 1.6 and 1.9 logs lower than the control with *Listeria* alone. Preliminary results of a challenge study at 5°C indicated that CS-74 was also efficient (results not shown), but M-35 performance appeared rather variable. The sensorial analysis did not reveal important differences between controls and slices with low LAB inoculum during the 21 days. Slices with high inoculum of both LAB strains, and with low inoculum of CS-74 had a slight undefined odor noticed by the panelists. No color changes appeared in samples over the 21 days period. The pH of the inoculated smoked salmon flesh always remained constant, and never dropped below 6.0 over the entire challenge period.

DISCUSSION

The results, particularly the inactivation by proteolytic enzymes as well as the inhibition activity of the supernatants, indicate that both *Carnobacterium* strains isolated produce antilisterial peptidic substance. Previous published experiments report the occurrence of *Carnobacterium piscicola* and *C. divergens* strains producing bacteriocins in meat and seafoods (Pilet *et al.*, 1995; Stoffels *et al.*, 1992; Shillinger *et al.*, 1993; Buchanan and Baggi, 1997; Duffes *et al.*, 1999; Nilsson *et al.*, 1999), making this bacteria a very interesting candidate for food applications. Lactic acid bacteria, including *Carnobacterium* species, are commonly found in cold-smoked salmon and brined shrimp. They frequently constitute the dominant microflora of lightly preserved seafoods, and are often associated with product spoilage (Magnusson and Traustadottir, 1982; Hansen *et al.*, 1995; Paludan-Muller *et al.*, 1998).) It is reported that

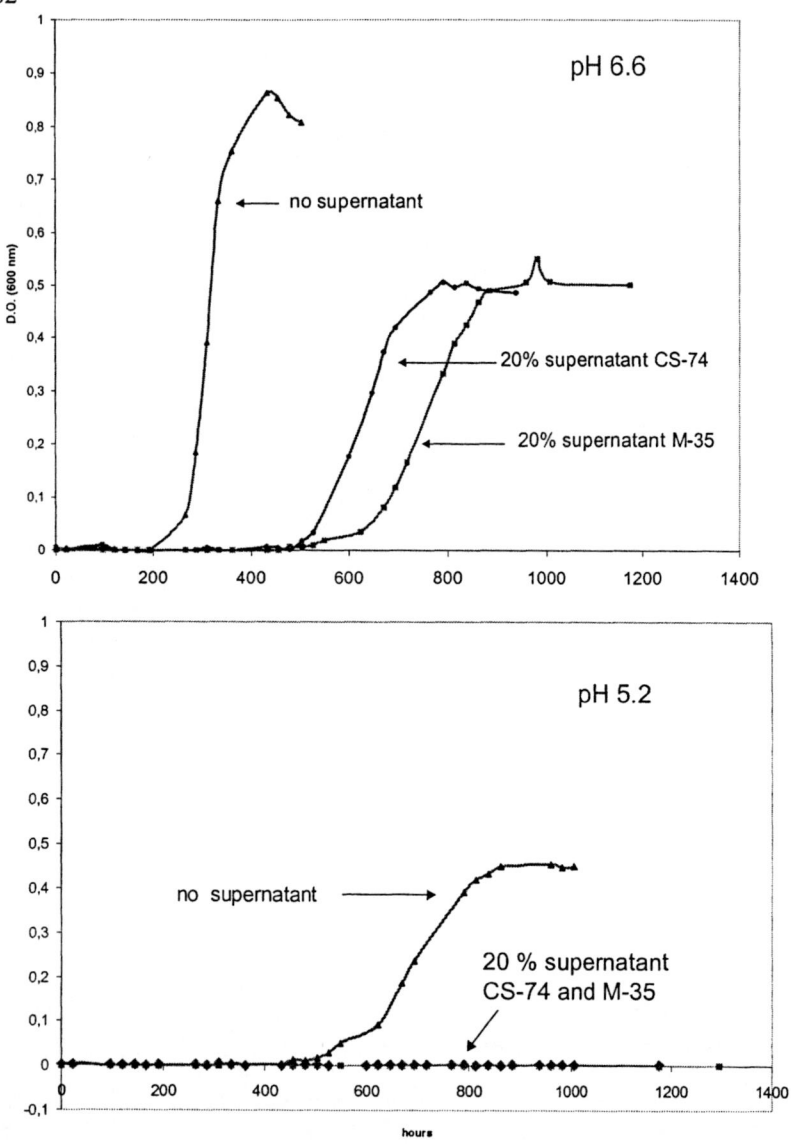

Figure 2. Changes in absorbance (600 nm) in APT-YE broth for growth at 5°C of Listeria monocytogenes at pH 6.6 and 5.2, with and without M-35 and CS-74 supernatant.

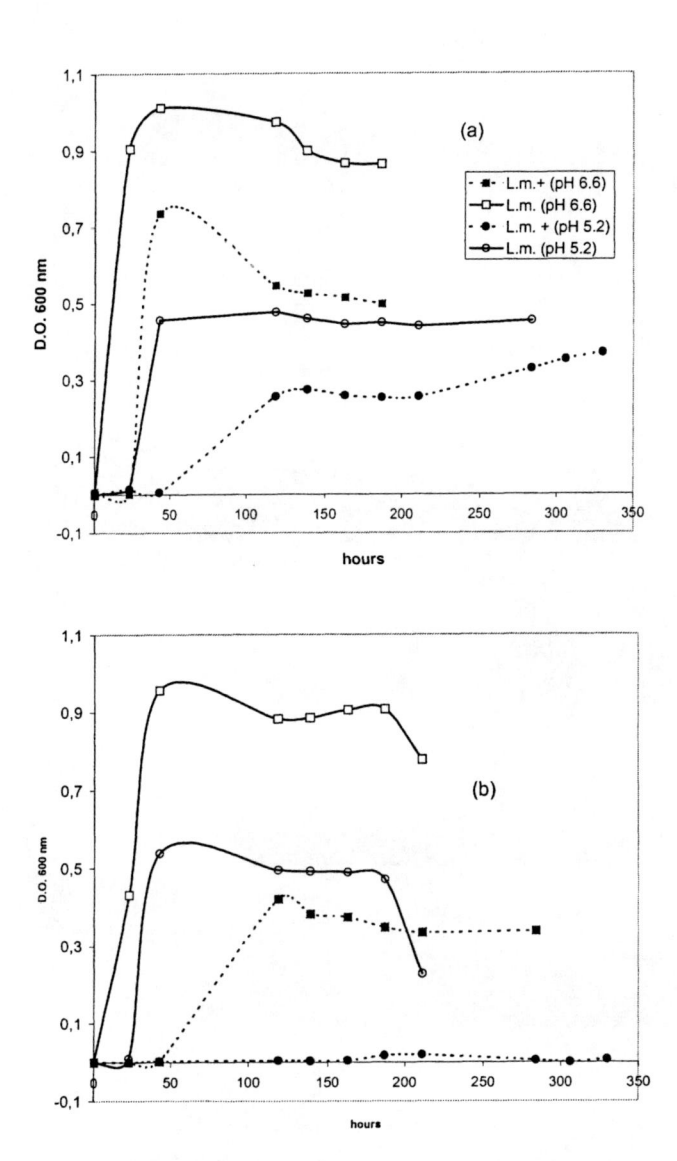

Figure 3. Changes in absorbance (600 nm) in APT-YE broth for growth at 30°C of Listeria monocytogenes 19115 (a) and 19112 (b), at pH 6.6 and 5.2, with and without CS-74 supernatant. (L.m.+: Listeria monocytogenes with CS-74 supernatant).

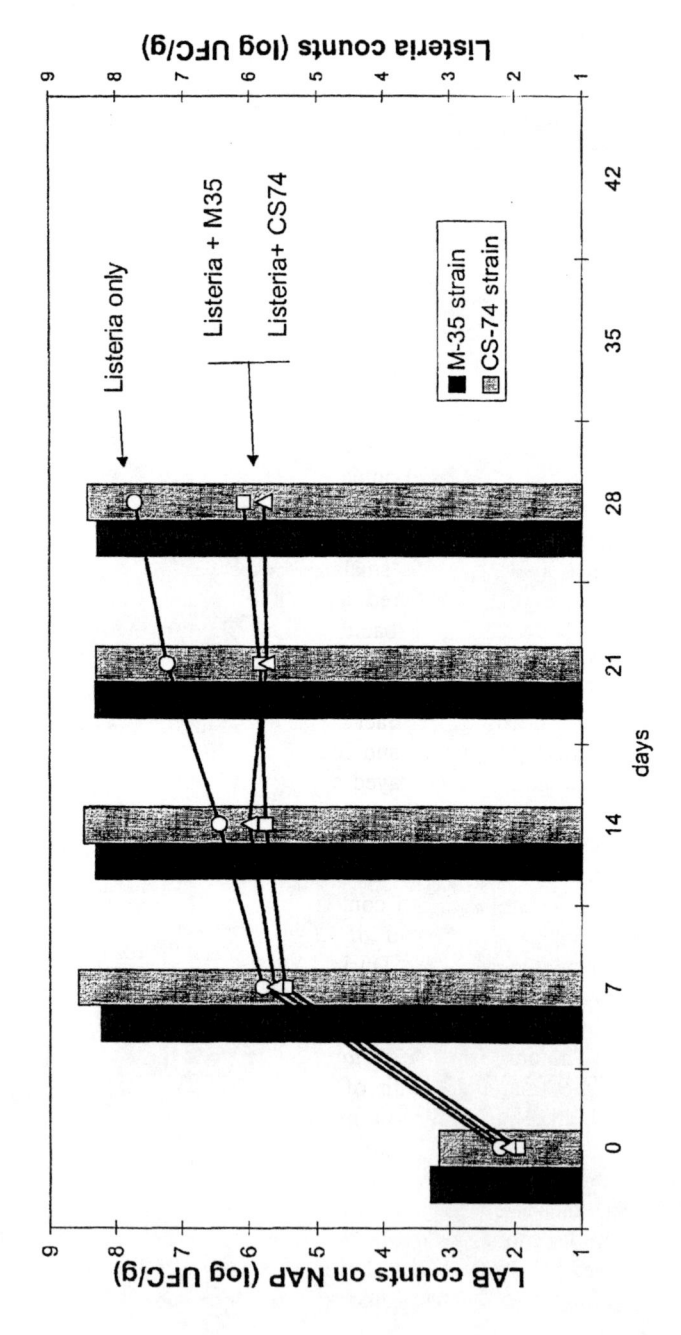

Figure 4. Evolution of total LAB (boxes) and Listeria counts (lines) during the challenge on smoked salmon slices at 10°C.

Carnobacterium species from smoked salmon generally do not spoil fish, while they are known for their growth and bacteriocin production capacities at refrigeration temperatures (Duffes *et al.*, 1999; Buchanan and Baggi, 1997). Indeed, during the challenged study on salmon slices, the spiked *Carnobacterium* strains appeared to have a very low spoilage potential at a moderate inoculum level, and did not result in a pH decrease in the product. The sensory evaluation scores were slightly lower only with a high LAB concentration at the end of the evaluation period. This property, as well as their salt tolerance and capacity to grow at refrigeration temperature, is of prime importance for the applicability of LAB as protective cultures on lightly preserved ready-to-eat products.

The inhibiting activity of the killed LAB cells on *Listeria* growth could be attributed to the adhesion of a proportion of the antimicrobial protein to the inoculated cell fragments. It has been noted that the plate counts of the *Listeria* cultures in presence of the killed LAB cells needed a longer incubation time to produce colonies, indicating that a stress factor occurred.

The LAB isolates also produced small inhibition zones on agar plates if catalase was omitted in the defererred antagonism assay, an indication of hydrogen peroxide production by the bacteria in the plates. In tubes, however, the peroxide eventually remaining in the supernatant probably did not contribute significantly to the inhibition, because of the instability of H_2O_2 over the long storage period, and the added yeast extract in the formulation of APT broth, that is reported to inactivate H_2O_2 (Jaroni and Brashears, 2000). On the other hand, production of peroxide could have played a minor role in the inhibition of *L. monocytogenes* in situ during the challenge on smoked salmon.

The lower *Listeria* counts measured in inoculated slices are not related to competition for nutrients, because total indigenous lactic acid bacteria counts on NAP agar performed on uninoculated controls showed a relatively high LAB concentration (10^7 cells/g) at the end of the experiment, only one log of magnitude under inoculated packages. During the challenge, the delay between the inoculation and the observed detrimental effect on *Listeria monocytogenes* is attributed to the time necessary to reach a bacteriocin concentration required to inhibit the bacteria. The number of *Listeria* cells per gram in the salmon slices was quite high, approximately a tenth of the LAB concentration. A higher LAB/*Listeria* ratio is likely to produce a more drastic antagonism, reflecting a more realistic situation where inoculated lactic acid bacteria should exceed considerably the quantity of *Listeria* cells in a package.

In conclusion, the two *Carnobacterium* strains isolated are potential candidates for application with lightly preserved seafoods. Further experimentations are needed to define the cultural conditions to optimize the bacteriocin production kinetics and the methodology of application in the food matrix. The direct addition of pure bacteriocin in the food products is a possibility, but the introduction of protective LAB strains or the incorporation of semi-purified bacteriocin could represent an alternative.

ACKNOWLEDGEMENTS

The technical assistance of Josée Blais, Annie Raté and Jeanne d'Arc Rioux (MAPAQ, Gaspé) is gratefully acknowledged.

REFERENCES

Bousfield, I.J.; Keddie, R.M.; Dando, T.R. and Shaw, S. 1985. Simple rapid method of cell wall analysis as an aid in the identification of aerobic coryneform bacteria. In *Chemical methods in bacterial systematics*. M. Goodfellow and D.E. Minnikin (eds.) Academic Press, London, UK, pp. 221-236.

Buchanan, R.L. and Baggi, L.K. 1997. Microbial competition; effect of culture conditions on the suppression of Listeria monocytogenes Scott A by Carnobacterium piscicola. *J. Food Protec.* 60: 254-261.

Duffes, F.; Corre, C.; Leroi, F.; Dousset, X. and Boyaval, P. 1999. Inhibition of Listeria monocytogenes by *in situ* produced and semi-purified bacteriocins of Carnobacterium spp. on vacuum-packed, refrigerated cold-smoked salmon. *J. Food Protec.* 62: 1394-1403.

Hansen, L.T.; Gill, T. and Huss, H.H. 1995. Effects of salt and storage temperature on chemical, microbiological and sensory changes in cold-smoked salmon. *Food Res. Int.* 28: 123-130.

Jaroni, D. and Brashears, M.M. 2000. Production of hydrogen peroxyde by Lactobacillus delbrueckii subsp. lactis as influenced by media used for propagation of cells. *J. Food Sci.* 65: 1033-1036.

Magnusson, H. and Traustadottir, K. 1982. The microbial flora of vacuum packed smoked herring fillets. *J. Food Technol.* 17: 695-702.

Mauguin, S. and Novel, G. 1994. Characterization of lactic acid bacteria isolated from seafood. *J. Appl. Bacteriol.* 76: 616-625.

Nilsson, L.; Gram, L. and Huss, H.H. 1999. Growth control of Listeria monocytogenes on cold-smoked salmon using a competitive lactic acid bacteria flora. *J. Food Protec.* 62: 336-342.

Paludan-Muller, C.; Dalgaard, P.; Huss, H.H. and Gram, L. 1998. Evaluation of the role of Carnobacterium piscicola in spoilage of vacuum- and modified-atmosphere- packed cold-smoked salmon stored at 5°C. *Int. J. Food Microbiol.* 39: 155-166.

Pilet, M.F.; Dousset, X.; Barré, R.; Novel, G.; Desmazeaud, M. and Piard, J.C. 1995. Evidence of two bacteriocins produced by Carnobacterium piscicola and Carnobacterium divergens isolated from fish and active against Listeria monocytogenes. *J. Food Protec.* 58: 256-262.

Schillinger, U.; Stiles, M.E. and Holzapfel, W.H. 1993. Bacteriocin production by Carnobacterium piscicola LV 61. *Int. J. Food Microbiol.* 20: 131-146.

Stoffels, G.; Nes, I.F. and Guomundsdottir, A. 1992. Isolation and properties of a bacteriocin-producing Carnobacterium piscicola isolated from fish. *J. Appl. Bacteriol.* 73: 309-316.

Tagg, J.R.; Dajani, A.S. and Wannamaker, L.W. 1976. Bacteriocins of Gram-positive bacteria. *Bacteriol. Rev.* 40: 722-756.

Turtell, A. and Delves-Broughton, J. 1998. International acceptance of nisin as food preservative. *Int. Dairy Fed. Bull.* 329: 20-23.

Wessels, S. and Huss, H.H. 1996. Suitability of Lactococcus lactis subsp. lactis ATCC 11454 as a protective culture for lightly preserved fish products. *Food Microbiol.* 13: 323-332.

CHAPTER 28

AQUA FEED: RESEARCH CHALLENGES AND FUTURE TRENDS

Fernando Luis García-Carreño

CIBNOR. PO Box 128. La Paz, BCS. México

Management for sustainability in aquaculture involves several aspects, including aqua feed production and feeding strategies. This contribution deals with studies done in several laboratories, both in America and Europe, relating to the quality of the protein ingredients used to produce feed for aquafarming. The modeling of the digestive systems of fish and crustaceans and the use of *in vitro* techniques for understanding the biochemical physiology of organisms intended for aquafarming and evaluating the presence of inhibitors of digestive enzymes in feed ingredients is at the Heart of this issue. Also, the quality of currently-used raw materials, processes of preparation, products available, and the use of alternate ingredients are discussed.

INTRODUCTION

According to the FAO, a major challenge for aquaculture is to maintain, and where sustainable, enhance the contributions made to global fish supplies. In 1996, aquaculture contributed 26.5% of the world's fish production, becoming the fastest growing food production sector in agriculture. Moreover, aquaculture makes a major contribution to global food security, and opportunities still exist to expand its role.

Management for sustainability in aquaculture involves several aspects, including aqua feed production and feeding strategies. The expected increase in aquaculture productivity will require adequate feed, environment care, quality of the product, and profit. Production of feed for aquatic organisms is a current challenge, both scientifically and technically. Aqua feed is made from trash fish, fishmeal, and some slaughterhouse leftovers, as the main source of protein for feeding carnivores, omnivores, and even herbivores. Such protein feed practices generate many criticisms that have to be solved to support sustainable aquafarming activities in the 21st Century.

Usually, a high concentration of protein is used in feed. Up to 7 kg of caught fish is needed to produce one kg of farmed organisms. There are severe discrepancies in the lowest feed protein concentrations that yield acceptable performance (usually called the optimum) of farmed organisms. All of this is related to the quality of protein ingredients used to fabricate feed. It is a common practice to produce fish or slaughterhouse leftover meals by a high temperature process during the drying of the product before packing. This process affects the nutritional quality of the product by reducing the bioavailability of amino acids and increasing ash content. Additionally, a possible embargo of feed produced from animal-derived ingredients is shadowing the encouraging panorama of aquafarm feeding technologies.

This chapter deals with studies done in several laboratories, both in America and Europe, relating to the quality of the protein ingredients used to produce feed for aquafarming. The modeling of the digestive systems of fish and crustaceans and the use of *in vitro* techniques for understanding the biochemical physiology of organisms intended for aquafarming and evaluating the presence of inhibitors of digestive enzymes in feed ingredients is at the heart of this issue. Also, the quality of currently-used raw materials, processes of preparation, products available, and the use of alternate ingredients are discussed.

The current status and challenges of aqua feeds

Feeds are used for a variety of organisms, both terrestrial and aquatic. The biggest demanders for feed are the poultry and pig industries with 32 and 31% of the total, respectively. Dairy and beef industries demand 17 and 11%, respectively. Aqua feeds constitute about 3% of the total. Aquaculture competes for feed ingredients with traditional farming. Aqua feed production in 1994 was 4,250,000 tons (Smith and Guerin 1995), with Asia consuming about 60% of the total and North America about 10%. Most of the feed produced is intended for salmonids and shrimps (27 and 25%, respectively). The feeding habits of aqua farmed organisms are diverse. Some, the photosynthetic ones, only need sunlight and minerals. But most of them, omnivorous and carnivorous, need a food supply, even herbivores are fed formulated feeds, including animal protein ingredients to increase the yield and reduce time of nurture.

Fish meal is the most widely used protein ingredient in feed formulations, with more than a million tons produced in 1994. Poultry is the largest demanding sector (Pike, 1997). Aqua feeds use 17% of total fish meal production. The need for fish meal in 2010 is estimated to be more than a million and a half tons, about a 50% increase in 16 years. Aqua feeds will demand 23% of the total production in 2010, an increase of 6% in 16 years.

However, the total production of fish meal may not satisfy the demand, because the total production has remained unchanged between 1961 and 1995. Worsening the situation is the fact that most fish meal producers are burning the fish proteins because of obsolete methods of fishing, managing, and industrialization, especially during the drying process. The process, from a nutritional point of view yields a poor product. Moreover, environmental issues and rational use of marine harvests are of interest. Such issues are common in environmental and political arenas. It has been said that fish feed is exhausting the oceans. Thus, new findings on unintended impacts of fish farming that put both oceans and the aquaculture industry at risk.

Recent developments in aqua feed

Aqua feeds are under the spot light for several reasons. Increasing demand, the search for improved nutritional characteristics, abuse of resources (mainly from ignorance), operational problems (like inefficient processing and the presence of natural anti-nutritional compounds in raw material), and political pressures are challenging issues, both scientifically and technologically.

On the other hand, aquaculture stands for opportunities for development if scientifically sound techniques for production of food are used. Aqua feeds lie at the intersection of fisheries industrialization and aquafarming. Opportunities are open to both the developed and developing countries.

Several techniques have been developed to study the efficiency of formulated feed for particular organisms. Two variables are most important, digestibility of the protein and the food-biomass conversion ratio. There are *in vivo* techniques, which are tedious and expensive because they involve dozens of organisms, replicate and feeding, sampling, and analyzing residual feed and feces. Alternative *in vitro* techniques are also available. A technique has been developed to evaluate protein digestibility for salmonids (Dimes and Haard, 1994). An *in vitro* technique to evaluate protein digestibility for farmed shrimp has also been developed at the CIBNOR. The technique evaluates the degree of hydrolysis of protein by enzymes from the digestive system of the studied organism, and involves the evaluation of the progress of protein hydrolysis using a pHstat. A correlation between the quality of the protein source and digestibility, in which the poorest the quality of the ingredient, the lowest the DH was found (García-Carreño *et al.*, 1997; Ezquerra *et al.*, 1997). We also found a correlation of 0.77 between digestibility obtained in the *in vivo* assay and the degree of hydrolysis in the *in vitro* assays. This allows the use of *in vitro* techniques to evaluate the digestibility of ingredients, feeds, and the effect of processing on the quality of the product (Ezquerra et al., 1997). Some plant protein ingredients possess antinutritional compounds such as lectins, toxic amino acids, allergens, and proteinase inhibitors. Such inhibitors can reduce the proteolytic activity of enzymes from the digestive gland of shrimp. Inhibitors in several legume seeds reduced up to 60% the total proteolytic activity from the digestive proteinases of white and brown shrimp (*Penaeus vannamei* and *P. californiensis*). The effect is species specific, the enzymes from the brown shrimp being most affected (García Carreño et al., 1997). Fortunately, there are procedures to reduce inhibitory activity of protease inhibitors. An increase in the degree of hydrolysis of protein in legume seeds is shown when treating the seed protein extracts are heated at 85°C. The response to the process is also species specific, both for the legume seeds and shrimp.

At CIBNOR, there is interest in demonstrating that better protein ingredients will yield improved performances in farmed organisms. We have evaluated the supplementation of a commercial shrimp feed with commercial fish protein hydrolyzate, krill protein hydrolyzate, and a squid concentrate produced in the lab (Córdova Murueta and García Carreño, 2001, 2002). The final weight and food conversion ratio of groups of shrimp fed with the supplemented feed at concentrations of 3, 9, and 15%. Final weight was higher in all treatments, except in squid fed at 15%, when compared to the control group that was fed with commercial feed. Food conversion rates were improved in all the treatments, except for fish hydrolyzate at 15% and squid at 9 and 15%. All treatments at the lowest supplementation level of 3% improved the performance of shrimp.

Variables of protein digestibility, both *in vivo* and *in vitro*, like apparent digestibility coefficient (ADC), regardless of the protein ingredient, were significantly improved in all the experimental groups fed with supplemented feeds (Córdova and García Carreño, 2002). Shrimp digested the supplemented feeds better. In the *in vitro* approach to analyzing the degree of hydrolysis by digestive enzymes indicated that shrimp fed with supplemented feeds exhibited a higher degree of hydrolysis than that of the control group. Also, the degree of hydrolysis of supplemented feeds showed better digestion when using commercial enzymes. Thus, feeds were substantially improved when supplemented with better protein ingredients

When feeding piracanjuba, a fish from Brazilian rivers, the final weight was higher in specimens fed with an experimental feed than the specimens fed on a commercial feed. The weight of fatty tissue was higher in the experimental group. This is important because, at the time of assay, these fish were preparing for migration upstream to reproduce, and fatty tissue is the reserve of energy for the migration (García Carreño et al., 2002). For the piracanjuba, the digestibility of casein by enzymes from this fish fed on commercial and experimental feeds was significantly different. The degree of hydrolysis, when using the four enzyme cocktail of Hsu *et al.* (1977) was lowest. The degree of hydrolysis, when using the enzymes from piracanjuba fed the commercial feed, was lower than when using enzymes from piracanjuba fed with experimental feed. The quality of feed, beyond nutritional factors, definitely affects variables in the metabolism of farmed organisms. Modification of digestibility of food and feed by modulator-supplemented feeds is expected in future research.

Because the availability of animal protein ingredients is not enough to satisfy demand and health-associated problems, the use of alternative protein ingredients must be assessed. One alternative is plant protein ingredients, but the presence of anti-physiological compounds, such as digestive enzyme inhibitors, are among drawbacks that need to be considered. Even some animal protein ingredients possess enzyme inhibitors. A cheap, fast, and easy methodology was developed to evaluate the effect of inhibitors present in protein ingredients on the digestive proteases of aquafarmed organisms (Alarcón *et al.*, 2001). Extracts of digestive enzymes from the organism and plant ingredients under study were tested for their inhibitory effects. The enzymes and the test ingredient were mixed and the mixtures assayed for proteinase activity, using casein as a substrate. The absorbance was transformed to a percentage when the control of no inhibition was set at zero inhibition. Also, the protein, enzyme, and inhibitor composition were evaluated using a substrate-SDS-PAGE. Both animal and plant protein ingredients yielded some degree of inhibition for digestive proteinases of farmed organisms. The Mediterranean seabream, *Sparus aurata* was inhibited up to 19% of the total proteinase activity by a commercial fish meal. Ovalbumin inhibits more than half of the total activity. Plant ingredients are good inhibitors for *S. aurata* enzymes, which means that anti-physiological compounds in feed can eliminate a good deal of the total enzyme secreted to digest food protein. Commercial microencapsulated feed for *S. aurata* fingerlings were fabricated using ovalbumin as binder. The feed inhibited 22% of the total proteinase activity of the fish, one of the reasons why the use of microcapsules to fully substitute live food has failed. The problem is that food technologists have to deal with ingredients that have naturally occurring inhibitors.

Enzyme extract from four shrimp species were inhibited to varying extents by both animal and plant protein ingredient extracts. To understand how digestive enzymes were affected, the effect of inhibitors on enzyme extracts of three fish was calculated. The response to inhibition factors was species specific, and followed particular kinetics and extents. By using a technique that combined the evaluation of the degree of hydrolysis, electrophoresis, and lane densitometry to follow the reduction in size of the main proteins in ingredients and feed, it is possible to calculate a coefficient of protein degradation (CPD).

The algorithm allowed calculation of the maximum concentration of a particular proteinase inhibitor containing ingredient in a feed with a negligible effect on digestive enzymes.

$$CPD = \sum_{i=1}^{n}\left[\frac{A_i(t=0) - A_i(t=90\text{min})}{A_i(t=0)} \times 100\right]\frac{A_i(t=0)}{\sum_{i=1}^{n} A_i(t=0)}$$

Trends

Efforts in the near future should be directed at:

1) Reducing the amount of fishery resources used by producing improved protein ingredients. 2) Improving the performance of farmed organisms by using better feed, including growth factors and nutraceuticals. 3) Using aqua feed to support sustainability in aqua farming.

All these goals are possible, if a strong science of aqua feed production is developed.

Farmers have to be educated to understand that the health of the farm depends on a deep knowledge of organisms and inputs to the farm. One should also recognize that most aquaculture problems result from trying to push a balanced ecosystem too far, in the interests of higher production or higher profits.

REFERENCES

Alarcón, J.; García-Carreño, F. and Navarrete del Toro, M. 2001. Effect of plant protease inhibitors on digestive proteases in two fish species with aquafarming potential. Pacific Fisheries Technologists. 53 Annual Meeting, La Paz, BCS. México, PFT.

Córdova Murueta, J. and García Carreño, FL. 2001. The effect on growth and protein digestibility of shrimp *Penaeus stylirrostris* fed with feeds supplemented with squid (*Dosidicus gigas*) meal dried by two different processes. *J. Aquatic Food Prod. Technol.* 10: 35-47.

Córdova-Murueta, J. and García-Carreño, FL. 2002. Nutritive value of squid and hydrolyzed protein supplement in shrimp feed. *Aquaculture* 210: 371-384.

Dimes, L. and Haard, F. 1994. Estimation of protein digestibility-I development of an in vitro method for estimating protein digestibility in salmonids (Salmo gairdneri). *Comp. Biochem. Physiol.* 108A(2/3): 349-362.

Ezquerra, M. and García-Carreño, FL. 1997. pH-stat method to predict protein digestibility in white shrimp (Penaeus vannamei). *Aquaculture* 157: 251-262.

García-Carreño, FL.; Navarrete, M. and Ezquerra, M. 1997. Digestive shrimp proteases for evaluation of protein digestibility in vitro. I: Effect of protease inhibitors in protein ingredients. *J. Marine Biotechnol.* 5: 36-40.

García-Carreño, FL.; Albuquerque-Cavalcanti, C.; Navarrete del Toro, MA. and Zaniboni-Filho, E. 2002. Digestive proteinases of Brycon orbignyanus (Characidae, Teleostei): Characteristics and effects of protein quality. *Comp. Biochem. Physiol.* In press.

Hsu, H.; Savak, D.; Satterlee, L. and Miller, G. 1977. A multienzymatic technique for estimating protein digestibility. *J. Food Sci.* 42: 1269-1273.

Pike, I. 1997. Future supplies of fish meal and fish oil: quality requirements for aquaculture with particular reference to shrimp. In *Feed Ingredients Asia '97*. Uxbridge, Turret-RAI, pp. 26.

Smith, P. and Guerin, M. 1995. Aquatic feed industry and its role in sustainable aquaculture. In *Feed Ingredients Asia '95 Conference*. Uxbridge, Turret-RAI, pp. 1-11.

INDEX

acid
 acetic, 235
 docosahexaenoic, 4
 docosapentaenoic, 4
 eicosapentaenoic, 4
 thiobarbituric, 71, 139
acid protease activity, 95
acidic marinade, 309
aerobic plate counts, 286, 318
air-blast freezing, 139
alkaline protease activity, 95
ameripure process, 307
annual equivalent cost of capital, 262
antifreeze proteins, 64
antifungal activity, 189
antilisterial activity, 359
antimicrobial activity of chitosan, 187
antimicrobial polypeptides, 62, 65
apparent digestibility coefficient, 372
aqua feed, 371
aromascan analysis, 160
Atlantic cod, 62
Atlantic halibut, 62
Atlantic wolfish, 62
automatic identification system, 39, 40, 47
bacterial counts, 358
bacteriocin, 357
bar code, 39
batch deep-fat frying, 137
biogenic amine(s), 155, 160
biomarkers and traceability, 37
biometric procedures, 53
biomolecules from fish, 64
biopesticide, 189
biotechnology, 200
blue fish, 156

blue mussels, 319
breaded shrimp, 130, 197
breaded butterfly jumbo shrimp, 129
brine salting, 111, 117
brined shrimp, 358
brown shrimp, 374
cadaverine, 156, 162
cadmium sequestration, 217
capillary electrophoresis, 160
carnobacterium piscicola, 363
carotenoid pigments, 200, 201, 260
carotenoprotein, 259, 266
carp, 99, 148
cell hybridization, 8
central composite rotatable design, 206
changes in texture, 146
chelatin of toxic metals, 204
chemical analysis, 341
chitin, 4, 197, 198, 201, 223, 259
chitobiose, 276, 275
chitosan, 1, 4, 187, 209, 223, 233, 251, 259
chitosan coating, 227
chitosan sorption of heavy metals, 210
chitosan flakes, 203
chitosan film, 223
chitosan oligomer, 1
chitosan modifications, 253
chitotriose, 276, 275
chondroitins, 4
cod, 226
cold-smoked salmon, 281, 283, 296, 337, 350, 358
color measurement, 116
cooling and freezing treatment, 309
cooking drip, 99

correspondence factorial analysis, 340
crab, 196, 204
crab chitosan, 204
crab shell chitin, 226
creamed phase, 239
cross-linked palmitoyl chitosan, 252
cryogenic equipment, 113
cultured fish, 5, 317
cultured blue mussels, 321
D-glucosamine, 223
data analysis, 331
decolorization, 225
degree of hydrolysis, 372
deheaded gutted mackerel, 71
demineralization, 225
deproteinization, 225
depuration, 306
descriptive analysis, 159
digestive enzymes, 66
drip loss, 146
drop volume, 100
drug release, 252
dry salting, 117
drying-smoking, 115
e-commerce, 39
electrofusion, 8
electroplating facilities, 203
electronic nose, 158
electronic device manufacturing, 203
electrostatic smoking, 111, 112
emulsion stabilizing properties, 233
emulsion preparation, 235
environmental pollutants, 2
enzymatic activity, 91
enzymes, 94
extracellular ice crystals, 148
extractable protein nitrogen, 71
fish allergy, 156

fish farming, 61, 371
fishmeal, 369
flavors, 201
flavor extracts, 66
fluorometric method, 161
food safety and quality, 7
food quality, 81
free amino acids, 158
free fatty acids, 71, 96, 139
free fatty acids release, 142
freeze-dried shrimp, 196
freezing and thawing, 98
fresh salmon, 155
fresh fish chain, 35
frying parameters, 129
fungicides, 187
gel formation, 180
gel melting, 180
genetically modified foods, 40
global trade item number, 39
global location number, 39
glucosamine, 1,4, 204, 223, 275
glucosamine oligomers, 276
hazard analysis critical control point, 1, 156, 196
heat shock treatment, 308
heavy metal contaminated leachate, 203, 207
herring, 226
high pressure processing, 22, 91, 311
high pressure low temperature, 101
high pressure equipment, 331
high pressure thawing, 99
highly unsaturated fatty acids, 3
histamine analysis, 160
histamine, 156, 162
histidine decarboxylase, 156
hydrolysis of protein, 374
hydroperoxides, 142
hypocholesterolemic, 251

identification and authentication, 36
inactivation of microorganisms, 91
instantaneous pressure kill, 329
irradiation treatment, 310
isoelectric focusing, 36
isostatic pressure, 91
kinetics of the microbial destruction, 329
krill, 204
lactic acid bacteria, 281, 286, 296, 341, 357, 361
lactobacillus curvatus, 288
leachate, 203
lipid oxidation, 95, 142
listeria monocytogenes, 281, 287, 295, 296, 359
lobster, 196, 204, 259
machine vision, 54
mahi-mahi, 155, 158, 161, 162
mango fruits, 188
marine algae, 317
marine oils, 3
mesophilic bacteria, 156
metal-plating facilities, 203
methylated agars, 187
microbial quality, 96, 319
microbial inactivation kinetics, 11
microbial analysis, 331, 341
microbial control, 335
microbial count, 11, 164
microbial quality, 100, 320
microbial quality of mussels, 322
microbial survival rate, 11
most probable number, 306
mucopolysaccharides, 4
multiple compound quality indices, 281
mussels, 319, 323, 355
myofibrillar proteins, 93
N-acetylglucosamine, 204, 275

N-acetyl-Dglycosamine, 223
non-protein nitrogenous, 3
non-scombroid fish, 156
non-thermal processing, 8
ocean ranched salmon, 113
omega-3 fatty acids, 5
optical character recognition, 53
optimization of PEF parameters, 14
oxygen barrier properties, 227
oysters, 305
palmitoyl chitosan, 252
particle size determination, 236
pasteurization, 307, 310
pathogenic bacteria, 306
pectin methyl esterase, 329
peeled and cooked shrimp, 196
PEF processing, 8
peroxide value, 227
pesticide residues, 187
phase change, 98
phenol deposits, 122
plate count agar, 224
polyunsaturated fatty acids, 1
predictive microbiology, 2
pressure shift freezing, 98, 139
pressure death time, 332
pressure sensitivity, 333
principal component analysis, 340
protease inhibitors, 374
protein denaturation, 93
protein emulsifiers, 234
protein hydrolysis, 373
protein hydrolyzates, 66
protein-polysaccharide, 233
protein load, 236, 244
proteoglycans, 4
psychrotrophic bacteria, 156, 180, 321
pulsed electric fields, 7, 8
putrescine, 156, 162

quality assurance, 40, 82
quality control, 40
quality index method, 82
quality indices, 341, 344
quality loop, 48
quality of fish, 82
radio frequency identification, 39
red algae, 179
red lipophilic dye, 239
reduce surface symbology, 39
relative rate of spoilage, 285
response surface methodology, 203
salmon, 161, 163
salmonids, 371
salting and smoking, 113
salting, 113
sarcoplasmic proteins, 92
sarcoplasmic, 92
sardines, 156
scallops, 196, 319, 321
scombroid poisoning, 156
seabream, 373
seafood proteins, 3, 92
seafood quality, 155
seafood quality preservation, 223
seafood spoilage predictor, 290
sensory evaluation, 72, 76, 115, 159
sensory analysis, 341
sensory rejection time, 340
shelf-life, 290, 339
shrimp cooked meats, 197
shrimp, 195, 204, 329, 333, 355
shrimp by-product enhancement, 200
shrimp enhancement protenial, 195
shrimp processing, 196
shrimp waste, 259
slicing and packaging, 115
small and medium enterprises, 39
smart codes, 54

smoked salmon quality, 111
smoked salmon, 112, 355
smoked mussels, 358
soil leachate, 217
sorption capacity, 216
spider crab, 259
spoilage microflora, 293
spoilage microorganisms, 285, 294, 339, 340, 342
squalene, 1
stability of emulsions, 236
stroma, 92
stroma proteins, 93
temperature of powder
solubilization, 180
thawing drip, 99
thermal processing, 7
thiobarbituric acid reactive substances, 227
total heterotrophic bacterial, 318
total volatile basic nitrogen, 156
total microbial counts, 159
total phenols, 115
total psychrotrophic microorganisms, 341
traceability, 31, 32, 48
traceability and quality assurance, 34
traceable unit, 33
trimethylamine, 339
trout, 329, 333
tryptic soy agar, 331
tuna, 155, 158, 163
turbot, 99
ungutted mackerel, 77
UV irradiation, 23
UV light processing, 23
vbrio bacteria, 305
volatile basic-nitrogen, 225, 339
whey protein isolate, 233
wheying off, 239, 240

wolffish cultivation, 61
yield measurement, 115

zero-valent metal regeneration, 209